Recent Advances in Mechanics of Non-Newtonian Fluids

Recent Advances in Mechanics of Non-Newtonian Fluids

Special Issue Editors

Wei-Tao Wu

Mehrdad Massoudi

MDPI • Basel • Beijing • Wuhan • Barcelona • Belgrade • Manchester • Tokyo • Cluj • Tianjin

MDPI

Special Issue Editors
Wei-Tao Wu
Nanjing University of Science and Technology
China

Mehrdad Massoudi
Carnegie Mellon University
USA

Editorial Office
MDPI
St. Alban-Anlage 66
4052 Basel, Switzerland

This is a reprint of articles from the Special Issue published online in the open access journal *Fluids* (ISSN 2311-5521) (available at: https://www.mdpi.com/journal/fluids/special_issues/non_newtonian_fluids).

For citation purposes, cite each article independently as indicated on the article page online and as indicated below:

LastName, A.A.; LastName, B.B.; LastName, C.C. Article Title. *Journal Name* **Year**, *Article Number*, Page Range.

ISBN 978-3-03928-308-8 (Pbk)
ISBN 978-3-03928-309-5 (PDF)

Contents

About the Special Issue Editors vii

Wei-Tao Wu and Mehrdad Massoudi
Recent Advances in Mechanics of Non-Newtonian Fluids
Reprinted from: *Fluids* **2020**, *5*, 10, doi:10.3390/fluids5010010 1

Lorenzo Fusi, Angiolo Farina, Fabio Rosso and Kumbakonam Rajagopal
Thin-Film Flow of an Inhomogeneous Fluid with Density-Dependent Viscosity
Reprinted from: *Fluids* **2019**, *4*, 30, doi:10.3390/fluids4010030 5

Naser Hamedi and Lars-Göran Westerberg
On the Interaction of Side-By-Side Circular Cylinders in Viscoplastic Fluids
Reprinted from: *Fluids* **2019**, *4*, 93, doi:10.3390/fluids4020093 21

D. Andrew S. Rees and Andrew P. Bassom
The Effect of Internal and External Heating on the Free Convective Flow of a Bingham Fluid in a Vertical Porous Channel
Reprinted from: *Fluids* **2019**, *4*, 95, doi:10.3390/fluids4020095 35

Bruno M.M. Pereira, Gonçalo A.S. Dias, Filipe S. Cal and Kumbakonam R. Rajagopal and Juha H. Videman
Lubrication Approximation for Fluids with Shear-Dependent Viscosity
Reprinted from: *Fluids* **2020**, *5*, 10, doi:10.3390/fluids5010010 51

Masoud Jabbari, Jim McDonough, Evan Mitsoulis and Jesper Henri Hattel
Application of a Projection Method for Simulating Flow of a Shear-Thinning Fluid
Reprinted from: *Fluids* **2019**, *4*, 124, doi:10.3390/fluids4030124 69

U. S. Mahabaleshwar, P. N. Vinay Kumar, K. R. Nagaraju, Gabriella Bognár and S. N. Ravichandra Nayakar
A New Exact Solution for the Flow of a Fluid through Porous Media for a Variety of Boundary Conditions
Reprinted from: *Fluids* **2019**, *4*, 125, doi:10.3390/fluids4030125 85

Ângela Ribau, Luís L. Ferrás, Maria L. Morgado, Magda Rebelo and Alexandre Afonso
Semi-Analytical Solutions for the Poiseuille–Couette Flow of a Generalised Phan-Thien–Tanner Fluid
Reprinted from: *Fluids* **2019**, *4*, 129, doi:10.3390/fluids4030129 107

Evgenii S. Baranovskii, Anastasia A. Domnich and Mikhail A. Artemov
Optimal Boundary Control of Non-Isothermal Viscous Fluid Flow
Reprinted from: *Fluids* **2019**, *4*, 133, doi:10.3390/fluids4030133 119

Francesco Farsaci, Ester Tellone, Antonio Galtieri and Silvana Ficarra
A New Model for Thermodynamic Characterization of Hemoglobin
Reprinted from: *Fluids* **2019**, *4*, 135, doi:10.3390/fluids4030135 133

Hossein Asadi, Amir Mahdi Akbarzadeh, Mohammad Taeibi-Rahni, Khodayar Javadi and Goodarz Ahmadi
Investigation of Hydrodynamically Dominated Membrane Rupture, Using Smoothed Particle Hydrodynamics–Finite Element Method
Reprinted from: *Fluids* **2019**, *4*, 149, doi:10.3390/fluids4030149 151

Eilis Rosenbaum, Mehrdad Massoudi, and Kaushik Dayal
The Influence of Bubbles on Foamed Cement Viscosity Using an Extended Stokesian Dynamics Approach
Reprinted from: *Fluids* **2019**, *4*, 166, doi:10.3390/fluids4030166 . . . . **169**

Yahaya D. Baba, Archibong Archibong-Eso, Aliyu M. Aliyu, Olawale T. Fajemidupe, Joseph X. F. Ribeiro, Liyun Lao and Hoi Yeung
Slug Translational Velocity for Highly Viscous Oil and Gas Flows in Horizontal Pipes
Reprinted from: *Fluids* **2019**, *4*, 170, doi:10.3390/fluids4030170 . . . . **183**

Emad Jafari Nodoushan, Taeil Yi, Young Ju Lee and Namwon Kim
Wormlike Micellar Solutions, Beyond the Chemical Enhanced Oil Recovery Restrictions
Reprinted from: *Fluids* **2019**, *4*, 173, doi:10.3390/fluids4030173 . . . . **203**

Chengcheng Tao, Wei-Tao Wu and Mehrdad Massoudi
Natural Convection in a Non-Newtonian Fluid: Effects of Particle Concentration
Reprinted from: *Fluids* **2019**, *4*, 192, doi:10.3390/fluids4040192 . . . . **227**

About the Special Issue Editors

Wei-Tao Wu, Professor, PhD., School of Mechanical Engineering; Nanjing University of Science and Technology, China. Dr. Wei-Tao Wu is a professor in the School of Mechanical Engineering at Nanjing University of Science and Technology. He received his Ph.D. degree from the Department of Mechanical Engineering of Carnegie Mellon University, USA, and his B.S. degree from Xi'an Jiaotong University, China. His research interests include multiphase flows, non-Newtonian fluids, bio-fluid mechanics and aerodynamics. His work covers both mathematical modeling and computational simulation. In the last five years, Dr. Wu has authored or co-authored over 40 peer-reviewed journal papers. He has also served as an editorial board member of Fluids and is a member of the Physics and Aerodynamics Committee of Chinese Aerodynamics Research Society.

Mehrdad Massoudi, PhD. Adjunct Professor. Department of Biomedical Engineering, Carnegie Mellon University, Pittsburgh, PA 15213. Dr. Massoudi received his BS, MS, and PhD in 1979, 1982, and 1986 from University of Pittsburgh. His research interests are in the areas of mathematical modeling of non-linear materials, non-Newtonian fluid mechanics, multiphase flows, and granular materials. He is an ASME Fellow and the Editor-in-Chief of Fluids.

fluids

MDPI

Editorial

Recent Advances in Mechanics of Non-Newtonian Fluids

Wei-Tao Wu [1] and Mehrdad Massoudi [2,*]

[1] School of Mechanical Engineering, Nanjing University of Science & Technology, Nanjing 210094, China; weitaowwtw@njust.edu.cn
[2] U. S. Department of Energy, National Energy Technology Laboratory (NETL), Pittsburgh, PA 15236, USA
* Correspondence: mehrdad.massoudi@netl.doe.gov

Received: 2 January 2020; Accepted: 5 January 2020; Published: 8 January 2020

Keywords: lubrication; suspensions; viscoplastic fluids; cement; biofluids; oil recovery; porous media

---

Flow of non-Newtonian (non-linear) fluids occurs not only in nature, for example, mud slides and avalanches, but also in many industrial processes involving chemicals (polymers), biological materials (blood), food (honey, ketchup, yogurt), pharmaceutical and personal care items (shampoo, creams), etc. In general, these fluids exhibit certain distinct features such as shear-rate dependency of the viscosity (related to shear-thinning or shear-thickening aspects of the fluid), normal stress effects (related to die-swell and rod-climbing), creep or relaxation (viscoelasticity), yield stress effects (viscoplasticity), history effects (time dependent response), etc. There are many different models which can be used for different fluids under different conditions. For excellent overall discussion of these fluids, see Barnes, et al. [1], Larson [2], Tanner [3], Schowalter [4], Carreau, et al. [5], Macosko [6], and Bird, et al. [7]; for a more mathematical approach, see Deville and Gatski [8], Coleman, et al. [9], and Huilgol and Phan-Thien [10]; for a computational approach, see Crochet, et al. [11] and Owen and Phillips [12]; for a historical perspective, see Tanner and Walters [13]; and for a general introduction to measurement techniques, etc., see Coussot [14] and Walters [15].

This special issue of Fluids is dedicated to recent advances in the mathematical and physical modeling of non-linear fluids, with specific applications in lubrication, suspensions, viscoplastic fluids, cement, biofluids, oil recovery, porous media, and relevant numerical issues.

Formulating and solving flows of inhomogeneous fluids presents special difficulties, especially in the mathematical and numerical scheme. Fusi, et al. [16] consider the pressure-driven thin film flow of fluids whose viscosity depends on the density of the fluid. They use a thermodynamical framework to obtain the constitutive relation for the fluid, and by assuming a small aspect ratio for the channel, they use the lubrication approximation. The non-linear equations are solved numerically and the evolutions of the density in the fluid are plotted.

Hamedi and Westerberg [17] look at the two-dimensional flow of a Bingham fluid and its interactions with side-by-side cylinders. They study this problem numerically using the Open Source CFD Code OpenFOAM and discuss the influence of the gap between the cylinders on the drag force and the shape of the unyielded regions. Their results are presented for various Reynolds and Bingham numbers.

Rees and Bassom [18] study the free convection flow of a non-Newtonian fluid modeled as a Bingham fluid in a vertical porous channel. The heat is supplied both at the sidewalls and through an internal source of heat generation. Their results indicate that four different flow regimes arise: (1), one corresponding to complete stagnation, (2) one having one stagnation region, (3) and (4) two more regions with two stagnant regions. The authors also discuss how the locations of the yield surfaces evolve depending on the values of the Darcy–Rayleigh and Rees–Bingham numbers.

Pereira, et al. [19] rigorously derive the Reynolds lubrication approximation for steady flows of incompressible non-Newtonian fluids with shear dependent viscosity. They use a perturbation analysis and show that depending on the power-law character of the fluid, the derived equations can either be prescribed as a higher-order correction to the Reynolds equation or a new type of Reynolds equation for these non-linear fluids. They present the results for two flow conditions: flow in an eccentric journal bearing and flow between a rolling rigid cylinder and a plane.

Jabbari, et al. [20] use a first-order projection method to numerically solve the unsteady flow of (1) an incompressible Navier–Stokes fluid inside a three-dimensional lid-driven cavity, and (2) a power-law non-Newtonian fluid, for different cavity aspect ratio, and for three different Reynolds numbers, namely 100, 400, and 1000. They notice that as the power-law exponent n decreases (the fluid being more shear-thinning) the secondary flows become less pronounced.

Seeking and obtaining exact solutions for flows of fluids under different flow conditions and geometries is a great challenge to mathematicians and engineers. Mahabaleshwar, et al. [21] obtain an exact solution for the flow of a fluid through a semi-infinite porous medium with a slip boundary condition (using Navier's approach). They present their results considering the effects of slip parameters, the mass transpiration parameter, the Brinkman ratio, and the impact of shearing.

Ribau, et al. [22] study the Poiseuille–Couette flow of a viscoelastic fluid when the fluid is represented by a recently proposed model, generalizing the Phan–Thien–Tanner (PTT) constitutive equation. The new model has one or two new fitting parameters which can be used to fit the experimental data for a wider ranging application. The authors present their results for the velocity profiles for various parameters, including the new parameters in the generalized PTT model.

Baranovskii, et al. [23] study the non-linear equations describing the steady non-isothermal creeping flows of an incompressible viscous fluid and establish the existence of weak solutions. In their formulation, they assume that viscosity and thermal conductivity coefficients depend on the temperature; they also use mixed boundary conditions.

Farsaci, et al. [24] develop a thermodynamic model of hemoglobin; their model is based on experimental data and the physical point of view favoring the binding of oxygen to the protein. Their model, using non-equilibrium thermodynamics, introduces a phenomenological coefficient related to the displacement polarization current.

Asadi, et al. [25] numerically study the rupture of a membrane located between two fluids. They use the smoothed particle hydrodynamics (SPH) and the finite element method (FEM) to solve the problem and consider a range of pressure difference and membrane thicknesses. For the fluid phase, they use the Lagrangian form of the Navier–Stokes equations; these equations are coupled to the solid phase equations which use the von Mises elastic–plastic criterion.

Rosenbaum, et al. [26] study the influence of bubbles on the shear viscosity of foamed cement; the cement slurry is modeled as a fluid suspension and in their numerical simulations, they use an extended version of Stokesian Dynamics along with Fast Lubrication Dynamics. Bubbles are introduced/injected into the cement slurry to reduce the density, increase the viscosity, and improve the properties of cement. The simulated results for the viscosity show close agreement with traditional viscosity correlations for suspensions.

Baba, et al. [27] study the mixture flow of gases and liquids in pipelines; slug flow pattern has been shown to be the dominant flow regime for very viscous oils which are used in chemical and petroleum industries. They consider the impact of air and mineral oils, along with the influence of high liquid viscosity on the slug velocity; the authors propose a new empirical correlation for the slug velocity for highly viscous two-phase flows.

Viscoelastic surfactants such as Wormlike Micellar Solutions (WMS) have certain positive characteristics which can be used to overcome some of the difficulties encountered in chemical enhanced oil recovery (CEOR) techniques. Jafari Nodoushan, et al. [28] review the major approaches and challenges encountered in using these chemical agents. They indicate that the shear-thinning property of WMS along with their breakage under high shear stresses can lead to a better injectivity and trapping of oil in low-permeability zones.

The buoyancy driven flow (natural convection) of a suspension composed of rigid particles and a fluid between two vertical walls is studied by Tao, et al. [29]. They model the suspension as a non-linear fluid with a viscosity which depends on the shear rate and the particle concentration. To consider the motion of the particles, they use a convection–diffusion equation and perform a parametric study (after having non-dimensionalized the equations). They present their results for concentration, velocity, and temperature profiles in terms of various dimensionless numbers.

Acknowledgments: Finally, we would like to thank all the authors who contributed to this special issue. We are also grateful to all the anonymous reviewers for their help; without the help of qualified reviewers, it would not have been possible to organize this special issue. A personal note of appreciation and gratitude to Sonia Guan, the Managing Editor of Fluids, and the editorial staff at the Fluids Office; without their help and assistance, Fluids could not publish high quality papers in a short period of time.

Conflicts of Interest: The authors declare no conflict of interest.

References

1. Barnes, H.A.; Hutton, J.F.; Walters, K. *An Introduction to Rheology*; Elsevier: Amsterdam, The Netherlands, 1989; Volume 3.
2. Larson, R.G. *The Structure and Rheology of Complex Fluids*; Oxford University Press: New York, NY, USA, 1999; Volume 150.
3. Tanner, R.I. *Engineering Rheology*, 2nd ed.; Oxford University Press: Oxford, UK, 2000; Volume 52.
4. Schowalter, W.R. *Mechanics of Non-Newtonian Fluids*; Pergamon Press: Oxford, UK, 1978.
5. Carreau, P.J.; De Kee, D.C.R.; Chhabra, R.P. *Rheology of Polymeric Systems: Principles and Applications Hanser*; American Institute of Chemical Engineers: New York, NY, USA, 1997.
6. Macosko, C. *Rheology: Principles, Measurements and Applications*; Wiley-VCH Inc.: New York, NY, USA, 1994.
7. Bird, R.B.; Armstrong, R.C.; Hassager, O. *Dynamics of Polymeric Liquids. Volume 1: Fluid Mechanics*, 2nd ed.; Wiley: Hoboken, NJ, USA, 1987.
8. Deville, M.; Gatski, T.B. *Mathematical Modeling for Complex Fluids and Flows*; Springer Science & Business Media: Berlin/Heidelberg, Germany, 2012.
9. Coleman, B.D.; Markovitz, H.; Noll, W.; Markovitz, H.; Noll, W. *Viscometric Flows of Non-Newtonian Fluids: Theory and Experiment*; Cambridge University Press: Cambridge, UK, 1966; Volume 5.
10. Huilgol, R.R.; Phan-Thien, N. *Fluid Mechanics of Viscoelasticity: General Principles, Constitutive Modelling, Analytical and Numerical Techniques*; Elsevier: Amsterdam, The Netherlands, 1997; Volume 6.
11. Crochet, M.J.; Davies, A.R.; Walters, K. *Numerical Simulation of Non-Newtonian Flow*; Elsevier: Amsterdam, The Netherlands, 2012; Volume 1.
12. Owens, R.G.; Phillips, T.N. *Computational Rheology*; World Scientific: Singapore, 2002; Volume 14.
13. Tanner, R.I.; Walters, K. *Rheology: An Historical Perspective*; Elsevier: Amsterdam, The Netherlands, 1998; Volume 7.
14. Coussot, P. *Rheometry of Pastes, Suspensions, and Granular Materials: Applications in Industry and Environment*; John Wiley & Sons: Hoboken, NJ, USA, 2005.
15. Walters, K. *Rheometry, Industrial Applications*; Wiley: Hoboken, NJ, USA, 1980; Volume 1.
16. Fusi, L.; Farina, A.; Rosso, F.; Rajagopal, K. Thin-Film Flow of an Inhomogeneous Fluid with Density-Dependent Viscosity. *Fluids* **2019**, *4*, 30. [CrossRef]
17. Hamedi, N.; Westerberg, L.-G. On the Interaction of Side-By-Side Circular Cylinders in Viscoplastic Fluids. *Fluids* **2019**, *4*, 93. [CrossRef]
18. Rees, D.A.S.; Bassom, A.P. The effect of internal and external heating on the free convective flow of a Bingham fluid in a vertical porous channel. *Fluids* **2019**, *4*, 95. [CrossRef]
19. Pereira, B.M.M.; Dias, G.A.S.; Cal, F.S.; Rajagopal, K.R.; Videman, J.H. Lubrication Approximation for Fluids with Shear-Dependent Viscosity. *Fluids* **2019**, *4*, 98. [CrossRef]
20. Jabbari, M.; McDonough, J.; Mitsoulis, E.; Hattel, J.H. Application of a projection method for simulating flow of a shear-thinning fluid. *Fluids* **2019**, *4*, 124. [CrossRef]

21. Mahabaleshwar, U.S.; Vinay Kumar, P.N.; Nagaraju, K.R.; Bognár, G.; Nayakar, S.N. A New Exact Solution for the Flow of a Fluid through Porous Media for a Variety of Boundary Conditions. *Fluids* **2019**, *4*, 125. [CrossRef]
22. Ribau, Â.M.; Ferrás, L.L.; Morgado, M.L.; Rebelo, M.; Afonso, A.M. Semi-Analytical Solutions for the Poiseuille-Couette Flow of a Generalised Phan-Thien-Tanner Fluid. *Fluids* **2019**, *4*, 129. [CrossRef]
23. Baranovskii, E.S.; Domnich, A.A.; Artemov, M.A. Optimal Boundary Control of Non-Isothermal Viscous Fluid Flow. *Fluids* **2019**, *4*, 133. [CrossRef]
24. Farsaci, F.; Tellone, E.; Galtieri, A.; Ficarra, S. A New Model for Thermodynamic Characterization of Hemoglobin. *Fluids* **2019**, *4*, 135. [CrossRef]
25. Asadi, H.; Taeibi-Rahni, M.; Akbarzadeh, A.M.; Javadi, K.; Ahmadi, G. Investigation of Hydrodynamically Dominated Membrane Rupture, Using Smoothed Particle Hydrodynamics-Finite Element Method. *Fluids* **2019**, *4*, 149. [CrossRef]
26. Rosenbaum, E.; Massoudi, M.; Dayal, K. The Influence of Bubbles on Foamed Cement Viscosity Using an Extended Stokesian Dynamics Approach. *Fluids* **2019**, *4*, 166. [CrossRef]
27. Baba, Y.D.; Archibong-Eso, A.; Aliyu, A.M.; Fajemidupe, O.T.; Ribeiro, J.X.F.; Lao, L.; Yeung, H. Slug translational velocity for highly viscous oil and gas flows in horizontal pipes. *Fluids* **2019**, *4*, 170. [CrossRef]
28. Jafari Nodoushan, E.; Yi, T.; Lee, Y.J.; Kim, N. Wormlike Micellar Solutions, Beyond the Chemical Enhanced Oil Recovery Restrictions. *Fluids* **2019**, *4*, 173. [CrossRef]
29. Tao, C.; Wu, W.-T.; Massoudi, M. Natural Convection in a Non-Newtonian Fluid: Effects of Particle Concentration. *Fluids* **2019**, *4*, 192. [CrossRef]

fluids

MDPI

Article

Thin-Film Flow of an Inhomogeneous Fluid with Density-Dependent Viscosity

Lorenzo Fusi [1,*], Angiolo Farina [1], Fabio Rosso [1] and Kumbakonam Rajagopal [2]

[1] Dipartimento di Matematica e Informatica "U. Dini", Viale Morgagni 67/a, 50134 Firenze, Italy; angiolo.farina@unifi.it (A.F.); fabio.rosso@unifi.it (F.R.)

[2] Department of Mechanical Engineering, Texas A&M University, College Station, TX 77843, USA; krajagopal@tamu.edu

* Correspondence: lorenzo.fusi@unifi.it

Received: 10 January 2019; Accepted: 19 January 2019; Published: 18 February 2019

Abstract: In this paper, we study the pressure-driven thin film flow of an inhomogeneous incompressible fluid in which its viscosity depends on the density. The constitutive response of this class of fluids can be derived using a thermodynamical framework put into place to describe the dissipative response of materials where the materials' stored energy depends on the gradient of the density (Mechanics of Materials, 2006, 38, pp. 233–242). Assuming a small aspect ratio for the channel, we use the lubrication approximation and focus on the leading order problem. We show the mathematical problem reduce to a nonlinear first order partial differential equation (PDE) for the density in which the coefficients are integral operators. The problem is solved numerically and plots that describe the evolution of the density in the fluid domain are displayed. We also show that it was possible to determine an analytical solution of the problem when the boundary data are small perturbations of the homogeneous case. Finally, we use such an analytical solution to validate the numerical scheme.

Keywords: inhomogeneous fluids; non-newtonian fluids; lubrication approximation (76A05, 76D08, 76A20)

---

1. Introduction

Most bodies are inhomogeneous, but in many of them the inhomogeneity is mild, and they can be approximated as homogeneous bodies. In this study, we consider bodies wherein the inhomogeneity cannot be neglected. Some of these inhomogeneous bodies could be approximated as being incompressible. Before providing a rigorous definition of what is meant by an inhomogeneous fluid, we will define the concept in loose terms. The homogeneity of a body is not decided on the basis of the properties of a body in its present configuration but, rather, is determined by whether there is some configuration wherein its material points are indistinguishable in terms of their response. For example, a homogeneous fluid which is shear thinning, on being sheared could, in its current flowing state, have its property, say viscosity, vary at different particles; but this notwithstanding the fluid is homogeneous if in its reference state, and this might be close to its static state, all particles of the body would respond in an identical fashion. A rigorous definition of the notion of homogeneity can be found in Reference [1], but here we provide an elementary treatment of the issues. Two material points, P_1 and P_2, belonging to an abstract body, are said to be materially uniform within the context of a purely mechanical purview, if there exist neighborhoods of the particles which, when placed by different placers into a three dimensional Euclidean space, respond in an indistinguishable manner with regard to their mechanical response. If all the particles belonging to the body are pairwise materially uniform, the body is said to be materially uniform. A body is said to be homogeneous if all the material points belonging to the body are materially uniform with respect to a single placer.

A body that is not homogeneous is said to be inhomogeneous. Thus, based on the fact that properties of a body vary in a specific configuration, say the current deformed configuration, does not deem the body inhomogeneous. Whether a body is homogeneous or not cannot be decided within the context of an Eulerian specification. However, the statement that one ought to find if there exists some configuration in which all neighborhoods of material points respond in an identical fashion is an impractical definition, as we can never exhaust all possible configurations in which the body might be placed and, in this sense, never be in a position to decide whether the body is homogeneous. We were interested in studying bodies which, in their rest state, are inhomogeneous.

Anand and Rajagopal [2] have shown how inhomogeneous fluids with properties that vary mildly in the mean value may lead to differences in the response between the inhomogeneous body and its homogenized approximation, which can be larger than one order of magnitude. This means that one has to be very careful when approximating a fluid as being homogeneous because small variations of the properties can lead to significant errors in the computation of global and local quantities.

Inhomogeneous fluids can be described within the theory, developed by Rajagopal and co-workers [3], that appeals to the notion of a body having multiple natural configurations. This theory allows one to derive the constitutive equations of a large disparate class of dissipative materials [4–8]. The central idea of the theory is the fact that the body can exist in multiple stress free configurations, and its evolution is regulated by thermodynamical considerations. In particular, the evolution of the natural configuration is determined by the maximization of the rate of entropy production which, in the case of a pure mechanical setting (isothermal case), is equivalent to the maximization of the rate of dissipation.

Malek et al. [9] have used this thermodynamical framework to derive the response of inhomogeneous incompressible fluid-like bodies where the stored energy (henceforth quantities with superscript * denote a dimensional quantity) (Helmoltz potential ψ^*) depends on the density ρ^* and on the gradient of the density $(\nabla\rho)^*$ and where the rate of dissipation ξ^* is a function of the mean normal stress (which we shall denote by the pressure p^*), of the density, and of the symmetric part of the velocity gradient, namely the tensor $\mathbf{D}^*$, that is:

$$\psi^*(\rho^*, (\nabla\rho)^*) \qquad \xi^* = \xi^*(p^*, \rho^*, \mathbf{D}^*).$$

Using the maximization criterion under the constraint of incompressibility and imposing isotropy for the Helmoltz potential, one can derive the generic form of the Cauchy stress tensor that now depends on the pressure, the density and its gradient, and the symmetric part of the velocity gradient, i.e., (see Reference [9] for all the details):

$$\mathbf{T}^* = -p^*\mathbf{I} - \rho^*\left(\psi^*_{(\nabla\rho)^*} \otimes (\nabla\rho)^* - \frac{1}{3}\psi^*_{(\nabla\rho)^*} \cdot (\nabla\rho)^*\mathbf{I}\right) + 2\mu^*(p^*, \rho^*, \mathbf{D}^* \cdot \mathbf{D}^*)\mathbf{D}^*, \tag{1}$$

where $\psi^*_{(\nabla\rho)^*}$ means differentiation with respect to $(\nabla\rho)^*$ and where μ^* has the dimension of a viscosity. The specification of the stored energy function ψ^* and of the viscosity μ^*, i.e., the way energy is stored in the continuum and the way the system dissipates energy, allows one to determine the response of the material. Examples of inhomogeneous fluids described in Reference [9] are studied in Reference [9,10].

In this paper, we are interested in materials described by a constitutive equation of the form given by Equation (1). In particular, we consider a fluid-like material in which ψ^* did not depend on $\nabla\rho^*$ and $\mu^* = \mu^*(\rho^*)$. In this specific case, Equation (1) reduced to:

$$\mathbf{T}^* = -p^*\mathbf{I} + 2\mu^*(\rho^*)\mathbf{D}^*. \tag{2}$$

Hence, we consider an inhomogeneous incompressible fluid where the viscosity changes with the density ρ^* and where volumes were preserved, i.e., $\mathrm{tr}\,(\mathbf{D}^*) = 0$. In practice, we are interested in materials with mass distribution that is not uniform in the reference configuration but with motion that could be reasonably approximated as isochoric (i.e., volume-preserving). In other words, the density did not vary as long as the particle is fixed.

The paper investigates the pressure-driven flow between parallel plates of a material whose constitutive equation is described by Equation (2) (see Figure 1). The height of the channel is assumed to be far smaller than its length so that the lubrication approximation applies because of the smallness of the aspect ratio. The leading order momentum equation can be solved analytically, producing a velocity field with components that depends on the density in a functional way. Substitution into the mass balance equation finally provides a very particular integro-differential equation for the density that is solved numerically. A numerical code is implemented for such an equation, and the evolution of the density is plotted as a function of time and space. To investigate the validity of the numerical scheme, we finally consider the case in which the density (and, hence, viscosity) is a small perturbation of the homogeneous case (Navier-Stokes system). In this particular situation, we are able to find an explicit analytical solution that could be compared with the one obtained numerically. The comparison shows good agreement, verifying the validity of the numerical scheme.

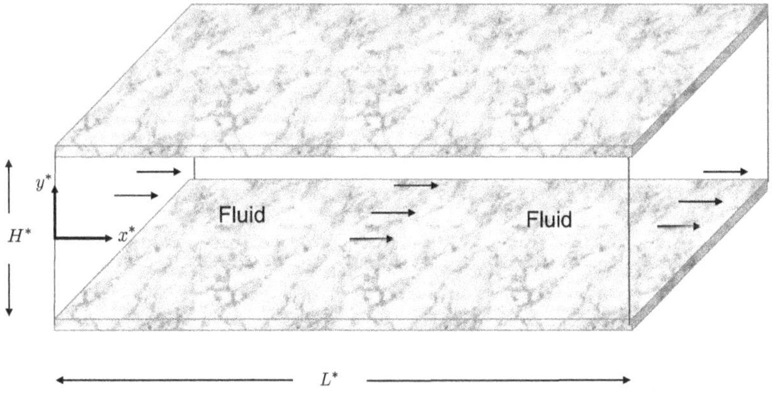

Figure 1. Sketch of the system.

2. The Mathematical Model

It is important to recognize that we are not dealing with a fixed set of particles in the flow domain at any instant of time, as particles are constantly entering and leaving the domain. As we are dealing with an inhomogeneous body, we could then have particles in the flow domain that have different response characteristics associated with them and, more importantly, the governing equations would apply to a different set of particles. That is, in order to study inhomogeneous bodies it is necessary to use the governing equations in the Lagrangian form and follow the partlcle and not use an Eulerian approach as the same point in space will be occupied by different particles. However, using a Lagrangian approach is very cumbersome and hence we make an approximation that will allow us to use an Eulerian approach. However, it is important to recognize that this approximation is only valid for a very small class of flows and also for materials whose inhomogeneity is mild. We make the following assumption concerning the particles that enter the flow domain that ensures we can indeed apply the Eulearian form of the governing equations for the different particles that are occupying the flow domain, as though they are the same set of particles. The assumption that we make is rather strong, but it is applicable for a reasonable class of problems. We assume that, at any instant of time, the properties associated with the particles that enter the flow domain at a fixed point at the boundary are identical, that assures that they have the same response characteristics. We further assume that they also satisfy the same boundary condition at $x^* = 0$ at all times. This is a critical assumption since this allows our domain, in which we did not have a fixed set of particles (as different particles entered the domain at $x^* = 0$), to be equivalent to a fixed set of particles, as those that enter the fluid domain satisfy the same conditions at $x^* = 0$. This allows us to approximate the problem in which we, in reality, had a different set of particles, as though they were the same set of particles, using an Eulerian framework.

Let us consider a fluid in which its Cauchy stress is given by Equation (2), where $\mu^*(\rho^*)$ is a prescribed positive smooth function. The mass balance is expressed as:

$$\frac{\partial \rho^*}{\partial t^*} + \mathbf{v}^* \cdot \nabla^* \rho^* + \rho^* \nabla^* \cdot \mathbf{v}^* = 0, \tag{3}$$

where $\mathbf{v}^*(\mathbf{x}^*, t^*)$ is the velocity field. Denoting the material derivative as:

$$\frac{d(.)}{dt^*} = \frac{\partial(.)}{\partial t^*} + \mathbf{v}^* \cdot \nabla^*(.)$$

Equation (3) can be rewritten as:

$$\frac{d\rho^*}{dt^*} + \rho^* \nabla^* \cdot \mathbf{v}^* = 0. \tag{4}$$

We say that the motion is volume-preserving (isochoric) if:

$$\nabla^* \cdot \mathbf{v}^* = 0 \tag{5}$$

for all $t \in \mathbb{R}$ and $\mathbf{x}^* \in \Omega^* \subset \mathbb{R}^3$. For an incompressible continuum, Equation (4) yields:

$$\frac{d\rho^*}{dt^*} = 0, \tag{6}$$

i.e., the density of a fixed particle does not change, even though the density can change from particle to particle in the body.

Using the standard procedure in continuum mechanics, we denote the Lagrangian density by $\hat{\rho}^*$, the coordinate of the particles in the reference configuration by $\mathbf{X}^*$, and the motion by χ^*. If the body is inhomogeneous, then the density varies in the reference configuration, i.e., $\hat{\rho}^* = \hat{\rho}^*(\mathbf{X}^*)$ and also in the current configuration, since:

$$\rho^*(\mathbf{x}^*, t) = \hat{\rho}^*(\chi^{*^{-1}}(\mathbf{x}^*, t)).$$

Following an Eulerian approach $\rho^*(\mathbf{x}^*, t^*)$ is an unknown in the problem, even though $\hat{\rho}^*(\mathbf{X}^*)$ is given. Indeed, the field $\rho^*(\mathbf{x}^*, t)$ is determined only when the mapping of $\chi^*(\mathbf{X}^*, t^*)$ has been explicitly found, solving the equations of motion. Hence, according to the Eulerian approach, we have to solve Equation (3) coupled with Equation (5) and with the balance of linear momentum which, in the absence of body forces, is given by:

$$\rho^* \frac{dv^*}{dt^*} = -\nabla p^* + \nabla^* \cdot \left[\mu^*(\rho^*)\mathbf{D}^*\right]. \tag{7}$$

Equations (5)–(7) constitute the mathematical formulation of the model to which one must add appropriate initial and boundary conditions. In particular, we consider a velocity field of the form $\mathbf{v}^*(x^*, y^*) = (u^*(x^*, y^*), v^*(x^*, y^*))$ and assume that the fluid is confined in a channel of width $2H^*$ and length L^* (see Figure 1).

We notice that the problem as specified by Equations (5)–(7) is given in terms of Eulerian quantities, while the inhomogenerity, which is a consequence of the density varying in the reference configuration, is a Lagrangian specification.

Concerning the boundary conditions for the velocity, because of symmetry w.r.t. the $y^* = 0$ plane, we limit ourselves to study the upper part of the channel. We shall write:

$$u^*(x^*, H^*, t^*) = v^*(x^*, H^*, t^*) = 0 \quad \text{(No slip)},$$

$$u^*_{y^*}(x^*,0,t^*) = v^*(x^*,0,t^*) = 0 \quad \text{(Symmetry)}, \tag{8}$$

where $u^*_{y^*}$ is the derivative w.r.t. y^* of the longitudinal component of the velocity. The pressure drop that drives the flow is taken as:

$$p^*(0^*,y^*,t^*) = \Delta p^* > 0 \qquad\qquad p^*(L^*,y^*,t^*) = 0,$$

where, for simplicity, we have rescaled the outlet pressure to zero and where Δp^* may depend on time. We also assign the initial condition for the velocity as $\mathbf{v}^*(x^*,y^*,0) = \mathbf{v}^*_s(x^*,y^*)$.

With regard to the density, we impose the boundary condition:

$$\rho^*(0,y^*,t^*) = \rho^*_\ell(y^*,t^*),$$

and the initial condition:

$$\rho^*(x^*,y^*,0) = \rho^*_s(x^*,y^*),$$

where ρ^*_ℓ (density at the lateral inlet) and ρ^*_s (initial density) are given functions. It is convenient to introduce:

$$\rho^*(x^*,0,t^*) = \rho^*_m(x^*,t^*).$$

Looking at the symmetry condition expressed by Equation (8) and Equation (3), we see that ρ^*_m must satisfy the following first order partial differential equation (PDE):

$$\frac{\partial \rho^*_m(x^*,t^*)}{\partial t^*} + u^*(x^*,0,t^*)\frac{\partial \rho^*_m(x^*,t^*)}{\partial x^*} = 0, \tag{9}$$

where the boundary and initial conditions are obtained from the compatibility conditions:

$$\rho^*_m(0,t^*) = \rho^*_\ell(0,t^*) \qquad\qquad \rho^*_m(x^*,0) = \rho^*_s(x^*,0). \tag{10}$$

Therefore ρ_m is an auxiliary unknown of the problem that can be determined solving Equations (9) and (10).

We suppose that the aspect ratio of the domain of the fluid is sufficiently small, so that we write:

$$\varepsilon = \frac{H^*}{L^*} \ll 1.$$

We rescale the variables as follows:

$$x^* = L^*x, \qquad y^* = \varepsilon L^* y, \qquad v^*_1 = U^* v_1, \quad v^*_2 = \varepsilon U^* v_2,$$

$$t^* = t^*_o t, \qquad p^* = p^*_o p, \qquad \rho^* = \rho^*_o \rho, \qquad \mu^*(\rho^*) = \mu^*_o \mu(\rho),$$

$$\rho^*_m = \rho^*_o \rho_m, \qquad \rho^*_\ell = \rho^*_o \rho_\ell, \qquad \rho^*_s = \rho^*_o \rho_s,$$

where we choose:

$$t^*_o = \frac{L^*}{U^*}, \qquad p^*_o = \frac{\mu^*_o U^* L^*}{H^{*2}},$$

and where ρ^*_o, $\mu^*_o > 0$ are characteristic values for the density and for the viscosity, respectively, and where U^* is the mean velocity in the channel. The time t^*_o represents the characteristic transit time (i.e., the time taken by a fluid particle to cross the entire channel), while p^*_o comes from the classical Poiseuille law. After little algebra, the system Equations (5)–(7) can be rewritten in the following non-dimensional form:

$$
\begin{cases}
u_x + v_y = 0, \\
\rho_t + u\rho_x + v\rho_y = 0, \\
\varepsilon \mathrm{Re}\left(u_t + uu_x + vu_y\right) = -p_x + \varepsilon^2\left(2\mu u_x\right)_x + \left(\mu u_y + \varepsilon^2 \mu v_x\right)_y, \\
\varepsilon^3 \mathrm{Re}\left(v_t + uv_x + vv_y\right) = -p_y + \left(\mu u_y + \varepsilon^2 \mu v_x\right)_x + \varepsilon^2\left(2\mu v_y\right)_y,
\end{cases}
\tag{11}
$$

where:

$$
\mathrm{Re} = \frac{\rho_o^* U^* H^*}{\mu_o^*}
$$

is the Reynolds number.

3. Leading Order

Assuming $\mathrm{Re} \leqslant O(1)$ and neglecting all the terms containing ε, we look for the zero order approximation of the general problem (Equation (11)):

$$
\begin{cases}
u_x + v_y = 0, \\
\rho_t + u\rho_x + v\rho_y = 0, \\
0 = -p_x + \left(\mu u_y\right)_y, \\
0 = -p_y.
\end{cases}
\tag{12}
$$

It immediately follows that $p = p(x,t)$ and, integrating $(12)_3$ with the symmetry condition $u_y(x,0,t) = 0$ and no-slip conditions $u(x,1,t) = 0$, we obtain:

$$
u(x,y,t) = -p_x \int_y^1 \frac{s}{\mu\left(\rho(x,s,t)\right)} ds.
$$

Now, exploiting the balance of mass $u_x = -v_y$, we find:

$$
v_y(x,y,t) = p_{xx} \int_y^1 \frac{s}{\mu\left(\rho(x,s,t)\right)} ds + p_x \int_y^1 \frac{d}{dx}\left[\frac{s}{\mu\left(\rho(x,s,t)\right)}\right] ds.
$$

Integrating again between y and 1 and exploiting $v(x,1,t) = 0$, we obtain:

$$
v(x,y,t) = -\frac{d}{dx}\left[p_x \int_y^1 dz \int_z^1 \frac{s}{\mu\left(\rho(x,s,t)\right)} ds\right].
$$

Next, imposing the symmetry condition $v(x,0,t) = 0$, we get:

$$
0 = \frac{d}{dx}\left[p_x \int_0^1 dz \int_z^1 \frac{s}{\mu\left(\rho(x,s,t)\right)} ds\right],
\tag{13}
$$

namely:

$$p_x = \frac{C}{\int_0^1 dz \int_z^1 \frac{s}{\mu\Big(\rho(x,s,t)\Big)} ds}, \tag{14}$$

where C may depend only on time. Exploiting the boundary conditions (we have set $\Delta p = \Delta p^*/p_o^*$, $p(0,t) = \Delta p$, and $p(1,t) = 0$), we integrate (14) between $x = 0$ and $x = 1$ and get:

$$C = -\frac{\Delta p}{\int_0^1 \frac{dx}{\int_0^1 dz \int_z^1 \frac{s}{\mu\Big(\rho(x,s,t)\Big)} ds}}.$$

So, in conclusion,

$$p_x = -\frac{\Delta p}{\int_0^1 \frac{dx}{\int_0^1 dz \int_z^1 \frac{s}{\mu\Big(\rho(x,s,t)\Big)} ds}} \cdot \frac{1}{\int_0^1 dz \int_z^1 \frac{s}{\mu\Big(\rho(x,s,t)\Big)} ds}.$$

Substituting p_x into the expressions for u and v, we find (notice that $v(x,y,t)$ can be rewritten as $v(x,y,t) = \int_y^1 u_x(x,z,t)dz$):

$$u(x,y,t) = \frac{\Delta p}{\int_0^1 \frac{dx}{\int_0^1 dz \int_z^1 \frac{s}{\mu} ds}} \cdot \frac{\int_y^1 \frac{s}{\mu} ds}{\int_0^1 dz \int_z^1 \frac{s}{\mu} ds}, \tag{15}$$

$$v(x,y,t) = \frac{d}{dx}\left[\frac{\Delta p}{\int_0^1 \frac{dx}{\int_0^1 dz \int_z^1 \frac{s}{\mu} ds}} \cdot \frac{\int_y^1 dz \int_z^1 \frac{s}{\mu} ds}{\int_0^1 dz \int_z^1 \frac{s}{\mu} ds}\right], \tag{16}$$

where $\mu = \mu(\rho(x,s,t))$. Relations (15), (16) represent the formal expressions for the velocity components.

One immediately sees that the velocity components are operators (and not functions) of the density ρ. With an abuse of notation, we shall write $u = u(\rho)$ and $v = v(\rho)$, recalling that the dependence on ρ is through an integral. In conclusion, we have ended up with a non-standard equation for ρ, i.e.,

$$\rho_t + u(\rho)\rho_x + v(\rho)\rho_y = 0,$$

with $u = u(\rho)$ and $v = v(\rho)$ given by (15), (16). This equation can be obviously solved only numerically.

Recalling the discussion of Section 2, the problem to be solved is the following:

$$
\begin{cases}
\dfrac{\partial \rho_m}{\partial t} + \left(u(\rho)\Big|_{y=0} \right) \dfrac{\partial \rho_m}{\partial x} = 0, \\
\rho_m(x,0) = \rho_s(x,0), \\
\rho_m(0,t) = \rho_\ell(0,t), \\
\dfrac{\partial \rho}{\partial t} + u(\rho)\dfrac{\partial \rho}{\partial x} + v(\rho)\dfrac{\partial \rho}{\partial y} = 0, \\
\rho(x,0,t) = \rho_m(x,t), \\
\rho(0,y,t) = \rho_\ell(y,t), \\
\rho(x,y,0) = \rho_s(x,y),
\end{cases}
\tag{17}
$$

with the compatibility condition:

$$
\rho_s(0,y) = \rho_\ell(y,0).
$$

In conclusion, we have to solve a system of two first order PDEs for the unknowns $\rho_m(x,t)$ and $\rho(x,y,t)$.

Remark 1. *Suppose $\rho_\ell(0,t) = \rho_s(x,0) = 1$. Then $\rho_m(x,t) \equiv 1$ is the only solution of (17)$_1$ and problem (17) reduces to:*

$$
\begin{cases}
\dfrac{\partial \rho}{\partial t} + u(\rho)\dfrac{\partial \rho}{\partial x} + v(\rho)\dfrac{\partial \rho}{\partial y} = 0, \\
\rho(x,0,t) = 1, \\
\rho(0,y,t) = \rho_\ell(y,t), \\
\rho(x,y,0) = \rho_s(x,y).
\end{cases}
\tag{18}
$$

Remark 2. *If, on the other hand, $\rho_\ell(y,t) = \rho_s(x,y) = 1$, then the system (17) is automatically satisfied by $\rho_m(x,t) = \rho(x,y,t) \equiv 1$. In this case, μ is a constant and, taking $\mu = 1$ from Equations (15) and (16):*

$$
u = \Delta p \left(\frac{1-y^2}{2} \right), \qquad v = 0,
$$

is the classical parabolic profile.

4. Numerics

The method adopted to solve Equation (17) is the following: We pick a function $\rho^{(0)}(x,y,t)$ representing a guess for the solution in some time interval, $[0,t]$ and we evaluate the coefficients $u(\rho^{(0)})$, $v(\rho^{(0)})$, and $u(\rho^{(0)})|_{y=0}$, which appear in Equation (17)$_{1,4}$. Then we solve Equation (17)$_{1-3}$ and use the solution as the boundary condition in Equation (17)$_5$. Finally, we solve the problem (17)$_{4-7}$, getting a solution $\rho^{(1)}(x,y,t)$ that clearly depends on the particular choice of $\rho^{(0)}$. With this procedure, we define the map:

$$
\Psi : \rho^{(0)}(x,y,t) \longrightarrow \rho^{(1)}(x,y,t).
$$

We then use $\rho^{(1)}(x,y,t)$ as the new guess and determine $\rho^{(2)}(x,y,t) = \Psi(\rho^{(1)})$. Proceeding in this way, we build a sequence of functions:

$$\left\{\rho^{(n)}(x,y,t)\right\} \qquad n \in \mathbb{N}.$$

If the map Ψ has a unique fixed point in some function space, then the unique fixed point is the solution to Equation (17). Here, we do not focus on the proof of the existence and uniqueness of a fixed point, but we use such an iterative procedure to evaluate the solution numerically. In particular, the iterative procedure will be stopped when the quantity:

$$\mathsf{tol} = \|\rho^{(n+1)} - \rho^{(n)}\| = \sup|\rho^{(n+1)} - \rho^{(n)}|, \tag{19}$$

is less than a fixed tolerance. The numerical scheme is implemented through a first-order upwind scheme for both Equations $(17)_1$ and $(17)_4$. A schematic representation of the stencils for the finite difference schemes for ρ_m and ρ is shown in Figure 2.

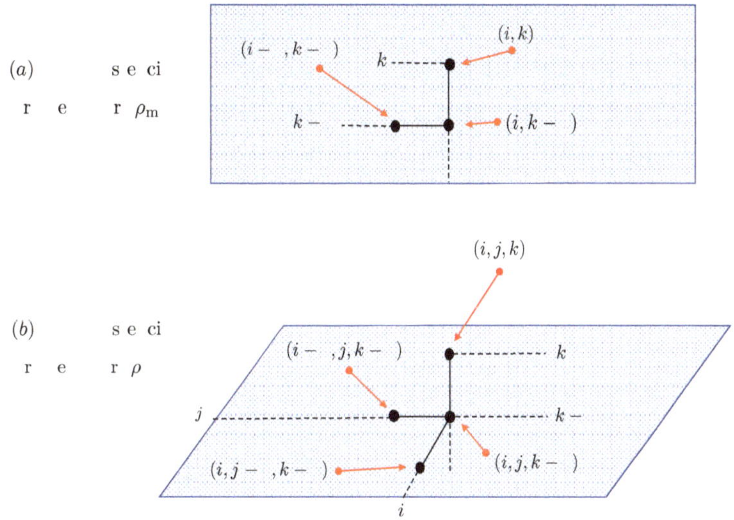

Figure 2. Stencil for the 2D and 3D problems.

Let us focus first on Equation $(17)_{1-3}$, Figure 2a. We build the grid:

$$x_i = i\Delta x \qquad t_k = k\Delta t \qquad i = 0,\ldots,n_x, \quad k = 0,\ldots,n_t.$$

and we approximate the derivatives in the following way:

$$\frac{\partial \rho_m}{\partial t} \approx \frac{\rho_{m_{i,k}} - \rho_{m_{i,k-1}}}{\Delta t} \qquad \frac{\partial \rho_m}{\partial x} \approx \frac{\rho_{m_{i,k-1}} - \rho_{m_{i-1,k-1}}}{\Delta x}.$$

The first-order upwind explicit scheme is given by:

$$\rho_{m_{i,k}} = \rho_{m_{i,k-1}} - u_{m_{i,k-1}}\frac{\Delta t}{\Delta x}\left(\rho_{m_{i,k-1}} - \rho_{m_{i-1,k-1}}\right), \tag{20}$$

where:

$$u_{m_{i,k-1}} = u(\rho^{(n)}(x_i, t_{k-1})),$$

and where $\rho^{(n)}$ is the guess at the nth iteration. The solution of Equation (20) provides the numerical solution of Equation $(17)_{1-3}$ at the nth iteration. This solution is then used as a boundary condition for Equation $(17)_4$. Let us now consider Equation $(17)_{4-7}$. We build the 3D grid:

$$x_i = i\Delta x \qquad y_j = j\Delta y \qquad t_k = k\Delta t$$

$$i = 0, \ldots, n_x, \quad j = 0, \ldots, n_y, \quad k = 0, \ldots, n_t,$$

and we approximate the derivatives:

$$\frac{\partial \rho}{\partial t} \approx \frac{\rho_{i,j}^{k} - \rho_{i,j}^{k-1}}{\Delta t} \qquad \frac{\partial \rho}{\partial x} \approx \frac{\rho_{i,j}^{k-1} - \rho_{i-1,j}^{k-1}}{\Delta x} \qquad \frac{\partial \rho}{\partial y} \approx \frac{\rho_{i,j}^{k-1} - \rho_{i,j-1}^{k-1}}{\Delta y}.$$

The first-order upwind explicit scheme is given by:

$$\rho_{i,j}^{k} = \rho_{i,j}^{k-1} - u_{i,j}^{k-1}\frac{\Delta t}{\Delta x}\left(\rho_{i,j}^{k-1} - \rho_{i-1,j}^{k-1}\right) - v_{i,j}^{k-1}\frac{\Delta t}{\Delta y}\left(\rho_{i,j}^{k-1} - \rho_{i,j-1}^{k-1}\right), \tag{21}$$

where

$$u_{i,j}^{k-1} = u(\rho^{(n)}(x_i, y_j, t_{k-1})) \qquad v_{i,j}^{k-1} = v(\rho^{(n)}(x_i, y_j, t_{k-1})),$$

and where $\rho^{(n)}$ is again the guess at the nth iteration. The solution of Equation (21) provides the solution at the $(n+1)$th iteration, that is $\rho^{(n+1)}$. We have set the tolerance with Equation (19) to be 10^{-5}, so that, when $\mathsf{tol} < 10^{-5}$, the iterations are stopped and the resulting solution is deemed as the solution to the problem.

5. Examples

To illustrate the behavior of the solutions of Equation (17), we have selected a pair of boundary conditions:

$$\text{(B1)} \begin{cases} \rho_s(x,y) = 1 + \frac{4}{5}\sin(6x^2) + \frac{4}{5}\cos(4y^2), \\ \rho_l(y,t) = 1 + \frac{4}{5}\cos(4y^2), \end{cases}$$

$$\text{(B2)} \begin{cases} \rho_s(x,y) = \exp\left[-\alpha\left(x - \frac{1}{2}\right)^2 - \alpha y^2\right] + \frac{x}{x^3+2}, & \alpha = 4, \\ \rho_l(y,t) = \exp\left[-\frac{\alpha}{4} - \alpha y^2\right], & \alpha = 4, \end{cases}$$

where the compatibility condition $\rho_\ell(y,0) = \rho_s(0,y)$ is always satisfied. These functions are taken arbitrarily and do not come from any particular application. They have been chosen just to to illustrate the variation of the density with time and space. In both cases, we have taken the viscosity function as:

$$\mu(\rho) = \exp\left[\beta(\rho - 1)\right] \qquad \beta = 1,$$

that is an exponential. The pressure drop is $\Delta p = 1$. We look for a solution in the time interval $t \in [0,1]$ and we solve the problem using the numerical scheme of Equations (20) and (21). In Figure 3 we have plotted the function ρ_s and $\rho(x,y,1)$ for the case (B1), while in Figure 4, the same functions for the case (B2). In both cases, the evolution of the density in the fluid domain from time $t = 0$ to time $t = 1$ can be observed. Notice that the boundary conditions are satisfied. Moreover, we observe that, because of the symmetry of the problem w.r.t. the plane $y = 0$, we have $\rho_y(x,0,t) = 0$. In the central part of the channel (near $y = 0$), the density tends to assume the value it has at the inlet. This is due to the fact ρ satisfies a homogeneous transport equation in which the boundary data are "transported" along the characteristics.

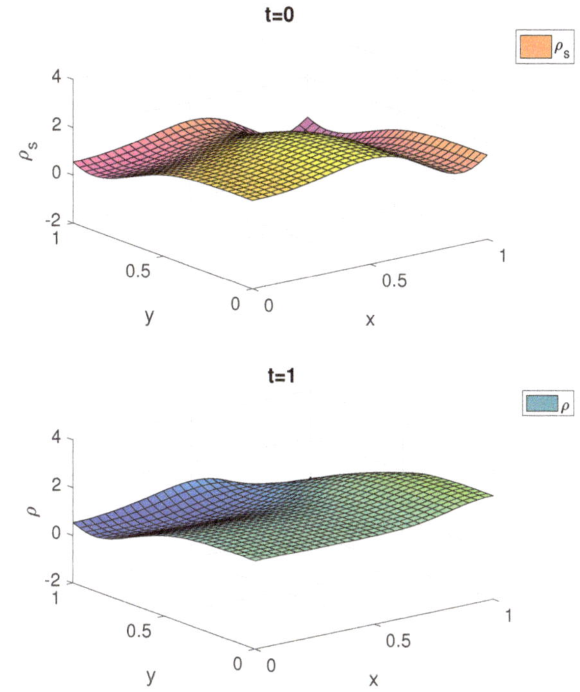

Figure 3. $\rho_s(x, y)$ and $\rho(x, y, 1)$ in the case (B1).

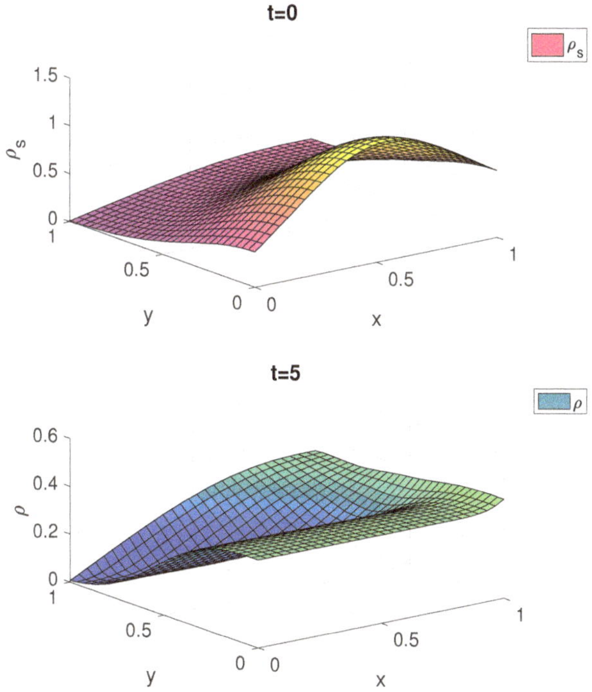

Figure 4. $\rho_s(x, y)$ and $\rho(x, y, 5)$ in the case (B2).

6. Analytical Solution

Recalling Remark 2, we know that $\rho = 1$ automatically satisfies the system (Equation (17)) and leads to the classical Navier-Stokes system. Suppose we take a "small perturbation" of this very special case, namely:

$$\rho = 1 + \delta\zeta\,(x, y, t)\,, \qquad \delta \ll 1, \tag{22}$$

Then, expanding the function μ around $\rho = 1$, we get (the prime denotes differentiation w.r.t. ρ):

$$\mu(\rho) = \mu(1) + \mu^{'}(1)(\rho - 1) + o(\rho - 1).$$

Since we are dealing with non-dimensional variables, we can take $\mu(1) = 1$, without loss of generality. Therefore, at the first order, we write:

$$\mu(\rho) = 1 + \delta\beta\zeta(x, y, t), \tag{23}$$

where $\beta = \mu^{'}(1)$ and where the $o(\delta)$ terms have been neglected. Let us now substitute Equations (22) and (23) into the system of Equation (15):

$$u = \frac{\Delta p}{\displaystyle\int_0^1 \frac{dx}{\displaystyle\int_0^1 dz \int_z^1 \frac{s}{1+\delta\beta\zeta} ds}} \cdot \frac{\displaystyle\int_y^1 \frac{s}{1+\delta\beta\zeta} ds}{\displaystyle\int_0^1 dz \int_z^1 \frac{s}{1+\delta\beta\zeta} ds}$$

Using Taylor expansion (because of the smallness of δ), we approximate:

$$\frac{1}{1+\delta\beta\zeta} \simeq 1 - \delta\beta\zeta, \tag{24}$$

so that:

$$u = \frac{\Delta p}{\displaystyle\int_0^1 \frac{dx}{\displaystyle\int_0^1 \int_z^1 s - \delta \int_0^1 \int_z^1 s\beta\zeta}} \cdot \frac{\displaystyle\int_y^1 s - \delta \int_y^1 s\beta\zeta}{\displaystyle\int_0^1 \int_z^1 s - \delta \int_0^1 \int_z^1 s\beta\zeta},$$

namely:

$$u = \frac{\Delta p \left(1 + 3\delta \displaystyle\int_0^1 \int_z^1 s\beta\zeta\right)}{1 + 3\delta \displaystyle\int_0^1 \int_0^1 \int_z^1 s\beta\zeta} \left[\frac{1-y^2}{2} - \delta \int_y^1 s\beta\zeta\right].$$

Using Equation (24) again, we finally find:

$$u = \Delta p \left(1 + 3\delta \int_0^1 \int_z^1 s\beta\zeta\right) \cdot \left(1 - 3\delta \int_0^1 \int_0^1 \int_z^1 s\beta\zeta\right) \left[\frac{1-y^2}{2} - \delta \int_y^1 s\beta\zeta\right].$$

Neglecting the $o(\delta)$ terms in the above, we find:

$$u = \Delta p \left(\frac{1-y^2}{2}\right) + \delta\Delta p \left\{ \left(\frac{1-y^2}{2}\right) \cdot \left[3 \int_0^1 \int_z^1 s\beta\zeta - 3 \int_0^1 \int_0^1 \int_z^1 s\beta\zeta\right] - \int_y^1 s\beta\zeta \right\}.$$

Finally, recalling that:

$$\int_y^1 \frac{1-z^2}{2} = \frac{(y-1)^2(y+2)}{6},$$

we make use of Equation (16) to find:

$$v = \delta \Delta p \left\{ \left(\frac{(y-1)^2(y+2)}{6} \right) \cdot \right.$$

$$\left. \left[3\int_0^1 \int_z^1 s\beta\zeta - 3\int_0^1 \int_0^1 \int_z^1 s\beta\zeta \right] - \int_y^1 \int_z^1 s\beta\zeta \right\}.$$

In conclusion, we have obtained the expansion of the velocity field at the first order in δ:

$$u = u_o + \delta u_1 \qquad v = \delta v_1, \tag{25}$$

with:

$$u_o = \Delta p \left(\frac{1-y^2}{2} \right),$$

and:

$$u_1 = \Delta p \left\{ \left(\frac{1-y^2}{2} \right) \left[3\int_0^1 \int_z^1 s\beta\zeta - 3\int_0^1 \int_0^1 \int_z^1 s\beta\zeta \right] - \int_y^1 s\beta\zeta \right\},$$

$$v_1 = \Delta p \left\{ \left(\frac{(y-1)^2(y+2)}{6} \right) \cdot \right.$$

$$\left. \left[3\int_0^1 \int_z^1 s\beta\zeta - 3\int_0^1 \int_0^1 \int_z^1 s\beta\zeta \right] - \int_y^1 \int_z^1 s\beta\zeta \right\}.$$

Next, substitute Equation (22) and Equation (25) into Equation (17). The zero order problem, i.e., the one in which all the terms containing δ are neglected, is the classical Navier-Stokes system with constant density. The first order problem becomes:

$$\begin{cases} \dfrac{\partial \zeta_m}{\partial t} + \dfrac{\Delta p}{2} \dfrac{\partial \zeta_m}{\partial x} = 0, \\ \zeta_m(x,0) = \zeta_s(x,0), \\ \zeta_m(0,t) = \zeta_\ell(0,t), \\ \dfrac{\partial \zeta}{\partial t} + \Delta p \left(\dfrac{1-y^2}{2} \right) \dfrac{\partial \zeta}{\partial x} = 0, \\ \zeta(x,0,t) = \zeta_m(x,t), \\ \zeta(0,y,t) = \zeta_\ell(y,t), \\ \zeta(x,y,0) = \zeta_s(x,y), \end{cases} \tag{26}$$

where $\zeta_\ell(y,t)$ and $\zeta_s(x,y)$ are obtained, expanding around δ the boundary data $\rho_\ell(y,t)$ and $\rho_s(x,y)$, respectively, namely:

$$\rho_\ell(y,t) = 1 + \delta\zeta_\ell(y,t) \qquad \rho_s(x,y) = 1 + \delta\zeta_s(x,y).$$

The compatibility of $\rho_\ell(y,0) = \rho_s(0,y)$ implies the condition $\zeta_\ell(y,0) = \zeta_s(0,y)$. From Equation $(26)_4$, we notice that y plays the role of a parameter for the equation for ζ. Equation $(26)_{1-3}$ can be solved by means of the method of characteristics. Indeed, extending $\zeta_s(x,y)$ for $x \leqslant 0$ in the following way:

$$\bar{\zeta}_s(x,y) = \begin{cases} \zeta_s(x,y), & x \geqslant 0, \\ \zeta_\ell\left(y, -\dfrac{2x}{\Delta p(1-y^2)}\right), & x \leqslant 0, \end{cases} \tag{27}$$

we find that the solution ζ_m is given by:

$$\zeta_m(x,t) = \bar{\zeta}_s\left(x - \frac{\Delta p}{2}t, 0\right). \tag{28}$$

Therefore we are left with the problem:

$$\begin{cases} \dfrac{\partial \zeta}{\partial t} + \Delta p\left(\dfrac{1-y^2}{2}\right)\dfrac{\partial \zeta}{\partial x} = 0, \\ \zeta(x,0,t) = \bar{\zeta}_s\left(x - \dfrac{\Delta p}{2}t, 0\right), \\ \zeta(0,y,t) = \zeta_\ell(y,t), \\ \zeta(x,y,0) = \zeta_s(x,y). \end{cases} \tag{29}$$

Still, exploiting the method of characteristics, we find that the solution of Equation (29) is given by:

$$\zeta(x,y,t) = \bar{\zeta}_s\left(x - \frac{\Delta p}{2}(1-y^2)t, y\right). \tag{30}$$

To verify the validity of the solution, we need to merely substitute Equations (28) and (30) into Equation (26).

Remark 3. *From Equation (27), we notice that the extension $\bar{\zeta}_s(x,y)$ of the function $\zeta_s(x,y)$ to the strip:*

$$\Omega = \Big\{(x,y): \quad x \in (-\infty, 1], \quad y \in [0,1]\Big\},$$

is "physically meaningful" only if $\zeta_\ell(y,t)$ is a bounded function of time t. Indeed, the function $\bar{\zeta}_s(x,y)$ is defined up to $y = 1$, implying that, for $x < 0$, we have to evaluate the limit $\lim_{t\to\infty} \zeta_\ell(1,t)$, which, of course, has to be bounded. On the other hand, if ζ_ℓ does not depend on time, i.e., the density at the inlet does not change with time, the function $\bar{\zeta}_s(x,y)$ becomes:

$$\bar{\zeta}_s(x,y) = \begin{cases} \zeta_s(x,y), & x \geqslant 0, \\ \zeta_\ell(y), & x \leqslant 0 \end{cases} \tag{31}$$

and the solution of the problem is still given by (28), (30).

To prove that the numerical scheme is consistent, we compare the solution obtained using the perturbation with the one obtained with the numerical procedure. Towards this aim, we consider the following pair of boundary conditions:

$$\begin{cases} \rho_s(x,y) = 1 + \delta\left[\sin(6x^2) + \dfrac{4}{5}\cos(4y^2)\right], \\ \rho_l(y,t) = 1 + \delta\left[\cos(4y^2)\right]. \end{cases} \tag{32}$$

We solve the problem using the perturbation of Equation (30) and with the numerical scheme of Equation (21), getting two solutions, $\rho_p(x,y,t)$ and $\rho_u(x,y,t)$, where ρ_p is the analytical solution using the perturbation and ρ_u is the solution obtained with the numerical scheme. We plot the difference of $\theta = \rho_p - \rho_u$ for different values of δ. In other words, we show that the perturbation is advantageous when the boundary data are small perturbations of the homogeneous case. Notice that, when $\delta = 4/5$, the numerical solution is exactly the one plotted in Figure 3. In Figures 5 and 6, the function θ (absolute error) is plotted for decreasing values of δ. As one can notice, the error decreases for decreasing δ, as expected.

The good agreement in the comparison suggests that the perturbation is not only useful as an approximation for small density variation around the constant value (homogeneous system), but it also provides some faith that the numerical method is reliable and can be used to infer the density distribution with general boundary data.

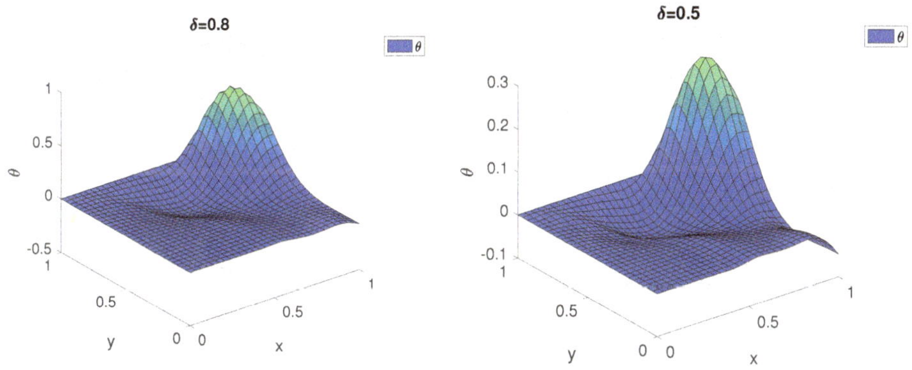

(a) θ with $\delta = 0.8$

(b) θ with $\delta = 0.5$

Figure 5. θ for decreasing δ.

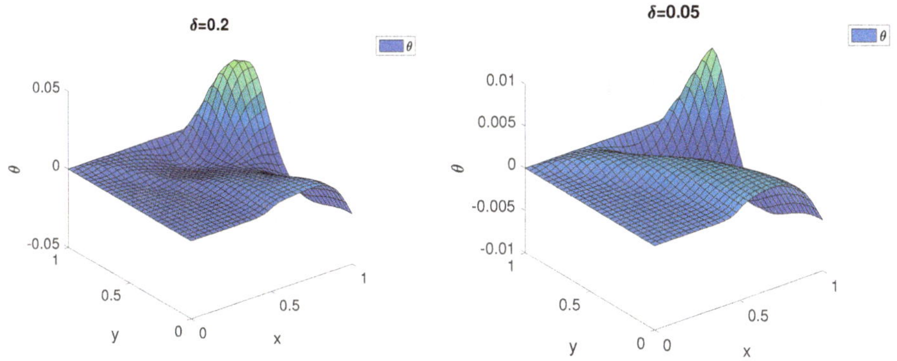

(a) θ with $\delta = 0.2$

(b) θ with $\delta = 0.05$

Figure 6. θ for decreasing δ.

7. Concluding Remarks

We investigated the thin film flow of an inhomogeneous fluid in which its viscosity was a function of the density. The constitutive equation for the class of fluids considered can be derived by assuming that the stored energy that depends on the gradient of the density [9]. We considered a channel flow driven by a given pressure gradient, and we assumed that the aspect ratio between the width and the length of the channel was such that we can use the lubrication approximation. We then focused on the leading order problem. The mathematical problem reduced to a nonlinear first order PDE for the density, where the coefficients are integral operators. The problem was solved numerically and plots that describe the evolution of the density in the fluid domain were displayed.

Author Contributions: Writing—original draft, L.F., A.F., F.R. and K.R.

Funding: This research received no external funding.

Conflicts of Interest: The authors declare no conflict of interest.

References

1. Truesdell, C.; Noll, W. *The Non-Linear Field Theories of Mechanics*; Springer-Verlag: Berlin, Germany, 1965.
2. Anand, M.; Rajagopal, K. A note on the flows of inhomogeneous fluids with shear-dependent viscosities. *Arch. Mech.* **2005**, *57*, 417–428.
3. Rajagopal, K. *Multiple Natural Configuration in Continuum Mechanics*; Technical Report; University of Pittsburgh: Pittsburgh, PA, USA, 1995.
4. Rajagopal, K.; Srinivasa, A. On the thermomechanics of shape memory wires. *Zeitschrift fur angewandte Mathematik und Physik ZAMP* **1999**, *50*, 459–496.
5. Rajagopal, K.; Srinivasa, A. A thermodynamic frame work for rate type fluid models. *J. Non-Newton. Fluid Mech.* **2000**, *88*, 207–227. [CrossRef]
6. Rajagopal, K.; Srinivasa, A. Modeling anisotropic fluids within the frame- work of bodies with multiple natural configurations. *J. Non-Newt. Fluid Mech.* **2001**, *99*, 109–124. [CrossRef]
7. Rao, I.; Rajagopal, K. A study of strain-induced crystallization of polymers. *Int. J. Solids Struct.* **2001**, *38*, 1149–1167. [CrossRef]
8. Rao, I.; Rajagopal, K. A thermodynamic framework for the study of crystallization in polymers. *Zeitschrift fur angewandte Mathematik und Physik ZAMP* **2002**, *53*, 365–406. [CrossRef]
9. Málek, J.; Rajagopal, K. On the modeling of inhomogeneous incompressible fluid-like bodies. *Mech. Mater.* **2006**, *38*, 233–242. [CrossRef]
10. Massoudi M.; Vaidya A. Unsteady flows of inhomogeneous incompressible fluids. *Int. J. Non-Linear. Mech.* **2011**, *46*, 738–741. [CrossRef]

fluids

MDPI

Article

On the Interaction of Side-By-Side Circular Cylinders in Viscoplastic Fluids

Naser Hamedi * and **Lars-Göran Westerberg**

Department of Engineering Sciences and Mathematics, Division of Fluid and Experimental Mechanics, Luleå University of Technology, SE-971 87 Luleå, Sweden; Lars-Goran.Westerberg@ltu.se

* Correspondence: naser.hamedi@ltu.se

Received: 12 April 2019; Accepted: 17 May 2019; Published: 21 May 2019

Abstract: In this paper, the static interaction of a train of three cylinders in a Bingham fluid is studied numerically using Computational Fluid Dynamics. The variation of drag forces for the cylinders in several configurations is investigated. Positions of the particles in relation to the reference particle are recognized by the separation distance between the cylinders. A steady state field is considered, with Bingham numbers between 5 and 150. Several separation distances (d) were considered, such that $2.0D \leq d \leq 6.0D$ where D is the cylinder diameter. The Reynolds number was chosen in the range of $5 \leq Re \leq 40$. In particular, the effect of the separation distance, Reynolds number and Bingham number on the shape and size of the unyielded regions was investigated. The functional dependence of this region and the drag coefficient is explored. The present results reveal the significant influence of the gap between the cylinders on the drag force and the shape of the unyielded regions surrounding the cylinders. It was found that there are several configurations in which the drag forces over the first and the third cylinders are almost equal depending on variation of the *Bi*, *Re* and the separation distance.

Keywords: particle interaction; viscoplastic fluid; Bingham fluid; computational fluid dynamics

---

1. Introduction

The flow around bluff bodies have been a matter of interest for researchers for many years. The study of non-Newtonian fluids past particulate objects are rapidly increasing. Such flow scenarios can be found in nature, and not the least in many different engineering applications. In nature, the flow of crude oil passed rocks is one example. In the industry, a few applications are flow in porous media such as during composite material manufacturing, flow of food materials through shells, flow in tube heat exchanger, and lubricant flow in machine elements. In the rheological behavior of materials like concrete, tomato paste and many dairy products, a so-called yield stress at low shear rates is often observed. This phenomenon has been studied by a number of researchers; see Refs. [1–5] for some central contributions, and there is an ongoing discussion whether a *true* yield stress exists. Regarding the yield stress, two main hypotheses have been proposed "true" yield stress and "apparent" yield stress. The first hypothesis anticipates that under a certain threshold shear stress value, the fluid exhibits a solid-like behavior while exceeding this threshold, the fluid starts to flow. According to the second hypothesis, all solid and fluid materials can creep and ultimately flow, provided the time scale is long enough [6,7]. In other words, for the case of a vanishing shear rate, the fluid can still flow and with increased shear rate, the apparent viscosity quickly and asymptotically decreases. The Bingham [8] and Herschel–Bulkley [9] rheology models are both based on hypothesis one, i.e., a true yield stress. In the Bingham model, the shear stress τ is assumed to be linearly proportional to the shear rate $\dot{\gamma}$, after it exceeds a certain Bingham yield stress (τ_0) such that

$$\begin{cases} \dot{\gamma} = 0; & \tau \leq \tau_0 \\ \tau = \tau_0 + \mu\dot{\gamma}; & \tau > \tau_0 \end{cases} \tag{1}$$

where μ is the Newtonian viscosity. In order to consider the non-linear behavior of the fluid in the yielded part, the Herschel–Bulkley model is used according to:

$$\begin{cases} \dot{\gamma} = 0; & \tau \leq \tau_0 \\ \tau = \tau_0 + K\dot{\gamma}^n; & \tau > \tau_0 \end{cases} \tag{2}$$

Here K is the consistency index and n the Power-law index. It should be noted that in both the Bingham and Herschel–Bulkley models, the yield stress is defined by the von Mises criteria which states that the yielding of a material begins when the second deviatoric stress invariant tensor reaches a critical value. According to Equations (1) and (2), the Bingham and Herschel–Bulkley rheology models are discontinuous, as there is no transition from unyielded to yielded built into the models. Papanastasiou [10] proposed a modified Bingham model in which the viscosity function smoothly changes for the whole domain such that:

$$\tau = \left(\mu + \tau_0 \frac{1 - \exp\left(-m\dot{\gamma}\right)}{\dot{\gamma}}\right)\dot{\gamma} \tag{3}$$

where m is a sufficiently large number to guarantee large apparent viscosity (η) at vanishing strain rate. This model is an example of a so-called regularize model, which is the most widely used approach to avoid the discontinuity. Some related works in this area are found in Refs. [11–14].

There are numerous applications in nature and industries such as particulate foods in non-Newtonian liquids or crude oil flow with rocks. The first work on the study of the flow past a Newtonian fluid is reported by Stokes (1851), where Stokes' law [15] was presented to calculate the drag force on a spherical object. In fact, the bulk of the studies in this research field is concerned with the Newtonian flow past spheres; see, e.g., [16–19]. The first study of creeping flow of a Bingham fluid past an object is by Beris et al. [20], who presented a Finite Element (FEM)/Newtonian method to solve the flow field including the yield surfaces in the creeping flow over a sphere. They found a non-dimension value for the critical yield stress value for which below this value the material acts as a solid in the domain. In addition, a detailed flow pattern around the sphere as well as the results for the drag coefficient was presented.

The earliest result for the solution of viscoplastic fluids around a 2D cylinder has been reported by Adachi and Yoshioka [21]. Using the slip-line analysis and the minimum- and maximum stress principles, they calculated upper and lower bounds for the drag coefficient and compared these values with each other. Moreover, they calculated the fluid region and the rigid region around the cylinder—i.e., the corresponding yield surface. In another work by Mitsoulis [22], the creeping flow of a Bingham fluid past a circular cylinder and the wall effects were investigated by FEM simulations. The blockage ratio varied between 2–50 in their study. Based on numerical prediction, they determined the drag coefficient as a function of the Bingham number for different gap/cylinder ratios. In addition, they examined un-sheared regions around the cylinder for various Bingham numbers. Different studied have been performed for the flow of Newtonian fluids past dual cylinders. Sheldon and Green [23] studied numerical simulation of the Newtonian flow around two-dimensional tandem circular cylinders perpendicular to the flow direction for the case of $1 \leq Re \leq 20$ and surface-to-surface separation distance to cylinder diameter ratio between 0.1 and 30. They investigated the effect of the gap distance between the cylinders on the lift and drag coefficient. Significant change of the lift and drag coefficients was found for the range of the Reynolds numbers studies. Due to the reduced viscous forces at small gap distances, they found a remarkable reduced drag force comparing to the single cylinder. For the lift coefficient variation with the gap distance, they found a sigmoidal pattern.

On the topic of viscoplastic fluid flow past dual cylinders, Jossic and Magnin [24] studied the interaction of two parallel cylinders, both numerically and experimentally, where $1 \leq Bi \leq 40$ and compared the results with a singular cylinder. Using the correlation between the drag coefficient and the Bingham number, they estimated a distance where there is no interaction between the two cylinders. They also investigated the influence of the slip condition on the drag forces. In the recent research, Koblitz et al. [25] presented the results of DNS simulations of interacting circular cylinders in three different cases whereas a Bingham fluid was changed between 0–2000 and the separation distance was kept constant at 1% of the cylinder radius in which the flow type was so called squeeze flow. They showed that the asymptotic lubrication solution is used in presence of the macroscopic flow assumption.

In this paper, we describe numerical simulations of the viscoplastic flow around three two-dimensional side-by-side circular cylinders with gap spacing $3.0D \leq d \leq 5.0D$, D being the cylinder diameter, and a Bingham number $10 \leq Bi \leq 150$. The Reynolds number is $5 \leq Re \leq 40$. In Section 2, we describe the numerical method and computational domain used for the simulations. Sections 4 and 5 discusses the results and Section 6 is comprised of a summary and conclusions. Although there are some studies made on the experimental or numerical approach of Newtonian fluid past three stationary spheres in steady or turbulent regions [26,27], to the best of the authors' knowledge, no study has been performed on the investigation of the viscoplastic fluid over three or more fixed stationary objects so far. Three cylinder objects enable us to further investigate the interaction between the objects and compare the outcomes with the single cylinder ones. It is important and of particular interest to build an understanding of how a train of cylinders affects the flow. The evolution and distribution of the yielded and unyielded regions are important for an understanding of the flow evolution in such domains.

2. Governing Equations and Numerical Models

The description below is based on the assumptions that the flow is steady, laminar and without secondary motions, which is indeed the case considering the very low Reynolds number for the flow in the domain. Non-dimensionalising the governing equations using the diameter of the cylinder D as characteristic length scale and the inlet velocity u as characteristic velocity we get:

$$\nabla \cdot \mathbf{u} = 0 \tag{4}$$

$$(\mathbf{u} \cdot \nabla)\mathbf{u} = -\nabla p + \frac{1}{Re} \nabla \cdot \tau \tag{5}$$

As previously mentioned, in this work, the shear thinning fluid is modelled using a so-called regularized Bingham Papanastasiou viscosity model expression [10]. The parameter m is considered sufficiently large to guarantee large apparent viscosity (η) at vanishing strain rate. Figure 1 shows the shear stress curve versus shear rate for different values of m. For the current study, $m = 1000$ was considered in all simulations, which does not show significant results compared to the ideal Bingham model.

The shear rate ($\dot{\gamma}$), is calculated using the second invariant of the rate of strain tensor S_{ij} [28]:

$$I_2 = S_{ij} S_{ij} \tag{6}$$

$$\dot{\gamma} = \sqrt{\frac{I_2}{2}} \tag{7}$$

To distinguish the yielded region from the non-yielded one considering the yield stress definition, one can use the criterion in Equation 8. The material consequently flows when the magnitude of the stress tensor exceeds the yield stress value in accordance with:

$$\begin{cases} \text{Non}-\text{yielded}; & \tau \leq \tau_0 \\ \text{yielded}; & \tau > \tau_0 \end{cases} \tag{8}$$

In this paper, the OpenFOAM open source Computational Fluid Dynamics (CFD) software [29] was employed to model the viscoplastic fluid flow past the three tandem cylinders. The finite volume solver, which uses a non-staggered grid, calculates the mass and momentum equations in the discretized form, which guarantees the conservation of fluxes through the control volume. Diffusion and pressure gradient terms were evaluated by a second order interpolation. The PISO scheme was employed for the pressure-velocity coupling.

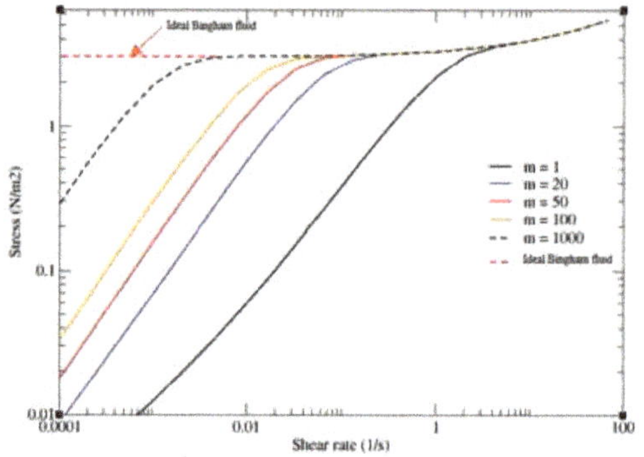

Figure 1. Shear stress versus shear rate for different values of m according to the Papanastasiou Bingham model.

3. Problem Setup

Three equally sized cylinders with diameter D are taken into account at several relative positions in a rectangular domain and Cartesian coordinates. Figure 2 shows the domain and cylinders positions. The positions are solely defined using separation distance d. Simulations are carried out for $d = 2.0D$, $3.0D$, $4.0D$, $5.0D$ and $6.0D$, respectively. The dimensions of the 2D computational domain are 15×40. The results of the grid independency of the computational domain are shown in Figure 2. The grid resolution of $h = D/20$ is used for the computation; see Figure 3.

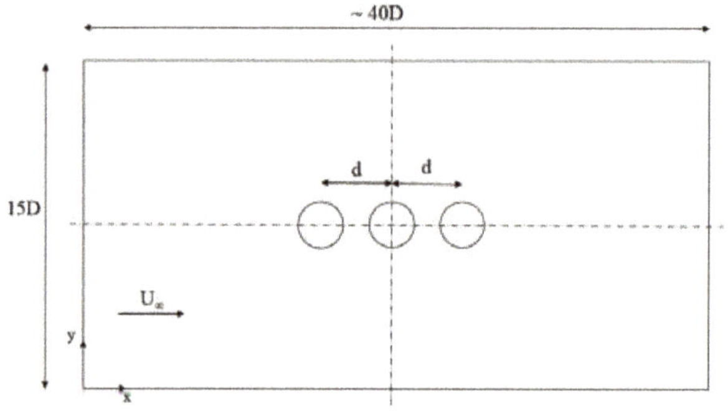

Figure 2. Problem setup and arrangement of the three cylinders.

A uniform velocity profile is applied at the inlet, a Dirichlet condition at the outlet, slip boundary conditions at the lateral boundaries, and a no-slip boundary condition on the cylinders surface.

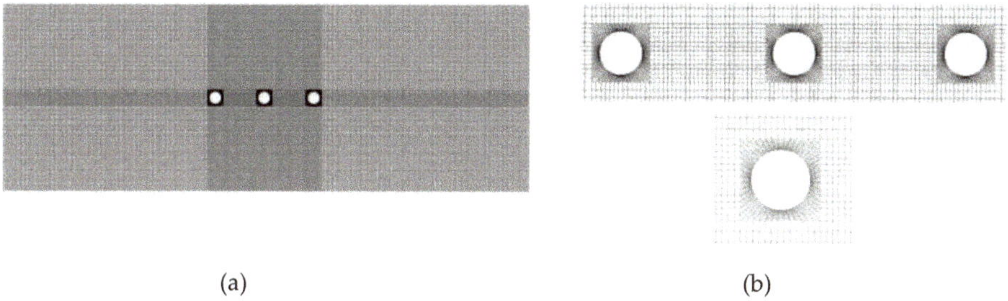

Figure 3. (**a**) Discretized model for the three cylinders with $d = 4.0D$, (**b**) a closer view of the grids for single cylinder and triple cylinders.

A grid sensitivity study of all computational domains used in this study was obtained. Various simulations of the Newtonian flow past the single cylinder and triple cylinders was performed and the drag coefficient was selected as a criterion to find a grid independent computational domain. It was found that there were only minor differences in the drag coefficient for the three finest resolution. As previously mentioned, the grid resolution of $D/20$ was used for all computational domains.

4. Results

4.1. Single Cylinder

For validating the non-Newtonian solver, the drag coefficient results of the single cylinder in viscoplastic media were compared with the results of the former researchers. Table 1 shows the results for viscoplastic fluid flow (Bingham number 10) for three Reynolds numbers—Re = 5, 20 and 40. The results agree well with previous studies.

Table 1. Drag variation for three Reynolds number at Bingham number 10 in the current study compared to the previous works.

Reference	$Re = 10$	$Re = 20$	$Re = 40$
Present work	33.30	17.11	8.85
Mossaz et al. [30]	34.79	17.21	9.43
Nirmalkar and Chhabra [31]	33.11	17.00	8.97
Takur et al. [32]	34.83	17.20	9.37

4.1.1. Drag Variation

Figure 4 shows the drag variation over a single cylinder where the Bingham number and Reynolds number are varied such that $5 \leq Re \leq 40$ and $5 \leq Bi \leq 150$. According to Figure 4—and as expected—the drag coefficient is increased with an increased Bingham number. However, the variation of the drag coefficient is more pronounced for the lower Reynolds numbers. In other words, for lower Reynolds numbers, the effect of the Bingham number is dominating compared to the range of higher Reynolds numbers.

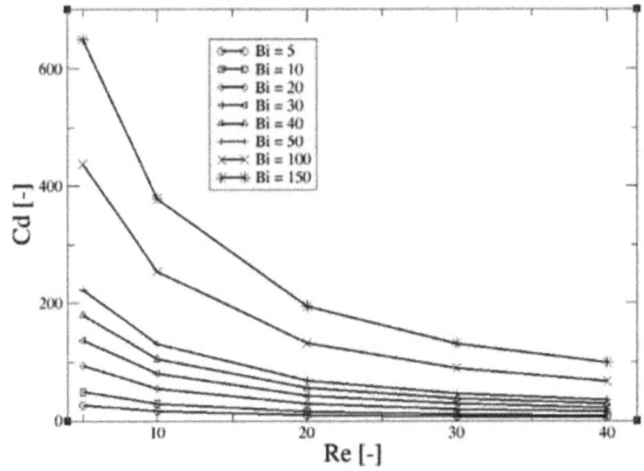

Figure 4. Variation of the drag coefficient for a single cylinder where the Reynolds number (*Re*) and Bingham number (*Bi*) are varied.

4.1.2. Non-Yielded Region Variation

Figure 5 shows the unyielded region (shaded) observed in the viscoplastic fluid past the single sphere. The unyielded region appeared around the cylinder is varied by changing the Reynolds and the Bingham numbers. Island and polar caps regions can be observed as well. Although increasing the Bingham number, increases the unyielded regions around the cylinder, the size and the shape of these regions are different and does not follow a similar pattern in different Reynolds numbers. When $Re > 10$, two small unyielded regions appear at top and bottom of the cylinder, which were not seen in the lower Reynolds number. In addition, while the Bingham number increases, the unyielded region gets smaller at flow direction, while at the vertical direction, it remains almost constant.

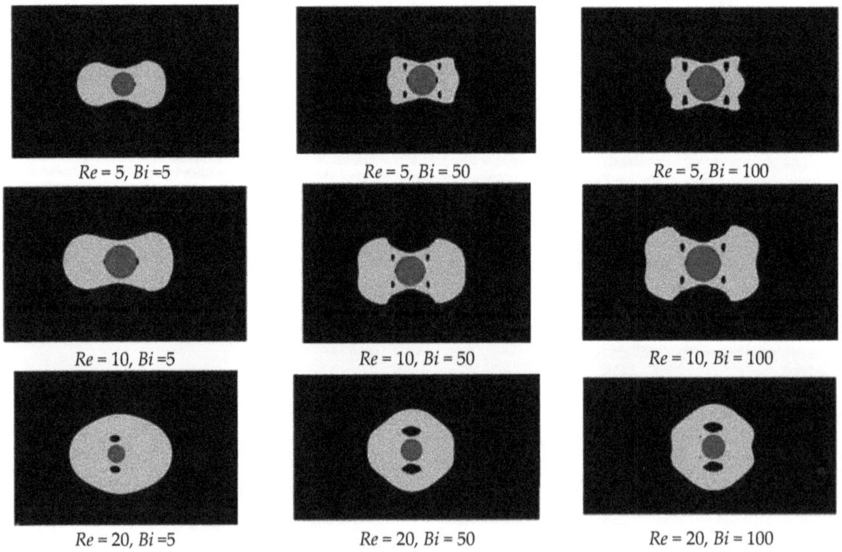

Figure 5. *Cont.*

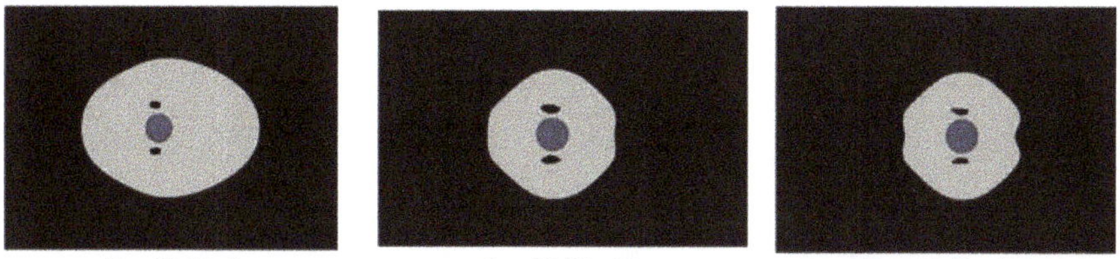

Re = 30, *Bi* =5 *Re* = 30, *Bi* = 50 *Re* = 30, *Bi* = 100

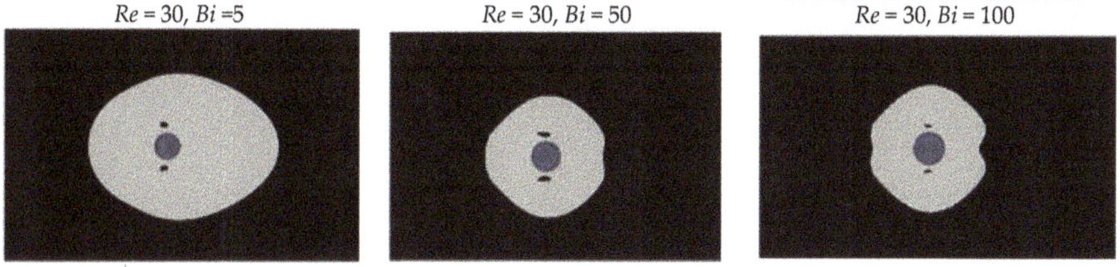

Re = 40, *Bi* =5 *Re* = 40, *Bi* = 50 *Re* = 40, *Bi* = 100

Figure 5. Simulation results to show the progressive growth of the unyielded region (shaded).

It should be noted that the small zig-zag lines divide the unyielded and yielded regions, relating to the use of higher values of m coefficient (m = 1000 s) in Equation (3), which generate non-smooth lines in the solution. These lines has been observed in previous research [33].

4.2. Three Cylinders

4.2.1. Drag Coefficients

Figure 6 shows the variation of the drag coefficient for the three cylinders, where the separation distances are changed to $2.0D \leq d \leq 6.0D$ and $5 \leq Bi \leq 100$. Like the drag alteration for the single cylinder, the drag coefficient is increased as the Bingham number increased. The cylinder in the middle position is also significantly affected by these parameters. However, the effect is varied by the variation of both Reynolds number and the Bingham number. This will be discussed later in this study.

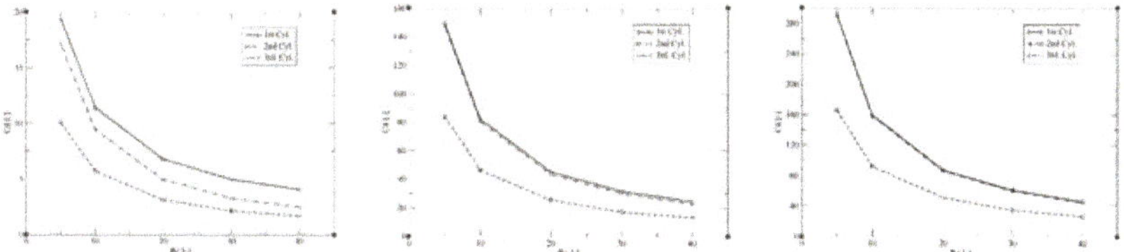

d = 2.0D, Bi = 5 d = 2.0D, Bi = 50 d = 2.0D, Bi = 100

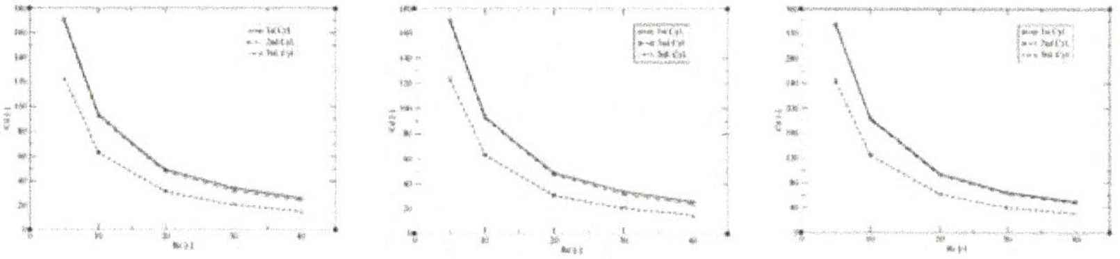

d = 3.0D, Bi = 5 d = 3.0D, Bi = 50 d = 3.0D, Bi = 100

Figure 6. *Cont.*

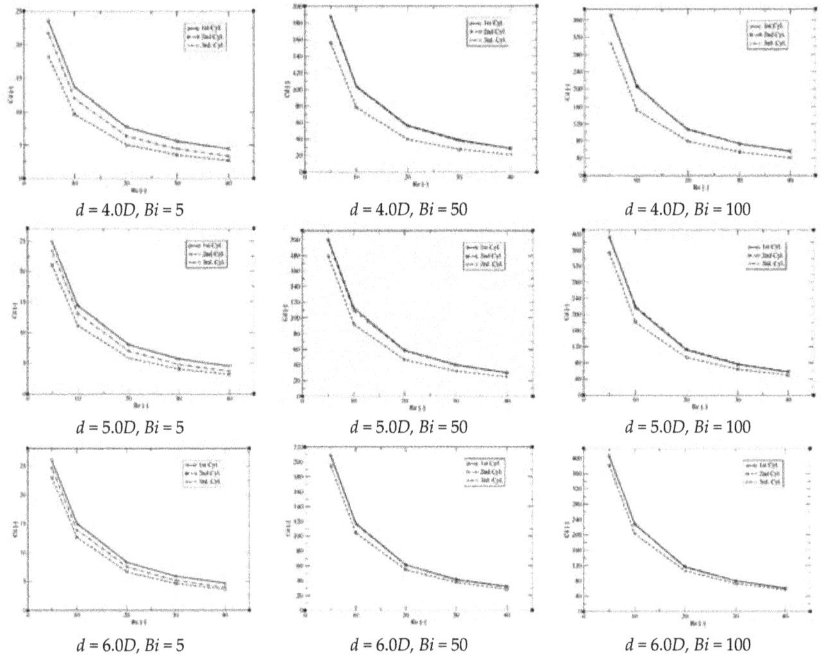

Figure 6. Drag variation over three cylinders for various separation distances and Bingham numbers.

4.2.2. Unyielded Region

Figures 7–10 show the growth of the unyielded region around three cylinders in tandem position for separation distances $d = 2.0D$, $3.0D$, $4.0D$ and $6.0D$, $Bi = 5$, 50 and 100 and $Re = 5$, 20 and 40. As shown in the figures, increasing the Bingham number or decreasing the Reynolds number, both enhances the yield regions around and between the cylinders.

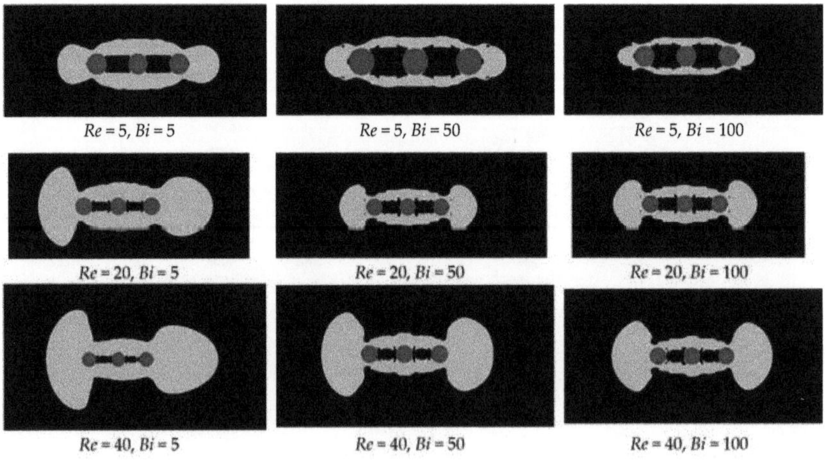

Figure 7. Progressive growth of the unyielded region (shaded) with the increasing Bingham number and decreasing Reynolds number where separation distance $d = 2.0D$.

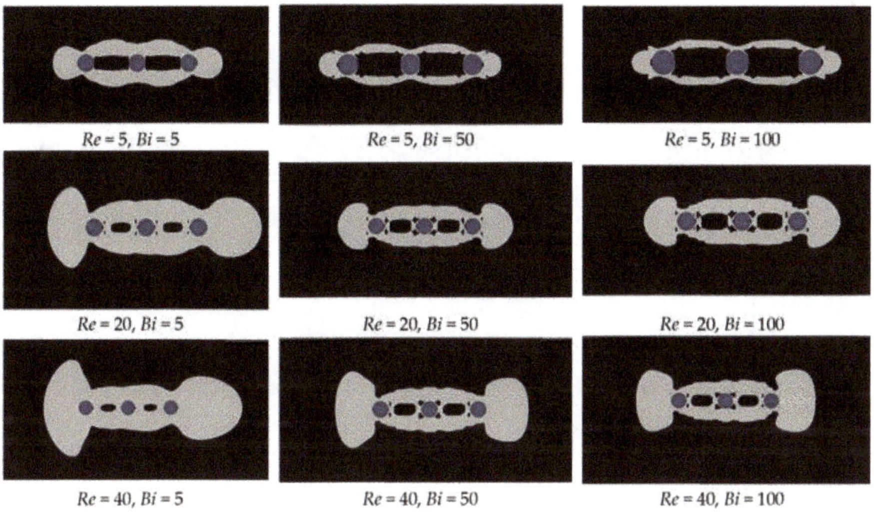

Figure 8. Progressive growth of the unyielded region (shaded) with the increasing Bingham number and decreasing Reynolds number where separation distance $d = 3.0D$.

From the figures, one can notice the separation distance influences on the breakage of the unyielded region between the cylinders. This phenomenon is observed for separation distance greater than $3.0D$. For example, for $d = 4.0D$, the continuous unyielded region, which was previously observed for all cases in $d = 2.0D$ and $d = 3.0D$, were broken and depending on the Bingham number and the Reynolds number, these region sizes are varied. The size of the unyielded region between the cylinders are more noticeable for cases with the lowest Reynolds number and the highest Bingham number where $d \geq 4.0D$. In these cases, the size of the un-sheared zone is significantly larger than the other cases (e.g., see cases with $Re = 5$, $Bi = 150$ in Figures 9 and 10).

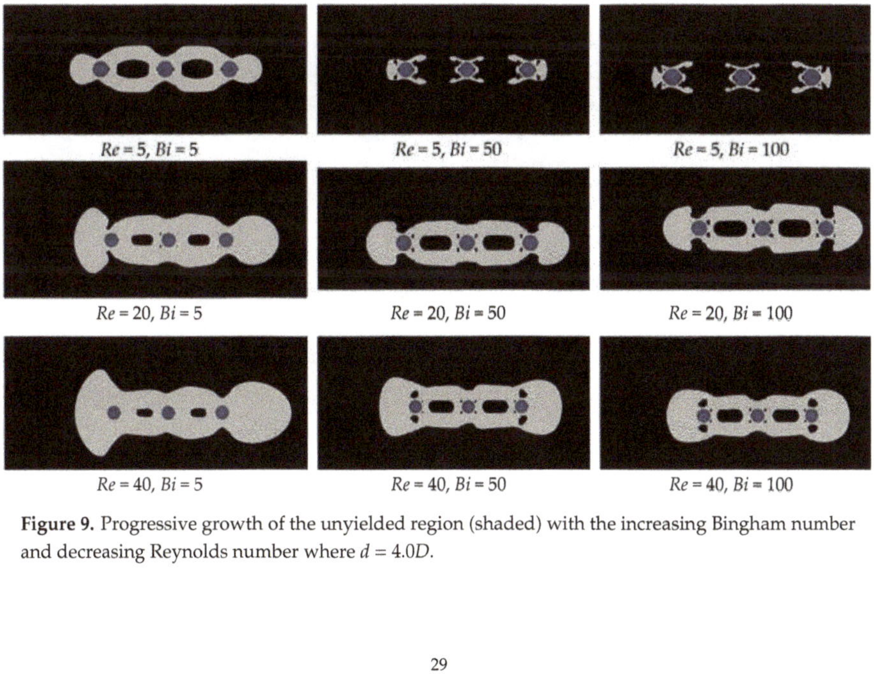

Figure 9. Progressive growth of the unyielded region (shaded) with the increasing Bingham number and decreasing Reynolds number where $d = 4.0D$.

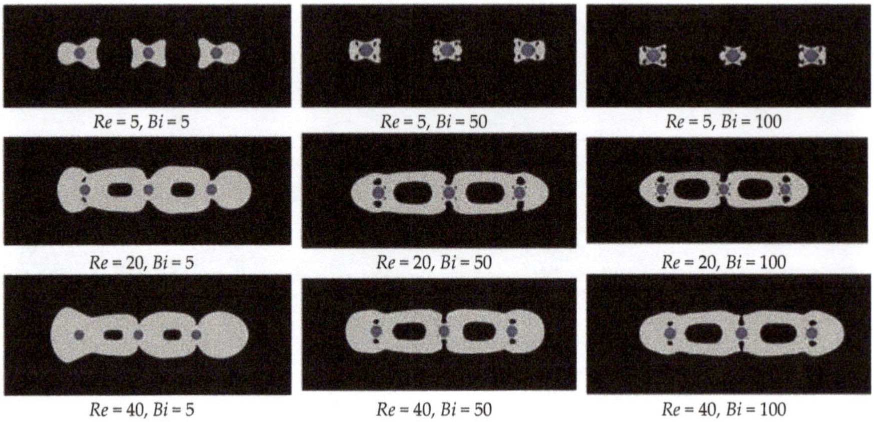

Figure 10. Progressive growth of the unyielded region (shaded) with the increasing Bingham number and decreasing Reynolds number where $d = 6.0D$.

4.2.3. Velocity Distribution

Figure 11 shows the velocity distribution of the three cylinders in separation distance $d = 4.0$, three Reynolds number $Re = 5$, 20 and 40, and three Bingham number $Bi = 5$, $Bi = 50$ and $Bi = 100$. As shown in the plots, the distribution of the velocity is different in various Reynolds numbers and Bingham numbers. As is shown in the figure, by increasing the Bingham number, the fluid shows a more symmetrical shape, which is related to the developing unyielded areas and the plug flow around the cylinders.

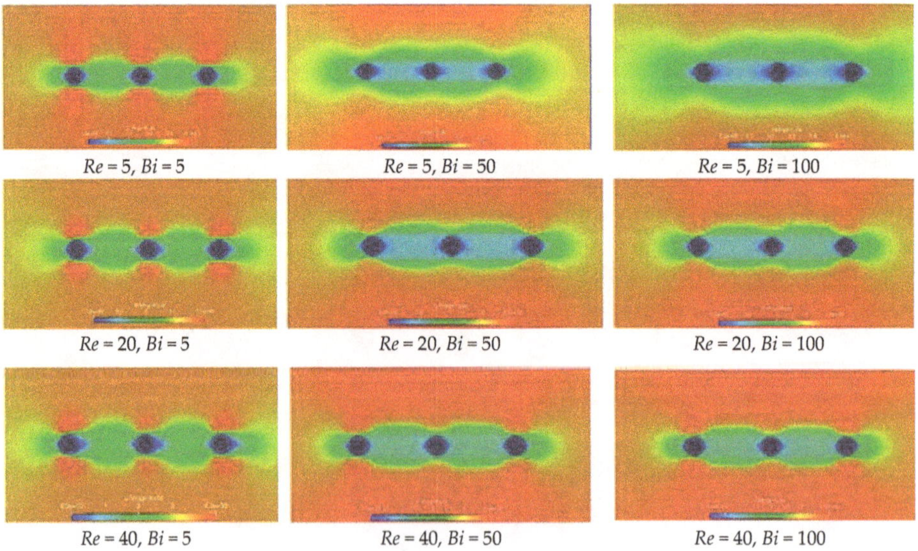

Figure 11. Velocity profile around three cylinder in three Bingham numbers (Bi = 5, 50 and 100) and three Reynolds number (Re = 5, 20 and 40) where $d = 4.0D$.

5. Discussion

The results in the previous section comprise of the outcome of viscoplastic fluid flow past single- and triple cylinders, respectively. The drag coefficient of the single cylinder was compared with previous results and the results were found to agree well. Concerning the drag coefficient for various Bingham numbers, it was shown that its variation is more pronounced for lower Reynolds numbers, which is due to the dominant viscous forces in the actual range of Reynolds numbers. Likewise, the unyielded region was reported for various Reynolds and Bingham numbers. Island and polar caps regions, which were reported in previous research [12,21,33,34], were observed in this study as well. The structure of the unyielded regions are very similar to the earlier findings cited above.

Regarding the viscoplastic fluid flow past triple cylinders (as mentioned in the Introduction), to the best of the authors' knowledge, this is the first study that examines the interaction between the cylinders and the unyielded regions around the objects in the presence of the viscoplastic fluid. The drag coefficient of three cylinders was depicted in various configurations. The higher the Reynolds number, the more difference between the drag coefficients of the cylinder in the middle position and the first/third position was observed. It should be noted that this pattern is repeated for all separation distances, but the rate of this difference is a matter of importance. In other words, for the smaller separation distances, this influence is more than the same effect for the bigger separation distances. It is worthwhile to note that due to the effect of the viscoplastic properties of the carrying fluid, by decreasing the Bingham number, the difference between the drag coefficient over the second and the first/third cylinders are decreased.

It is also interesting to observe how the third cylinder is affected by the flow comparing to the first cylinder. According to the graphs in Figure 7, the drag force over the cylinder in the middle position is always between the two other cylinders. However, the difference between the drag coefficients for the first and second cylinder is influenced by the separation distance, Bingham number, and the Reynolds number. As the Bingham number is increased, this difference is decreased. In contrast, by increasing the Reynolds number, this difference is increased due to the existing dominant viscous forces. In fact, except in cases where separation distance $d = 2.0D$, for other separation distances, there are some cases that the drag force over the first and the third cylinder is almost equal. For example, for $d = 3.0D$, cases where $Bi = 100$ and $Bi = 150$, the drag force for the cases with $Re = 5$ are equal. By increasing the separation distance, these cases can be found in lower Bigham numbers as well. For example, for $d = 6.0D$ and $Bi = 30$, the first and third cylinders encounter a similar drag force.

Further, it was found that the maximum difference between the drag forces of the first and third cylinder is about 38% where $d = 3.0D$, $Re = 5$ and $Bi = 150$. In this case, the difference between the drag force of the second and first cylinder was maximum too, whereas approximately a 59% difference was observed between these two drag forces.

As aforementioned, and likewise the cases for single cylinder, the Reynolds number has influence on the size and shape of the shaded region. Regarding the plots in Figures 7–10, one can mention that an increased Reynolds number has more effect on decreasing the size of the unyielded region around the cylinder rather than the size of the similar region in between the cylinders.

Increasing the Bingham number has another effect on the size and shape of the unyielded region in between the objects. It is noteworthy that in all cases, the height of the unyielded region is developed by increasing the Bingham number. In addition, some small islands are observed above the cylinders or around the cylinders when separation distance is increased to $d \geq 3.0D$ and Bingham number above 50.

6. Conclusions

The variation of drag coefficient over three fixed cylindrical particles for various configurations in tandem position was investigated. Simulations were performed for $10 \leq Bi \leq 150$, $5 \leq Re \leq 40$ and separation distances $2.0D \leq d \leq 6.0D$. The results had good agreement with previous studies for a single particle.

For a single particle, it was shown that due to the dominant viscous forces in the lower Reynolds number, the effect of the Bingham number is more pronounced. The unyielded region shape and size were investigated, and several patterns were found dependent on Reynolds number and the Bingham number. It was shown that the unyielded region gets smaller in the flow direction by increasing the Bingham number, while it remains almost constant at the vertical direction.

For multiple cylinders, it was found that the separation distance has a significant effect on the drag variation on the cylinders. Several conditions (in terms of Bingham number, separation distance and Reynolds number) were found in which the first and the third cylinder are affected with the same drag force. Concerning the drag force applied to the cylinder in the middle position, higher difference between the drag coefficients of this cylinder and the other two cylinders was found for the higher Reynolds number.

Author Contributions: Conceptualization, N.H. and L.-G.W.; Methodology, N.H.; Software, N.H.; Validation, N.H.; Formal Analysis, N.H.; Investigation, L.-G.W.; Resources, L.-G.W.; Data Curation, N.H.; Writing–Original Draft Preparation, N.H.; Writing–Review & Editing, N.H. and L.-G.W.; Visualization, N.H.; Project Administration, L.-G.W.; Funding Acquisition, L.-G.W.

Funding: This research was funded by Swedish KEMPE Foundations, Grant Number SMK-1739.

Acknowledgments: This research was conducted using the resources of High Performance Computer Center North (HPC2N), Sweden.

Conflicts of Interest: The authors declare no conflict of interest.

References

1. Dzuc, N.Q.; Boger, D.V. Yield Stress Measurement for Concentrated Suspensions. *J. Rheol.* **1983**, *27*, 321–349.
2. Wang, W.; Meng, B.; De Kee, D.; Khismatullin, D. Numerical investigation of plate edge and slot size effects in low yield stress measurements with a slotted plate device. *Rheol. Acta* **2012**, *51*, 151–162. [CrossRef]
3. Borgia, A.; Spera, F.J. Error analysis for reducing noisy wide-gap concentric cylinder rheometric data for nonlinear fluids: Theory and applications. *J. Rheol.* **1990**, *34*, 117–136. [CrossRef]
4. Heirman, G.; Vandewalle, L.; Gemert, D.V. Integration approach of the Couette inverse problem of powder type self-compacting concrete in a wide-gap concentric cylinder rheometer. *J. Non-Newtonian Fluids Mech.* **2008**, *150*, 93–103. [CrossRef]
5. James, P.W.; Jones, T.E.R.; Hughes, J.P. The determination of apparent viscosity using a wide gap, double concentric cylinder. *J. Non-Newton. Fluid Mech.* **2004**, *124*, 33–41. [CrossRef]
6. Barnes, H.; Walters, K. The yield stress myth? *Rheol. Acta* **1985**, *24*, 323–326. [CrossRef]
7. Barnes, H.A. The yield stress—A review or 'παντα ρει'—Everything flows? *J. Non-Newton. Fluid Mech.* **1999**, *81*, 133–178. [CrossRef]
8. Bingham, E.C. *An Investigation of the Laws of Plastic Flow*; US Government Printing Office: Pueblo, CO, USA, 1917; Volume 13.
9. Herschel, W.; Bulkley, R. Konsistenzmessungen von Gummi-Benzollösungen. *Colloid Polym. Sci.* **1926**, *39*, 291–300. [CrossRef]
10. Papanastasiou, T.C. Flows of Materials with Yield. *J. Rheol.* **1987**, *31*, 385–404. [CrossRef]
11. KÉ, D.D.; Turcotte, G. Viscosity of biomaterials. *Chem. Eng. Commun.* **1980**, *6*, 273–282. [CrossRef]
12. Mitsouls, E.; Abdali, S. Flow simulation of Herschel-Bulkley fluids through extrusion dies. *Can. J. Chem. Eng.* **1993**, *71*, 147–160. [CrossRef]
13. Zhu, H.; De Kee, D. A numerical study for the cessation of Couette flow of non-Newtonian fluids with a yield stress. *J. Non-Newton. Fluid Mech.* **2007**, *143*, 64–70. [CrossRef]
14. Zhu, H.; Kim, Y.D.; De Kee, D. Non-Newtonian fluids with a yield stress. *J. Non-Newton. Fluid Mech.* **2005**, *129*, 177–181. [CrossRef]
15. Stokes, S.G.G. *Mathematical and Physical Papers*; Cambridge University Press: Cambridge, UK, 1901.
16. Johnson, T.; Patel, V. Flow past a sphere up to a Reynolds number of 300. *J. Fluid Mech.* **1999**, *378*, 19–70. [CrossRef]
17. Kim, I.; Pearlstein, A.J. Stability of the flow past a sphere. *J. Fluid Mech.* **1990**, *211*, 73–93. [CrossRef]

18. Lee, S. A numerical study of the unsteady wake behind a sphere in a uniform flow at moderate Reynolds numbers. *Comp. Fluids* **2000**, *29*, 639–667. [CrossRef]
19. Fornberg, B. Steady viscous flow past a sphere at high Reynolds numbers. *J. Fluid Mech.* **1988**, *190*, 471–489. [CrossRef]
20. Beris, A.; Tsamopoulos, J.; Armstrong, R.; Brown, R. Creeping motion of a sphere through a Bingham plastic. *J. Fluid Mech.* **1985**, *158*, 219–244. [CrossRef]
21. Adachi, K.; Yoshioka, N. On creeping flow of a visco-plastic fluid past a circular cylinder. *Chem. Eng. Sci.* **1973**, *28*, 215–226. [CrossRef]
22. Mitsoulis, E. On creeping drag flow of a viscoplastic fluid past a circular cylinder: wall effects. *Chem. Eng. Sci.* **2004**, *59*, 789–800. [CrossRef]
23. Vakil, A.; Green, S.I. Two-dimensional side-by-side circular cylinders at moderate Reynolds numbers. *Comp. Fluids* **2011**, *51*, 136–144. [CrossRef]
24. Jossic, L.; Magnin, A. Drag of an isolated cylinder and interactions between two cylinders in yield stress fluids. *J. Non-Newton. Fluid Mech.* **2009**, *164*, 9–16. [CrossRef]
25. Koblitz, A.R.; Lovett, S.; Nikiforakis, N. Viscoplastic squeeze flow between two identical infinite circular cylinders. *Phys. Rev. Fluids* **2018**, *3*, 023301. [CrossRef]
26. Ozgoren, M. Flow structures around an equilateral triangle arrangement of three spheres. *Int. J. Multiph. Flow* **2013**, *53*, 54–64. [CrossRef]
27. Wu, Y. Numerical simulation of flows past multiple cylinders using the hybrid local domain free discretization and immersed boundary method. *Ocean Eng.* **2017**, *141*, 477–492. [CrossRef]
28. Bird, R.B.; Hassage, O. *Dynamic of Polymeric Fluids*, 2nd ed.; John Wiley and Sons Publication: Hoboken, NJ, USA, 1987.
29. Weller, H.G.; Tabor, G.; Jasak, H.; Fureby, C. A tensorial approach to computational continuum mechanics using object-oriented techniques. *Comp. Phys.* **1998**, *12*, 620–631. [CrossRef]
30. Mossaz, S.; Jay, P.; Magnin, A. Criteria for the appearance of recirculating and non-stationary regimes behind a cylinder in a viscoplastic fluid. *J. Non-Newton. Fluid Mech.* **2010**, *165*, 1525–1535. [CrossRef]
31. Nirmalkar, N.; Chhabra, R. Momentum and heat transfer from a heated circular cylinder in Bingham plastic fluids. *Int. J. Heat Mass Transfer* **2014**, *70*, 564–577. [CrossRef]
32. Thakur, P.; Mittal, S.; Tiwari, N.; Chhabra, R. The motion of a rotating circular cylinder in a stream of Bingham plastic fluid. *J. Non-Newton. Fluid Mech.* **2016**, *235*, 29–46. [CrossRef]
33. Zisis, T.; Mitsoulis, E. Viscoplastic flow around a cylinder kept between parallel plates. *J. Non-Newton. Fluid Mech.* **2002**, *105*, 1–20. [CrossRef]
34. Chatzimina, M.; Georgiou, G.C.; Argyropaidas, I.; Mitsoulis, E.; Huilgol, R. Cessation of Couette and Poiseuille flows of a Bingham plastic and finite stopping times. *J. Non-Newton. Fluid Mech.* **2005**, *129*, 117–127. [CrossRef]

fluids

MDPI

Article

The Effect of Internal and External Heating on the Free Convective Flow of a Bingham Fluid in a Vertical Porous Channel

D. Andrew S. Rees [1,*] and Andrew P. Bassom [2]

1 Department of Mechanical Engineering, University of Bath, Bath BA2 7AY, UK
2 School of Natural Sciences, University of Tasmania, Hobart TAS 7001, Australia; andrew.bassom@utas.edu.au
* Correspondence: D.A.S.Rees@bath.ac.uk

Received: 14 April 2019; Accepted: 16 May 2019; Published: 24 May 2019

Abstract: We study the steady free convective flow of a Bingham fluid in a porous channel where heat is supplied by both differential heating of the sidewalls and by means of a uniform internal heat generation. The detailed temperature profile is governing by an external and an internal Darcy-Rayleigh number. The presence of the Bingham fluid is characterised by means of a body force threshold as given by the Rees-Bingham number. The resulting flow field may then exhibit between two and four yield surfaces depending on the balance of magnitudes of the three nondimensional parameters. Some indication is given of how the locations of the yield surfaces evolve with the relative strength of the Darcy-Rayleigh numbers and the Rees-Bingham number. Finally, parameter space is delimited into those regions within which the different types of flow and stagnation patterns arise.

Keywords: porous media; convection; Bingham fluid; yield stress; channel flow

1. Introduction

This short paper considers the steady flow which is induced when a vertical porous channel is heated both externally and internally, and when that porous medium is saturated by a Bingham fluid. These fluids form, perhaps, the simplest model of a yield stress fluid where the fluid shears when the applied stresses are greater than a threshold value, the yield stress, but acts like a solid when the applied stresses are too small. Within the context of a porous medium the natural yield threshold is expressed in terms of a pressure gradient and, in the context of convective flows, this includes the buoyancy force. Pascal's piecewise-linear law (Pascal [1,2]) is itself the simplest relationship, and in this model there is no flow (as opposed to zero shear) when the threshold has not been met.

One of the earliest models for the flow of a yield stress fluid in a porous medium is by Gheorghitza [3] who, like Pascal, cites a number of authors reporting the presence of a yield threshold in filtration flows although Gheorghitza does not name the fluid as a Bingham fluid. Gheorghitza names this phenomenon as an *initial gradient*. Wu and Pruess [4] adds the practical observation that heavy oils in reservoirs and water within clay soils also exhibit a threshold gradient. Bingham fluids also arise elsewhere and these include drilling mud, ceramic pastes, yoghurt, mayonnaise, sewage sludges and magma.

Other models for the flow of a Bingham fluid in a porous medium also exist. One of the most frequently used is the Buckingham-Reiner model (Buckingham [5] and Reiner [6]) which, strictly speaking, corresponds to the Hagen-Poiseuille flow of a Bingham fluid, but it may be applied to a porous medium by assuming a unidirectional flow within a medium consisting of identical pores. More complicated scenarios may be envisaged, and it is worth mentioning that Nash and Rees [7] performed an analytical study of the effect of having distributions of pore diameters. In general this was found to 'soften' the initial stages of flow post-threshold, and each distribution of pores has its own analogue of a Buckingham-Reiner law. The work of Malvault et al. [8] relaxes the assumptions of [7] that the pores have a uniform cross-section.

The present paper considers what might be regarded as a very straightforward steady free convection problem in a vertical porous channel. The channel is heated via a uniform heat generation mechanism, but the walls are also held at different temperatures from one another. The resulting temperature distribution is simple to determine, and should the porous medium be saturated by a Newtonian fluid, most of the resulting analysis centres around the analytical determination of a suitable reference temperature. When the porous medium is saturated by a Bingham fluid then the locations of yield surfaces need to be found, and allowance has to be made for when a yield surface attaches to a bounding surface. In more detail, we find that four different flow regimes arise where the first corresponds to complete stagnation, the second to having one stagnant region, and two more to having two stagnant regions. Much depends on the competing effects of the internal and external heating mechanisms and how these interact with the yield threshold. This work is part of a study of different aspects of porous channel and boundary layer flows involving Bingham fluids; see Rees and Bassom [9–11]. These earlier works consider the unsteady evolution of yield surfaces in such flows.

2. Governing Equations

In the present paper we shall adopt Pascal's model (Pascal [1,2] to describe how a Bingham fluid flows within a homogeneous and isotropic porous medium. This is a piecewise linear relationship between the fluid velocity and the applied pressure gradient and it may be expressed as follows,

$$\overline{w} = \begin{cases} -\dfrac{K}{\mu}\left[1 - \dfrac{G}{|\overline{p}_{\overline{z}}|}\right]\overline{p}_{\overline{z}} & \text{when } |\overline{p}_{\overline{z}}| > G, \\ 0 & \text{otherwise,} \end{cases} \tag{1}$$

where the quantity, G, denotes the threshold pressure gradient above which the fluid yields and flows but below which fluid is stagnant. Here K is the permeability, μ is the plastic viscosity, $\overline{p}$ is the pressure, $\overline{z}$ the vertical coordinate and $\overline{w}$ the vertical velocity. Here we assume that the plastic viscosity is a constant which corresponds to a piecewise-linear stress/strain relationship, and therefore Equation (1) models a Bingham fluid, rather than a Herschel-Bulkley fluid. More complicated expressions than Equation (1) exist, such as those derived in Nash and Rees [7] for different tube bundle distributions. The simplest expression is the Buckingham-Reiner model [5,6] which softens the discontinuous gradient in Pascal's law. Other expressions derived in [7] provide for slower approaches to the eventual linear relationship between the flow and the pressure gradient and depend on the probability distribution associated with the pore diameters. In the present paper we restrict ourselves solely to Pascal's model; the adoption of anything more detailed will only result in small quantitative differences in the solutions presented below.

We are concerned with free convective flows for which buoyancy also acts as a body force. Therefore Equation (1) becomes,

$$\overline{w} = \begin{cases} -\dfrac{K}{\mu}\left[1 - \dfrac{G}{|\overline{p}_{\overline{z}} - \rho_f g\beta(T - T_c)|}\right]\Big(\overline{p}_{\overline{z}} - \rho_f g\beta(T - T_c)\Big) & \text{when } |\overline{p}_{\overline{z}} - \rho_f g\beta(T - T_c)| > G, \\ 0 & \text{otherwise.} \end{cases} \tag{2}$$

In this equation ρ_f is the reference density of the fluid, g gravity, β the coefficient of cubical expansion and T the temperature. In the above we have assumed that the Boussinesq approximation applies when writing down the buoyancy term, and have taken T_c to be the reference temperature which will also be the coldest temperature experienced by the medium. The full two dimensional heat transport equation is

$$\frac{(\rho C)_{\rm pm}}{(\rho C)_f}\frac{\partial T}{\partial \overline{t}} + \overline{u}\frac{\partial T}{\partial \overline{x}} + \overline{w}\frac{\partial T}{\partial \overline{z}} = \alpha\left(\frac{\partial^2 T}{\partial \overline{x}^2} + \frac{\partial^2 T}{\partial \overline{z}^2}\right) + \frac{q'''}{(\rho C)_f}, \tag{3}$$

where C is the heat capacity, $\alpha = k_{\rm pm}/(\rho C)_{\rm pm}$ is the thermal diffusivity and q''' is the uniform rate of heat generation. Both the temperature and the resulting velocity fields are independent of $\overline{z}$ and are functions solely of $\overline{x}$, the horizontal coordinate. Thus the heat transport equation reduces to the ordinary differential equation,

$$\frac{d^2T}{d\overline{x}^2} = q'''/k_{\rm pm}. \tag{4}$$

Finally, the boundary conditions for the temperature are that

$$T = T_c \quad \text{at} \quad \overline{x} = 0, \qquad \text{and} \qquad T = T_h \quad \text{at} \quad \overline{x} = d, \tag{5}$$

and where the channel itself lies in the range $0 \leq \overline{x} \leq d$.

The governing equations, namely Equations (2) and (4), and the boundary conditions, (5), may be nondimensionalised using the following scalings,

$$(\overline{x}, \overline{y}) = d(x, y), \quad \overline{w} = \frac{\alpha}{d}w, \quad \overline{p} = \frac{\alpha\mu}{K}p \quad \text{and} \quad T = T_c + (T_h - T_c)\theta, \tag{6}$$

where T_h is the temperature of the hot surface. We obtain,

$$w = \begin{cases} \text{Ra}\,\theta - p_z - \text{Rb}, & \text{Rb} < \text{Ra}\,\theta - p_z, \\ 0, & -\text{Rb} < \text{Ra}\,\theta - p_z < \text{Rb}, \\ \text{Ra}\,\theta - p_z + \text{Rb}, & \text{Ra}\,\theta - p_z < -\text{Rb}, \end{cases} \tag{7}$$

and

$$\theta_{xx} + \frac{\text{Ra}_i}{\text{Ra}} = 0, \tag{8}$$

subject to,

$$\theta = 0 \quad \text{at} \quad x = 0, \qquad \text{and} \qquad \theta = 1 \quad \text{at} \quad x = 1. \tag{9}$$

In the above the external and internal Darcy-Rayleigh numbers are given by

$$\text{Ra} = \frac{\rho g \beta (T_h - T_c) K d}{\mu\alpha} \quad \text{and} \quad \text{Ra}_i = \frac{\rho g \beta q''' K d^3}{\mu\alpha k_{\rm pm}}, \tag{10}$$

respectively, and the Rees-Bingham number,

$$\text{Rb} = \frac{KL}{\mu\alpha}G, \tag{11}$$

which may be interpreted as being a convective porous Bingham number or as a nondimensional threshold gradient for the fluid.

We note that it is possible to reduce the set of three nondimensional parameters to two and therefore it is also possible to give a comprehensive account of the present problem using any two independent ratios of the three. However, this degeneracy is removed once convection becomes nonlinear. The computations undertaken by Rees [12] consider convection in a rectangular porous cavity with the same external heating as here but without internal heating ($\text{Ra}_i = 0$). The upper and lower surfaces of the cavity are insulated. The flow in [12] is two-dimensional as compared with the present one-dimensional flow. It is clear from the results of [12] that solution curves (such as the variation of the Nusselt number) do not map onto a single curve when plotted against Ra/Rb, which they will do when the channel is infinitely tall. Therefore it was decided to retain the present three parameters because of the context provided by [12].

3. Numerical Solutions

The solution for θ may be written down easily:

$$\text{Ra}\,\theta = \text{Ra}\,x + \text{Ra}_i(x - x^2)/2, \tag{12}$$

and it is this which is used in Equation (7). We note that, when there is no internal heating ($\text{Ra}_i = 0$), then the temperature profile is odd about $x = 1/2$, and when there is no sidewall heating ($\text{Ra} = 0$) then the temperature profile is even about $x = 1/2$.

Later we shall see that the present system admits situations where either the whole channel is stagnant, or else that there exists either one or two stagnant regions bounded by flowing regions. An alternative viewpoint is that, when flow exists, then there may be two, three or four yield surfaces. One example of a situation in which there are four yield surfaces and two stagnant regions is shown in Figure 1 and this corresponds to the case, $\text{Ra} = 3$, $\text{Ra}_i = 50$ and $\text{Rb} = 1$. The temperature profile has no symmetry and the locations of the four yield surfaces are denoted by the values, x_1, x_2, x_3 and x_4. These values satisfy the following equations,

$$\text{Ra}\,x_1 + \tfrac{1}{2}\text{Ra}_i(x_1 - x_1^2) - p_z + \text{Rb} = 0, \tag{13}$$

$$\text{Ra}\,x_2 + \tfrac{1}{2}\text{Ra}_i(x_2 - x_2^2) - p_z - \text{Rb} = 0, \tag{14}$$

$$\text{Ra}\,x_3 + \tfrac{1}{2}\text{Ra}_i(x_3 - x_3^2) - p_z - \text{Rb} = 0, \tag{15}$$

$$\text{Ra}\,x_4 + \tfrac{1}{2}\text{Ra}_i(x_4 - x_4^2) - p_z + \text{Rb} = 0, \tag{16}$$

and are obtained by setting $w = 0$ into the first or third options in Equation (7). While the present flow is a free convection flow in an infinitely long channel, it also approximates very well the flow which will arise in a tall channel sufficiently far from the upper and lower boundaries, and therefore it is necessary to apply a zero upward flux condition. Thus we have,

$$\int_0^1 w\,dx = 0, \tag{17}$$

which is equivalent to,

$$\int_0^{x_1} w\,dx + \int_{x_2}^{x_3} w\,dx + \int_{x_4}^{1} w\,dx = 0. \tag{18}$$

After integration this translates into,

$$\begin{aligned}
&\Big(\tfrac{1}{2}\text{Ra}\,x_1^2 + \text{Ra}_i(\tfrac{1}{4}x_1^2 - \tfrac{1}{6}x_1^3) - p_z x_1 + \text{Rb}\,x_1\Big) - \Big(\tfrac{1}{2}\text{Ra}\,x_2^2 + \text{Ra}_i(\tfrac{1}{4}x_2^2 - \tfrac{1}{6}x_2^3) - p_z x_2 - \text{Rb}\,x_2\Big) \\
&+\Big(\tfrac{1}{2}\text{Ra}\,x_3^2 + \text{Ra}_i(\tfrac{1}{4}x_3^2 - \tfrac{1}{6}x_3^3) - p_z x_3 - \text{Rb}\,x_3\Big) - \Big(\tfrac{1}{2}\text{Ra}\,x_4^2 + \text{Ra}_i(\tfrac{1}{4}x_4^2 - \tfrac{1}{6}x_4^3) - p_z x_4 + \text{Rb}\,x_4\Big) \\
&+\Big(\tfrac{1}{2}\text{Ra}\,x_5^2 + \text{Ra}_i(\tfrac{1}{4}x_5^2 - \tfrac{1}{6}x_5^3) - p_z x_5 + \text{Rb}\,x_5\Big) = 0.
\end{aligned} \tag{19}$$

In the above we have used $x_5 = 1$ solely to show how the constant which arises when evaluating Equation (17) is related to the rest of the integrand.

Equations (13)–(16) and (19) form a set of five nonlinear equations for the four yield surfaces and the value of the vertical pressure gradient correction, p_z. This correction needs to be found because T_c was chosen to be the reference temperature and therefore the corresponding hydrostatic pressure gradient is incorrect if an overall zero mean flux is to be maintained. To illustrate this, if we choose to consider a Newtonian fluid ($\text{Rb} = 0$) in a channel with external heating only ($\text{Ra}_i = 0$), then $p_z = \frac{1}{2}\text{Ra}$ corresponds to a zero mean flow, and $w = \text{Ra}(x - \frac{1}{2})$. A similar analysis for solely internal heating yields $p_z = \frac{1}{12}\mathrm{Ra}_i$. Both of these values may be found in the final line of Equation (19) when Rb set to be equal to zero, noting again that $x_5 = 1$. For general cases these five equations were solved using a standard Newton-Raphson method and therefore our solutions are essentially exact.

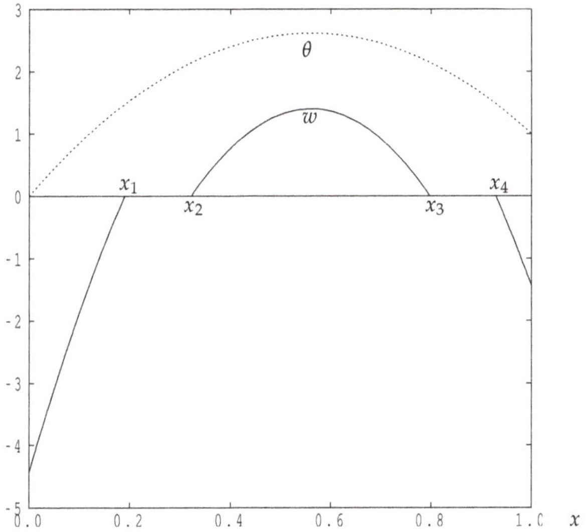

Figure 1. Showing the temperature and velocity profiles for the case $\mathrm{Ra} = 3$, $\mathrm{Ra}_i = 50$ and $\mathrm{Rb} = 1$. The locations of the yield surfaces are x_1, x_2, x_3 and x_4. The two stagnant regions lie in the ranges, $x_1 \le x \le x_2$ and $x_3 \le x \le x_4$.

We note that when Ra is increased gradually for the parameter set shown in Figure 1, then the rightmost yield surface at $x = x_4$ moves towards the right hand boundary at $x = 1$ and eventually reaches it. In our code this was modelled by simply replacing Equation (16) by $x_4 = 1$. This corresponds to the final integral in Equation (18) being zero, and is modelled correctly in Equation (19) by having the last two bracketed terms cancelling one another. A further increase in Ra eventually leads to the $x = x_3$ yield surface reaching $x = 1$, which means that the flow now has only one stagnant region. In our code this was modelled by replacing Equation (15) by $x_3 = 1$. Therefore, for practical reasons, we always began a computation with four yield surfaces and followed their trajectory as one of the governing parameters was varied.

4. Results and Discussion

4.1. Variation with Rb

Figure 2 shows how the location of the yield surfaces and the spatial extent of stagnant regions vary with Rb. Three cases are shown: one with purely external heating ($\mathrm{Ra} = 10$, $\mathrm{Ra}_i = 0$), one with purely internal heating ($\mathrm{Ra} = 0$, $\mathrm{Ra}_i = 120$), and an intermediate case which corresponds to the values of Ra and Ra_i used for the profile shown in Figure 1 (i.e., $\mathrm{Ra} = 10$, $\mathrm{Ra}_i = 120$).

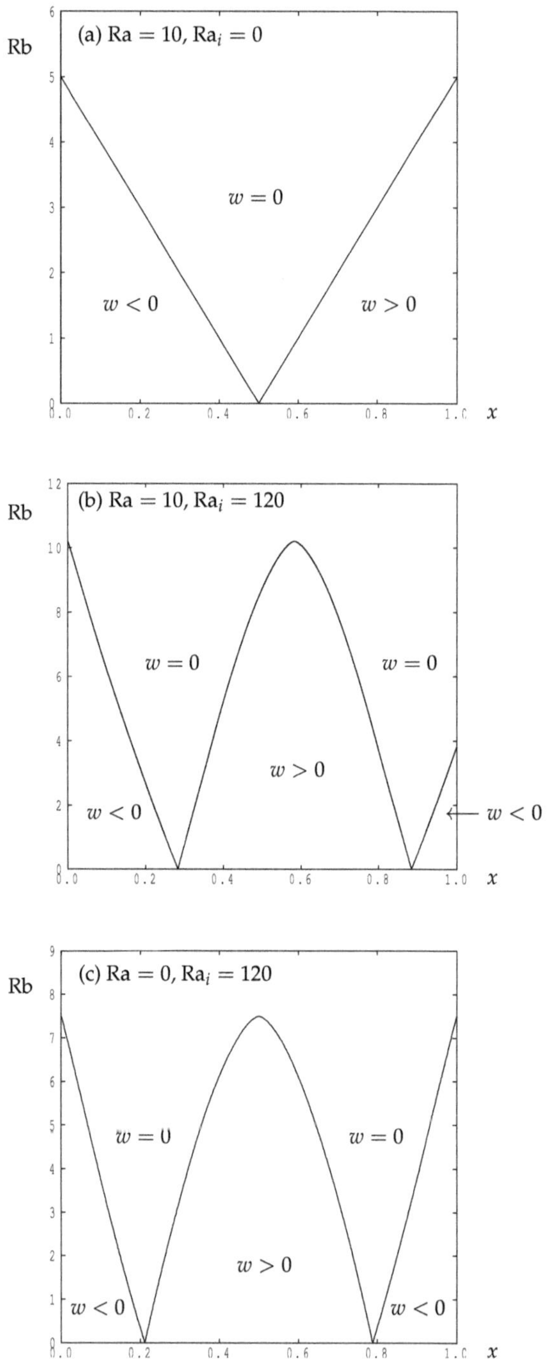

Figure 2. Displaying the evolution of the locations of the yield surfaces with Rb for three cases: (**a**) pure external heating, (**b**) a combination of internal and external heating and (**c**) pure internal heating. The corresponding values of Ra and Ra$_i$ are given in each frame.

When the heating is purely external then the system of equations for determining p_z and the yield surfaces may be solved analytically. We find again that $p_z = \frac{1}{2}\mathrm{Ra}$, and that the fluid velocity is given by,

$$w = \begin{cases} \mathrm{Ra}\,(x - \frac{1}{2}) + \mathrm{Rb}, & x < \frac{1}{2} - \frac{\mathrm{Rb}}{\mathrm{Ra}} \\ 0, & \frac{1}{2} - \frac{\mathrm{Rb}}{\mathrm{Ra}} < x < \frac{1}{2} + \frac{\mathrm{Rb}}{\mathrm{Ra}} \\ \mathrm{Ra}\,(x - \frac{1}{2}) - \mathrm{Rb}, & \frac{1}{2} + \frac{\mathrm{Rb}}{\mathrm{Ra}} < x. \end{cases} \tag{20}$$

It is clear from this expression that that there is flow only when $0 \leq \mathrm{Rb} \leq \frac{1}{2}\mathrm{Ra}$ and in this range of values of Rb the velocity profile is piecewise linear. The locations of the yield surfaces may be gleaned from Equation (20) and these linear functions of Rb are shown in Figure 2a. We have full stagnation when $\mathrm{Rb} > \frac{1}{2}\mathrm{Ra}$.

When the heating is purely internal then it is not possible to find an analytical expression for p_z or for the locations of the yield surfaces even though the symmetry of the system means that $x_1 + x_4 = x_2 + x_3 = 1$. Therefore we have resorted to purely numerical means to determine where the yield surfaces are as a function of Rb, and these are shown in Figure 2c. However, it is possible to find where the yield surfaces arise when Rb begins to increase from zero. Given that Ra = 0, Equations (13) and (14) with $p_z = \frac{1}{12}\mathrm{Ra}_i$ and $\mathrm{Rb} = 0$ give $x_1 = x_2 = \frac{1}{2}(1 - \sqrt{1/3}) = 0.211325$ while it is clear that $x_3 = x_4 = \frac{1}{2}(1 + \sqrt{1/3}) = 0.788675$; both these values may be seen in Figure 2c. It is also possible to determine the value of Rb above which the channel becomes fully stagnant. This is achieved by setting $x_1 = 0$, $x_2 = x_3 = \frac{1}{2}$ and $x_4 = 1$ into Equations (13)–(16), and then we find that stagnation corresponds to $\mathrm{Rb} > \frac{1}{16}\mathrm{Ra}_i$.

The intermediate range of cases is represented by the solutions shown in Figure 2b for which $\mathrm{Ra} = 10$ and $\mathrm{Ra}_i = 120$. Once Rb has risen above zero, two narrow stagnant regions appear but these are not symmetrically placed about $x = \frac{1}{2}$. The right hand yield surface attaches onto the $x = 1$ boundary when $\mathrm{Rb} \simeq 3.831$. Flow weakens as Rb increases further and full stagnation arises when $\mathrm{Rb} = \frac{245}{12} = 10.208333$. Once more this value may be found analytically by first substituting $x_1 = 0$ into Equation (13), which yields $p_z = \mathrm{Rb}$, and then by noting that Equations (14) and (15) must represent a double zero since $x_2 = x_3$. These two equations may be rearranged into the form,

$$\left[x - \left(\frac{1}{2} + \frac{\mathrm{Ra}}{\mathrm{Ra}_i}\right)\right]^2 = \left(\frac{1}{2} + \frac{\mathrm{Ra}}{\mathrm{Ra}_i}\right)^2 - \frac{4\mathrm{Rb}}{\mathrm{Ra}_i}. \tag{21}$$

It is clear that there will be two different solutions for this equations (i.e., x_2 and x_3) when the right hand side is positive, but none when it is negative. Therefore incipient stagnation corresponds to when the solutions are equal, for which the right hand side must be zero. Hence,

$$\mathrm{Rb} = \tfrac{1}{4}\mathrm{Ra}_i\left(\frac{1}{2} + \frac{\mathrm{Ra}}{\mathrm{Ra}_i}\right)^2 \tag{22}$$

represents the critical value of Rb in general. For the example shown in Figure 2b we have $\mathrm{Rb} = \frac{245}{12}$, as quoted above. Under these conditions x_1 is also zero and therefore full stagnation arises for larger values of Rb. The common values of x_2 and x_3 are now given by,

$$x = \frac{1}{2} + \frac{\mathrm{Ra}}{\mathrm{Ra}_i}, \tag{23}$$

and hence $x_2 = x_3 = \frac{7}{12}$ for the case shown in Figure 2b.

It is of interest to try to determine a general condition for stagnation to occur. The expression given in Equation (22) may be rearranged slightly to yield,

$$\frac{\mathrm{Rb}}{\mathrm{Ra}} = \frac{\mathrm{Ra}_i}{4\mathrm{Ra}}\left(\frac{1}{2} + \frac{\mathrm{Ra}}{\mathrm{Ra}_i}\right)^2. \tag{24}$$

However, this formula was derived by assuming that $x_2 = x_3$, i.e., that the coaslescence of two yield surfaces takes place in the interior of the domain, and therefore it is essential to check if the corresponding value of $x_2 = x_3$ does indeed lie in the interior. Equation (21), with a zero right hand side, tells us that $x_2 = x_3 = \frac{1}{2} + \frac{\text{Ra}}{\text{Ra}_i}$ and therefore the above analysis clearly applies only when $\frac{\text{Ra}}{\text{Ra}_i} \leq \frac{1}{2}$. It is straightforward to check that this criterion is also the criterion that the maximum value of θ given in Equation (12) is an internal maximum. Therefore we have a simple delineation between two separate regimes, one with an interior maximum for θ and one where that maximum lies on the right hand boundary.

An illustration of the approach to stagnation as Rb increases for cases where $\frac{\text{Ra}}{\text{Ra}_i} > \frac{1}{2}$ is shown in Figure 3 where Ra $= 10$ has been chosen and where the three separate values, 0, 10 and 20, have been taken for Ra_i. In all cases stagnation occurs when Rb $= 5$, and more generally this will be when

$$\frac{\text{Rb}}{\text{Ra}} = \frac{1}{2}. \tag{25}$$

In Figure 3 we see that the curve for $\text{Ra}_i = 20$, which is a transitional case because $\frac{\text{Ra}}{\text{Ra}_i} = \frac{1}{2}$, approaches $x = 1$ with a zero slope.

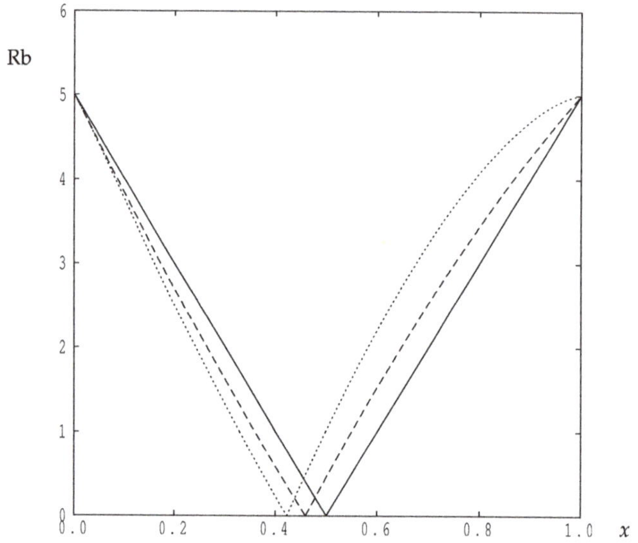

Figure 3. Displaying the evolution of the locations of the yield surfaces with Rb for Ra $= 10$ and for $\text{Ra}_i = 0$ (continuous), $\text{Ra}_i = 10$ (dashed) and $\text{Ra}_i = 20$ (dotted).

4.2. *Variation with Ra_i*

The detailed Figure 4 shows how the location of the stagnant regions changes as Ra_i/Rb increases for eight different values of Ra/Rb. When Ra $= 0$ the fluid remains stagnant until $\text{Ra}_i/\text{Rb} = 16$ as mentioned earlier. At larger values of Ra_i/Rb two stagnant regions appear symmetrically placed about the centre of the channel. The fluid flows upwards in the middle region and downwards in the outer two flowing regions. As Ra_i/Rb increases the stagnant regions becomer narrower and eventually become centred about $x = \frac{1}{2}(1 \pm \sqrt{1/3})$, as mentioned earlier.

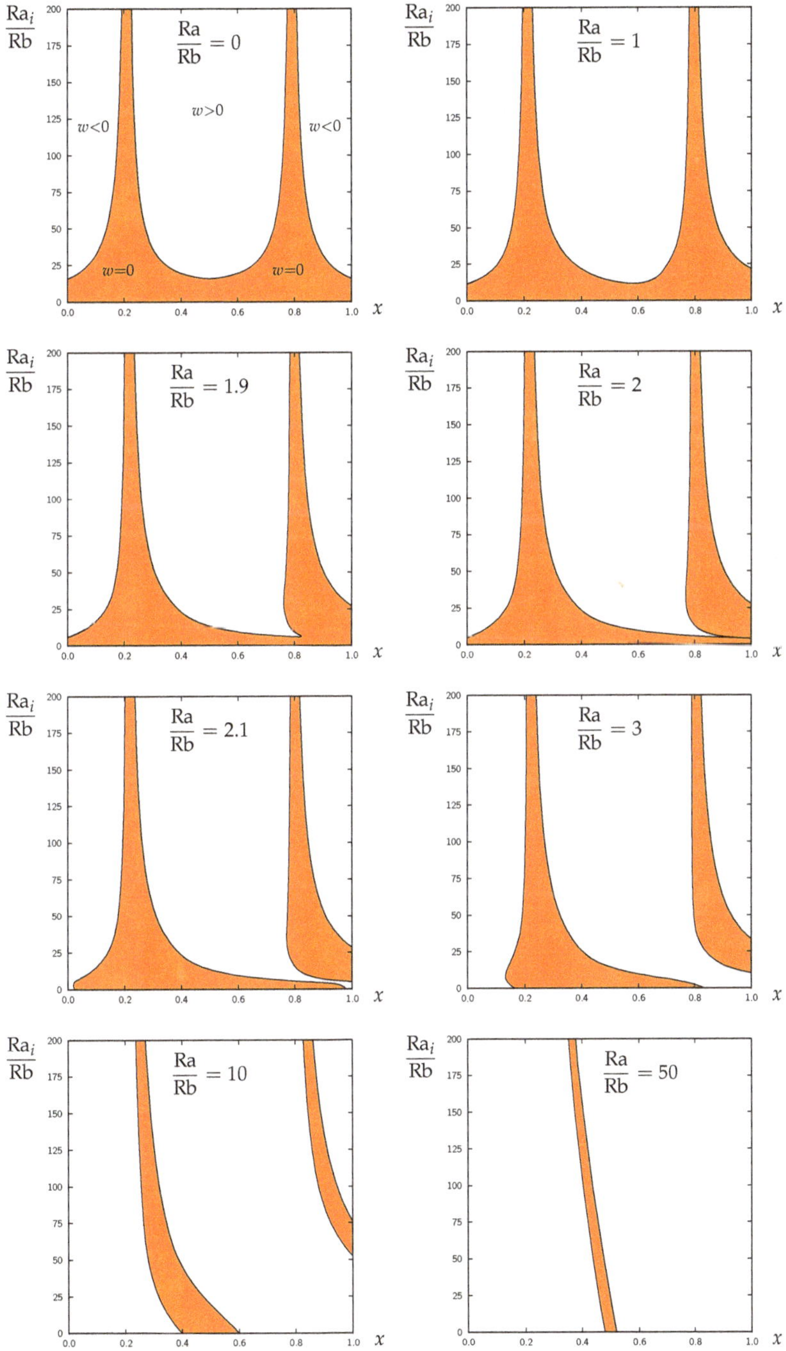

Figure 4. The evolution of the yield surfaces with Ra_i/Rb for the given values of Ra/Rb. The stagnant regions are shaded in orange. The flowing region adjacent to $x = 0$ always corresponds to downflow, $w < 0$.

When the external Darcy-Rayleigh number begins to rise from zero the pattern of flow and stagnation loses its symmetry. Flow begins at the right hand boundary at a larger value of $\mathrm{Ra}_i/\mathrm{Rb}$ than does flow at the left hand boundary and in the middle region both of which begin to flow simultaneously.

While $\mathrm{Ra}/\mathrm{Rb} < 2$, buoyancy forces remain too weak to cause flow when the heating is purely external, and therefore stagnation continues to be found at $\mathrm{Ra}_i/\mathrm{Rb} = 0$. But when Ra/Rb passes through 2, the shapes of the yield surfaces change dramatically, and evolve from one continuous region to two.

When $\mathrm{Ra}/\mathrm{Rb} = 2.1$ very narrow regions of flow occur at the two boundaries when $\mathrm{Ra}_i/\mathrm{Rb} = 0$ because buoyancy is only just in excess of what is required to overcome the yield threshold. But at larger values of Ra/Rb the two disconnected unyielded domains narrow, and, for $\mathrm{Ra}/\mathrm{Rb} = 3$ we see quite clearly the transition as $\mathrm{Ra}_i/\mathrm{Rb}$ increases from a flow pattern which is antisymmetric when $\mathrm{Ra}_i/\mathrm{Rb} = 0$ to one which is symmetric when $\mathrm{Ra}_i/\mathrm{Rb}$ becomes large. Even when $\mathrm{Ra}/\mathrm{Rb} = 50$, the transitions shown for smaller values of Ra/Rb also occur but do so at much larger values of $\mathrm{Ra}_i/\mathrm{Rb}$.

Figure 5 summarises all of our discussion about when stagnation occurs, but we have reinterpreted the data in terms of the variation of Rb/Ra with $\mathrm{Ra}_i/\mathrm{Ra}$. This figure delineates different regions of parameter space (ii) has a single stagnant region with flow either side, (iii) has two stagnant regions but the right hand one is attached to the right hand boundary, and (iv) has two stagnant and three flowing regions. These are indicated schematically on the figure itself for ease of interpretation.

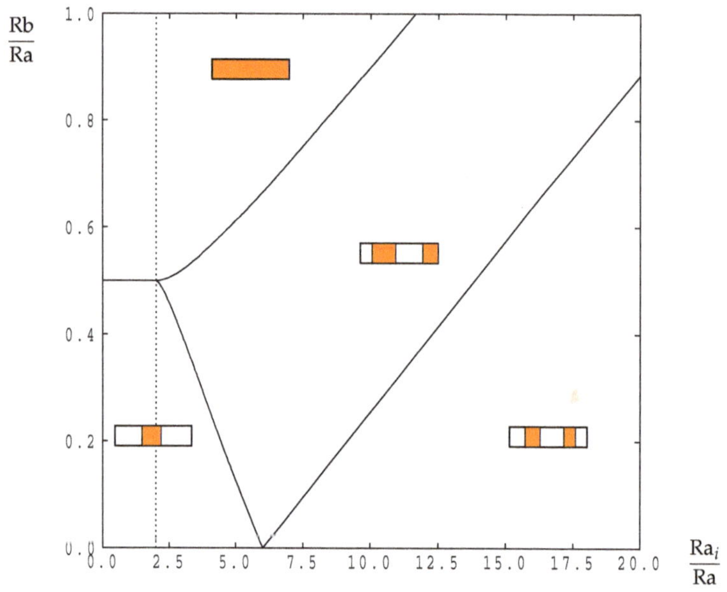

Figure 5. The loci of points in parameter space which delineate the various cases where the fluid flows and where it is stagnant. The different patterns of flow and stagnation are depicted by the insets. The region to the right of the dotted line corresponds to when θ displays an internal maximum value, while the region on the left corresponds to when the maximum is at $x = 1$.

4.3. *Detailed Analysis of Figure 5*

It is possible to explain analytically many of the features exhibited in Figure 5 and for these purposes it is convenient to define the ratios

$$\widetilde{\mathrm{Ra}}_i = \frac{\mathrm{Ra}_i}{\mathrm{Ra}} \quad \text{and} \quad \widetilde{\mathrm{Rb}} = \frac{\mathrm{Rb}}{\mathrm{Ra}}. \tag{26}$$

The dotted line separates the regions where the temperature field has an internal maximum ($\widetilde{\mathrm{Ra}}_i > 2$) and where it has a maximum at $x = 1$ ($\widetilde{\mathrm{Ra}}_i < 2$). The boundary of the fully stagnant region has already been shown to be given by Equation (24) when to the right of the dotted line and by Equation (25) when to the left—expressed in terms of the ratios given in Equation (26) these are

$$\widetilde{\mathrm{Rb}} = \frac{\widetilde{\mathrm{Ra}}_i}{4}\left(\frac{1}{2} + \frac{1}{\widetilde{\mathrm{Ra}}_i}\right)^2 \quad \text{and} \quad \widetilde{\mathrm{Rb}} = \frac{1}{2}, \tag{27}$$

respectively.

The other two lines in Figure 5 were obtained by suitably modified Newton-Raphson solvers and some of the numerical data corresponding to the yield surfaces just on the point of attaching to $x = 1$ in Figure 4. The middle region and the lower right hand region are distinguished by the location of the right hand stagnant region; in the former case it is attached to $x = 1$ but is not in the latter case.

We can derive expressions for the two boundaries in Figure 5 that meet at $\widetilde{\mathrm{Ra}}_i = 6$. The left one of these corresponds to the case when the single stagnant region that is present when $\widetilde{\mathrm{Ra}}_i$ and $\widetilde{\mathrm{Rb}}$ are both relatively small evolves into a pattern for which another stagnant zone begins to form at $x = 1$. The boundary is defined by $x_3 = 1$ and then combining Equations (14) and (15) gives that $x_2 = 2/\widetilde{\mathrm{Ra}}_i$. We can then determine x_1 in terms of x_2 and $\widetilde{\mathrm{Rb}}$ by eliminating the pressure term between Equations (13) and (14). Finally, the flux condition Equation (19) then simplifies so that if

$$X^2 \equiv \left(\tfrac{1}{2}\widetilde{\mathrm{Ra}}_i - 1\right)^2 + 4\widetilde{\mathrm{Ra}}_i\widetilde{\mathrm{Rb}}, \tag{28}$$

then X satisfies the cubic

$$2X^3 - \tfrac{3}{2}(2 + \widetilde{\mathrm{Ra}}_i)X^2 - \left(\tfrac{3}{8}\widetilde{\mathrm{Ra}}_i^3 - \tfrac{15}{4}\widetilde{\mathrm{Ra}}_i^2 + \tfrac{9}{2}\widetilde{\mathrm{Ra}}_i - 5\right) = 0. \tag{29}$$

This equation does not possess a simple analytic solution (although such an equation has solutions that can be written down albeit in very complicated form), but we can confirm some elementary results. First, when $\widetilde{\mathrm{Ra}}_i = 6$ we have $X = 2$ and then $\widetilde{\mathrm{Rb}} = 0$ as expected from Figure 5. Moreover, if $\widetilde{\mathrm{Ra}}_i = 6 - \delta$, with $0 < \delta \ll 1$, then $X = 2 + \delta + \cdots$ and Equation (28) gives $\widetilde{\mathrm{Rb}} = \frac{1}{8}\delta + \cdots$. This suggests that the boundary has slope, $-\frac{1}{8}$, at $\widetilde{\mathrm{Rb}} = 0$. Equation (29) also enables us to deduce the local behaviour near the other end of the boundary at $(\widetilde{\mathrm{Ra}}_i, \widetilde{\mathrm{Rb}}) = (2, \frac{1}{2})$ where it joins with the lines Equation (27). Near this point, if $\widetilde{\mathrm{Ra}}_i = 2 + \delta$ then Equation (29) leads to $X = 2 + \frac{1}{2}\delta - \frac{1}{2\sqrt{3}}\delta^{3/2} + \cdots$ and then to $\widetilde{\mathrm{Rb}} = \frac{1}{2} - \frac{1}{4\sqrt{3}}\delta^{3/2} + \cdots$. The presence of fractional powers is required to resolve an apparent contradiction at $O(\delta^3)$ in the expansion of Equation (29), but this matches perfectly the numerical data of the appropriate curve in Figure 5.

We now turn to the last boundary on Figure 5 that separates those flows with two stagnant regions with the right hand one attached to $x = 1$ and the situation when there are two stagnant and three flowing regions. We can pursue an analysis that in many ways parallels the argument just above. The boundary of interest arises when $x_4 = 1$; then Equations (13) and (16) lead quickly to the conclusion that $x_1 = 2/\widetilde{\mathrm{Ra}}_i$. Equation (16) yields the pressure gradient $p_z = 1 + \widetilde{\mathrm{Rb}}$ while Equation (14) shows that $x_2 + x_3 = 1 + (2/\widetilde{\mathrm{Ra}}_i)$. If these relationships are substituted into the flux condition and if we define

$$Z^2 \equiv \left(\tfrac{1}{2}\widetilde{\mathrm{Ra}}_i - 1\right)^2 + 4\widetilde{\mathrm{Ra}}_i\widetilde{\mathrm{Rb}}, \tag{30}$$

(cf. Equation (28)) then we find that $Z^3 = \frac{3}{2}\widetilde{\mathrm{Ra}}_i - 1$. Hence

$$4\widetilde{\mathrm{Ra}}_i\widetilde{\mathrm{Rb}} = \left(\tfrac{1}{2}\widetilde{\mathrm{Ra}}_i - 1\right)^2 - \left(\tfrac{3}{2}\widetilde{\mathrm{Ra}}_i - 1\right)^{2/3}, \tag{31}$$

is an analytical description of the boundary. It follows from this that, for small values of $(\widetilde{\mathrm{Ra}}_i - 6)$, we have $\widetilde{\mathrm{Rb}} \approx \frac{1}{16}(\widetilde{\mathrm{Ra}}_i - 6) + \cdots$.

We can also understand the behaviour of the system for small values of $\widetilde{\mathrm{Rb}}$ and illustrated in Figure 6. If we consider a fixed small value of $\widetilde{\mathrm{Rb}}$, which is taken to be 0.01 or 0.05 in Figure 6, then for values $\widetilde{\mathrm{Ra}}_i$ much less than 6 the flow contains a single stagnant zone. As $\widetilde{\mathrm{Ra}}_i$ approaches 6 so a second stagnant region forms on $x = 1$ and later it detaches so that the two stagnant zones separate three flowing regions. We can capture all this behaviour analytically by seeking solutions when $\widetilde{\mathrm{Ra}}_i = 6 + c\widetilde{\mathrm{Rb}}$ for values of $c = O(1)$. We suppose that the points x_1–x_4 are located at

$$x_1 = \tfrac{1}{3} + \tilde{x}_1\widetilde{\mathrm{Rb}} + \cdots, \quad x_2 = \tfrac{1}{3} + \tilde{x}_2\widetilde{\mathrm{Rb}} + \cdots, \quad x_3 = 1 + \tilde{x}_3\widetilde{\mathrm{Rb}} + \cdots, \quad x_4 = 1 + \tilde{x}_4\widetilde{\mathrm{Rb}} + \cdots. \tag{32}$$

If we substitute these expressions into Equations (13)–(16) and Equation (19) then at $O(\widetilde{\mathrm{Rb}})$ we obtain four linear equations for $\tilde{x}_1$–$\tilde{x}_4$ whose solution gives that

$$\begin{aligned} x_1 &= \tfrac{1}{3} + \left(-\tfrac{2}{3} - \tfrac{1}{72}c\right)\widetilde{\mathrm{Rb}} + \cdots, \quad & x_2 &= \tfrac{1}{3} + \left(\tfrac{1}{3} - \tfrac{1}{72}c\right)\widetilde{\mathrm{Rb}} + \cdots \\ x_3 &= 1 + \left(-\tfrac{1}{3} - \tfrac{1}{24}c\right)\widetilde{\mathrm{Rb}} + \cdots, \quad & x_4 &= 1 + \left(\tfrac{2}{3} - \tfrac{1}{24}c\right)\widetilde{\mathrm{Rb}} + \cdots. \end{aligned} \tag{33}$$

We remark that there may appear to be a something of a contradiction here because for sufficiently negative values of c these predictions for x_3 and x_4 do not lie in the region $0 \leq x \leq 1$. It turns out that the results quoted for x_1 and x_2 hold to the accuracy quoted irrespective of whether x_3 and/or x_4 lie inside the channel or not. We can then interpret the results encompassed by Equation (33) in conjunction with Figures 5 and 6 in the following way. For sufficiently large negative values of c only x_1 and x_2 lie within the channel so that there is a single isolated stagnant region corresponding to the lower left-hand part of Figure 5. As we slowly increase c then the first restructure of the flow occurs when the predicted value of x_3 becomes 1; so that $c = -8$. (We point out that at this stage $\widetilde{\mathrm{Rb}} = \frac{1}{8}(6 - \widetilde{\mathrm{Ra}}_i)$ confirming that the slope of the boundary here is $-\frac{1}{8}$.) For c slightly larger than this value, then the flow consists of one stagnant zone around $x = \frac{1}{3}$ while a second stagnant region next to the wall at $x = 1$. The signal for this second region detaching from the $x = 1$ is that $x_4 = 1$ which clearly occurs when $c = 16$. Now $\widetilde{\mathrm{Rb}} = \frac{1}{16}(\widetilde{\mathrm{Ra}}_i - 6)$ confirming that the slope of this bounding curve in $\widetilde{\mathrm{Ra}}_i/\widetilde{\mathrm{Rb}}$ parameter space is indeed $\frac{1}{16}$. For values of c greater than 16 the flow consists of two stagnant zones each away from either boundary.

This analysis is all consistent with the results shown in Figure 6 for two relatively small values of $\widetilde{\mathrm{Rb}}$. When $\widetilde{\mathrm{Ra}}_i \lesssim 6$ there is the single stagnant zone isolated from the walls of the channel. As $\widetilde{\mathrm{Ra}}_i$ grows so the centre of the stagnant region drifts towards $x = 1/3$ until at a value of $\widetilde{\mathrm{Ra}}_i$ slightly less than 6 there is evidence of the second stagnant zone forming on $x = 1$. As $\widetilde{\mathrm{Ra}}_i$ increases further, this second stagnant region soon detaches itself and we are left with three flowing regions separated by two stagnant zones (corresponding to region (iv) in Figure 5). We may use also Figure 5 in a qualitative manner by, for example, choosing a case for which $\mathrm{Rb}/\mathrm{Ra} = \frac{1}{4}$ while $\mathrm{Ra}_i/\mathrm{Ra} = 0$, i.e., pure external heating. This case lies in that part of the Figure for which there is a single stagnant region in the interior. Once more we see that as the strength of the internal heating is increased so a new stagnant region is induced at the right hand boundary, and then this ultimately detaches. We also see that it is impossible to have a single interior stagnant region when $\mathrm{Ra}_i/\mathrm{Ra} > 6$ irrespective of the value of Rb.

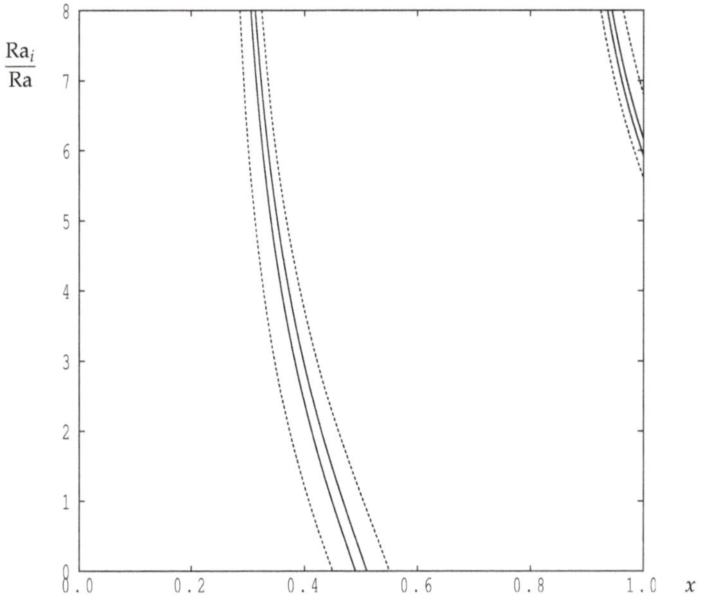

Figure 6. The evolution of the yield surfaces with $\mathrm{Ra}_i/\mathrm{Ra}$ when $\mathrm{Rb}/\mathrm{Ra} = 0.01$ (continuous line) and $\mathrm{Rb}/\mathrm{Ra} = 0.05$ (dotted line) to illustrate the small-Rb/Ra analysis of Section 4.3. The stagnant regions lie within the two narrow regions in each case.

5. Conclusions

In this short paper we have attempted to provide a comprehensive description of the properties of the flow in a vertical porous channel which is subjected to both internal heating (i.e., uniform heat generation) and external heating (i.e., a temperature difference across the layer) and where a Bingham fluid saturates the porous medium. We have adopted what is, perhaps, the simplest model for such flows namely the Pascal model. Some analytical results have been provided but the core results have been obtained using a multi-dimensional Newton-Raphson solver.

We have found the criterion for flow to arise and this is given either by Equations (24) or (25) depending on the value of $\mathrm{Ra}_i/\mathrm{Ra}$, and it is also shown as the uppermost curve in Figure 5. The number and locations of the resulting stagnant regions depend quite strongly on the balance between the values of Rb/Ra and $\mathrm{Ra}_i/\mathrm{Ra}$. We find that there is only one stagnant region when external heating dominates, but that a gradual transformation to a state where there are two stagnant regions arises as the strength of the internal heating increases.

The qualitative nature of our results will not change should a more accurate form of Pascal's law for the flow of a Bingham fluid in a porous medium be replaced by other models, such as the Buckingham-Reiner law which models porous media which are composed of identical unidirectional pores, or many of the models given in [7] which correspond to more general distributions of pores or channels.

Author Contributions: Formal analysis, A.P.B.; Software, D.A.S.R.; writing–original draft, D.A.S.R.; writing–review and editing, D.A.S.R. and A.P.B.

Funding: This research received no external funding.

Conflicts of Interest: The authors declare no conflict of interest.

Abbreviations

C	heat capacity ($L^2M/T^2\Theta$)
d	width of the channel (L)
g	gravity (L/T^2)
G	threshold gradient (M/T^2L^2)
k	thermal conductivity ($ML/T^3\Theta$)
K	permeability (L^2)
p	pressure (M/LT^2)
p_z	pressure gradient in the vertical direction (M/T^2L^2)
q'''	rate of internal heating (M/LT^3)
Ra	external Darcy-Rayleigh number (nondimensional)
Ra_i	internal Darcy-Rayleigh number (nondimensional)
Rb	Rees-Bingham number (nondimensional)
t	time (T)
T	temperature (dimensional) (Θ)
T_c	reference (cold) temperature (Θ)
T_h	temperature of heated surface (Θ)
w	vertical Darcy velocity (L/T)
x	horizontal coordinate (L)
z	vertical coordinate (L)
Greek letters	
α	thermal diffusivity (L^2/T)
β	coefficient of cubical expansion (Θ^{-1})
θ	temperature (nondimensional)
μ	dynamic viscosity (M/LT)
ρ	reference density (M/L^3)
Other symbols	
-	dimensional quantities
1, 2, 3, 4	indicating yield surfaces
f	fluid
pm	porous medium
z	z-derivative

References

1. Pascal, H. Influence du gradient de seuil sur des essais de remontée de pression et d'écoulement dans les puits. *Revue de l'Institut Francais du Petrole* **1979**, *34*, 387–404. [CrossRef]
2. Pascal, H. Nonsteady flow through porous media in the presence of a threshold gradient. *Acta Mech.* **1981**, *39*, 207–224. [CrossRef]
3. Gheorghitza, I. Motions with initial gradient. *Q. J. Mech. Appl. Math.* **1959**, *12*, 280–286. [CrossRef]
4. Wu, Y.S.; Pruess, K. Chapter 2: Flow of non-Newtonian fluids in a porous media. *Adv. Porous Media* **1996**, *3*, 87–184.
5. Buckingham, E. On Plastic Flow through Capillary Tubes. *Proc. Am. Soc. Test. Mater.* **1921**, *21*, 1154–1156.
6. Reiner, M. Ueber die Strömung einer elastischen Flüssigkeit durch eine Kapillare. Beitrag zur Theorie Viskositätsmessungen. *Colloid Polym. Sci.* **1926**, *39*, 80–87.
7. Nash, S.; Rees, D.A.S. The effect of microstructure on models for the flow of a Bingham fluid in porous media. *Transp. Porous Media* **2017**, *116*, 1073–1092. [CrossRef]
8. Malvault, G.; Ahmadi, A.; Omari, A. Numerical simulation of yield stress flow in capillary bundles: Influence of the form and the axial variation in the cross section. *Transp. Porous Media* **2017**, *120*, 255–270. [CrossRef]
9. Rees, D.A.S.; Bassom, A.P. Unsteady thermal boundary layer flows of a Bingham fluid in a porous medium. *Int. J. Heat Mass Transf.* **2015**, *82*, 460–467. [CrossRef]
10. Rees, D.A.S.; Bassom, A.P. Unsteady thermal boundary layer flows of a Bingham fluid in a porous medium following a sudden change in surface heat flux. *Int. J. Heat Mass Transf.* **2016**, *93*, 1100–1106. [CrossRef]

11. Rees, D.A.S.; Bassom, A.P. Unsteady free convection boundary layer flows of a Bingham fluid in cylindrical porous cavities. *Transp. Porous Media* **2019**, *127*, 711–728. [CrossRef]
12. Rees, D.A.S. The convection of a Bingham fluid in a differentially-heated porous cavity. *Int. J. Numer. Methods Heat Fluid Flow* **2016**, *26*, 879–896. [CrossRef]

fluids MDPI

Article

Lubrication Approximation for Fluids with Shear-Dependent Viscosity

Bruno M.M. Pereira [1], Gonçalo A.S. Dias [2], Filipe S. Cal [1] and Kumbakonam R. Rajagopal [3] and Juha H. Videman [2,*]

[1] Instituto Superior de Engenharia de Lisboa, Instituto Politécnico de Lisboa, Rua Conselheiro Emídio Navarro, 1959-007 Lisboa, Portugal; bmpereira@adm.isel.pt (B.M.M.P.); fcal@adm.isel.pt (F.S.C.)
[2] Center for Mathematical Analysis, Geometry and Dynamical Systems, Instituto Superior Técnico, Universidade de Lisboa, Av. Rovisco Pais 1, 1049-001 Lisboa, Portugal; goncalo.dias@tecnico.ulisboa.pt
[3] Department of Mechanical Engineering, Texas A&M University, College Station, TX 77843, USA; krajagopal@tamu.edu
* Correspondence: jvideman@math.tecnico.ulisboa.pt

Received: 30 April 2019; Accepted: 20 May 2019; Published: 28 May 2019

Abstract: We present dimensionally reduced Reynolds type equations for steady lubricating flows of incompressible non-Newtonian fluids with shear-dependent viscosity by employing a rigorous perturbation analysis on the governing equations of motion. Our analysis shows that, depending on the strength of the power-law character of the fluid, the novel equation can either present itself as a higher-order correction to the classical Reynolds equation or as a completely new nonlinear Reynolds type equation. Both equations are applied to two classic problems: the flow between a rolling rigid cylinder and a rigid plane and the flow in an eccentric journal bearing.

Keywords: power-law fluid; shear-dependent viscosity; Reynolds equation; lubrication approximation

1. Introduction

The Reynolds' lubrication approximation (Reynolds [1]) of the Navier-Stokes equation is a cornerstone of classical fluid mechanics. This approximation has tremendous value as it is relevant to many technological applications. However, many of the lubricating oils that are currently in use cannot be appropriately described by the Navier-Stokes constitutive theory. Many of the lubricating oils exhibit a variety of departures from Newtonian behavior, they shear thin, display stress-relaxation, instantaneous elasticity, non-linear creep, threshold for the strain rate before they start to flow, thixotropy, etc. Many constitutive models of the differential, rate and integral type have been developed to describe the non-Newtonian behavior exhibited by such fluids, and lubricating approximations have been derived for a variety of fluid flows governed by these equations. In this paper, we are only interested in a very special sub-class of non-Newtonian fluids, namely we are interested in developing a lubrication approximation for shear thinning fluids.

We do not aim to provide an exhaustive review of the numerous studies that have been carried out to simplify and approximate the governing equations that arise from assuming different non-Newtonian characteristics, a more complete list of the same can be found in the papers that we mention below. We merely cite some representative papers that consider the different departures from classical Newtonian behavior and in which the lubrication approximation has been developed. A discussion of several of the studies that have been carried out can be found in the book by Szeri [2], see also [3–5]. Early studies concerning the lubrication approximation for the flow of power-law constitutive relations were primarily concerned with purely one dimensional flows where inertial effects do not manifest themselves (see for instance Shukla et al. [6]); others concern non-inertial flows (see for example Park and Kwon [7]), and yet others concern the lubrication approximation for a power law fluid, under infinite wide gap approximation (see the study by Johnson and Mangkoesoebroto [8]). Bourgin [9], see also Kacou et al. [10], developed lubrication approximation for fluids of the differential type and Harnoy [11] studied the lubrication approximation for an elastic-viscous fluid in a short journal bearing. Ballal and Rivlin [12] studied the flow of a viscoelastic fluid in a journal bearing using a perturbation analysis which was shown to be incorrect by San Andres and Szeri [13]. Cal et al. [14] developed a lubrication approximation for viscoelastic fluids and showed that viscoelasticity can have pronounced effect for certain values of the film thickness and in the case of the journal bearing, the eccentricity. Also, in flows involving high pressures as encountered in elastohydrodynamics, the pressure dependence of viscosity has to be taken into account (see Barus [15], Bair [16]). Lubrication approximation has been developed in the case of fluids with pressure dependent viscosity by Rajagopal and Szeri [17] and Gustafsson et al. [18]. Finally, Fusi et al. [19] have studied the lubrication approximation for a Bingham fluid taking into account inertial effects; fluids like the Bingham fluid that have a threshold in the stress for flow to take place are best described by constitutive relations wherein the kinematics is described as a function of the stress rather than expressing the stress in terms of the kinematical variable in the traditional manner (see Rajagopal [20,21] for a discussion of such fluids as well as more general fluids that are described by implicit constitutive relations). As there is a threshold for the stress beyond which the fluid starts to flow, the governing equations are quite different from the lubrication approximation obtained in the case of the other studies that employ fluid models that do not have such a threshold for the flow to take place.

The current study that is being carried out takes into consideration nonlinearities both due to the shear dependent viscosity and inertia and the flow is two dimensional. Although the inertial effects are not omitted a priori they do not influence the flow characteristics at the orders of approximation considered in this work (see Nazarov and Videman [22] for the inertial correction to the Reynolds lubrication approximation). The power-law fluid model under consideration has two constants that determine its viscosity, a constant power-law exponent n and another constant α_0 that determines the departure of the viscosity from the Newtonian viscosity when the power-law exponent is 2 (see Equation (7)). A formal perturbation analysis is carried out, assuming two different possibilities for the material parameter α_0, namely that it is of the order $\mathcal{O}(\epsilon^3)$ and $\mathcal{O}(\epsilon^2)$, and new lubrication approximations are derived. In the former case, we simply obtain a higher-oder correction to the classical (linear) Reynolds equation but if α_0 is of $\mathcal{O}(\epsilon^2)$ the Reynolds type equation becomes fully nonlinear and must be solved together with an ODE for the main part of the flow velocity. Using these approximations, two problems are solved, the first being the fluid flow between a rolling rigid cylinder and a rigid plane, and the second being the problem of the fluid flowing in an eccentric journal bearing

2. Formulation of the Problem

Consider the following set of partial differential equations governing the isothermal flow of an incompressible, homogeneous power-law fluid

$$\rho\left(\frac{\partial \mathbf{v}}{\partial t}+[\nabla \mathbf{v}]\,\mathbf{v}\right)+\nabla p=\rho\,\mathbf{f}+\operatorname{div}\mathbf{S}\,, \tag{1}$$

$$\operatorname{div}\,\mathbf{v}=0\,, \tag{2}$$

where $\rho > 0$ is the constant density of the fluid, $\mathbf{v} = (u, v)$ is the velocity field, $\mathbf{S}$ is the deviatoric stress tensor ($\mathbf{T} = -p\mathbf{I} + \mathbf{S}$ is the Cauchy stress tensor), p is a scalar variable, often referred to as the mechanical pressure, associated with the incompressibility constraint (2), and $\mathbf{f}$ is an external force acting on the fluid. Whether p is the mean value of the stress depends on the constitutive relationship for the extra stress tensor $\mathbf{S}$. In general, p is not the mean value of the stress but for the model considered in this paper it is (see Rajagopal [23] for a detailed discussion of the notion of pressure, its use, misuse and abuse). The deviatoric stress tensor for a fluid with shear-dependent viscosity which we shall study is related to the symmetric part of the velocity gradient as follows:

$$\mathbf{S} = 2\mu_0 \left(1 + \alpha_0 \, |\mathbf{D}|^2\right)^{\frac{n-2}{2}} \mathbf{D}, \tag{3}$$

where $n \geq 1$ is the power-law exponent and α_0 a scalar coefficient related to the power-law character of the fluid, with dimensions of time squared ($\dim[\alpha_0] = T^2$). Moreover, $\mathbf{D}(\mathbf{v}) = \frac{1}{2}\left(\nabla\mathbf{v} + (\nabla\mathbf{v})^T\right)$ denotes the symmetric part of the velocity gradient $(\nabla\mathbf{v})_{ij} = \frac{\partial v_i}{\partial x_j}$, and $\mu_0 > 0$ is a constant.

2.1. Lubrication Approximation

Let us restrict our attention to steady two-dimensional thin flows without external forces ($\mathbf{f} = 0$). We will now introduce the following dimensionless (starred) quantities

$$(x^*, y^*) = L^{-1}(x, y), \quad \mathbf{v}^* = U^{-1}\mathbf{v}, \quad p^* = P^{-1}p, \quad \mathbf{S}^* = S^{-1}\mathbf{S}, \quad \alpha_0^* = U^2L^{-2}\alpha_0, \tag{4}$$

where L and U represent typical length and velocity scales and where the characteristic pressure P and the characteristic stress S are taken to be

$$P = S = \mu_0 \, UL^{-1}. \tag{5}$$

We will define the usual Reynolds number Re through

$$\mathrm{Re} = \rho \, UL\mu_0^{-1}, \tag{6}$$

and make the following assumptions that are appropriate for a class of lubrication problems:

- the flow takes place between two almost parallel surfaces situated at $y = 0$ and $y = H(x)$;
- the lubricating film is thin, that is, $H(x) = \epsilon \, h(x)$, where $\epsilon \ll 1$ denotes a small non-dimensional parameter;
- the flow is slow enough or the viscosity high enough so that $\mathrm{Re} = \mathcal{O}(\epsilon)$;
- the power-law parameter will have to be such that $\alpha_0 = \mathcal{O}(\epsilon^2)$ or $\alpha_0 = \mathcal{O}(\epsilon^3)$.

Next, we will drop the stars and introduce the fast variable $y \to \epsilon^{-1}y$ (the stretched normal coordinate). The equations then become

$$\mathrm{Re}\,\left([\nabla\mathbf{v}]\,\mathbf{v}\right) + \nabla p = 2\,\mathrm{div}\left\{\left(1 + \alpha_0 \, |\mathbf{D}|^2\right)^{\frac{n-2}{2}} \mathbf{D}\right\}, \tag{7}$$

$$\mathrm{div}\;\mathbf{v} = 0. \tag{8}$$

The adherence boundary conditions on rigid, impermeable boundaries read as

$$\mathbf{v} = \mathbf{U}^0, \quad \mathbf{U}^0 \cdot \mathbf{e}_y = 0, \quad \mathbf{U}^0 \cdot \mathbf{e}_x = U^0, \qquad \text{at } y = 0, \tag{9}$$

$$\mathbf{v} = \mathbf{U}^h, \quad \mathbf{U}^h \cdot \mathbf{e}_\nu = 0, \quad \mathbf{U}^0 \cdot \mathbf{e}_\tau = U^h, \qquad \text{at } y = h(x), \tag{10}$$

where $\mathbf{U}^0$ and $\mathbf{U}^h$ denote given constant velocities of the boundaries and $\mathbf{e}_\nu$ and $\mathbf{e}_\tau$ stand for the unit normal and tangent vectors at $y = h(x)$. These vectors are related to the unit vectors $\mathbf{e}_x$ and $\mathbf{e}_y$ through the formulae

$$\mathbf{e}_\nu = \frac{\nabla(\epsilon h(x) - y)}{\|\nabla(\epsilon h(x) - y)\|} = \frac{\frac{\partial}{\partial x}(\epsilon h(x) - y)\mathbf{e}_x + \frac{\partial}{\partial y}(\epsilon h(x) - y)\mathbf{e}_y}{\sqrt{1 + \epsilon^2(h'(x))^2}} = \frac{\epsilon h'(x)\mathbf{e}_x - \mathbf{e}_y}{\sqrt{1 + \epsilon^2(h'(x))^2}}, \tag{11}$$

$$\mathbf{e}_\tau = \frac{\mathbf{e}_x + \epsilon h'(x)\mathbf{e}_y}{\sqrt{1 + \epsilon^2(h'(x))^2}}. \tag{12}$$

2.2. Formal Asymptotic Analysis

2.2.1. Case $\alpha_0 = \mathcal{O}(\epsilon^3)$

We will assume the following ansatz

$$\mathbf{v}(\epsilon,\mathbf{x}) = \epsilon^0\mathbf{v}^{(1)}(\mathbf{x}) + \epsilon^1\mathbf{v}^{(2)}(\mathbf{x}) + \epsilon^2\mathbf{v}^{(3)}(\mathbf{x}) + \dots, \tag{13}$$

$$p(\epsilon,\mathbf{x}) = \epsilon^{-2}p^{(0)}(s) + \epsilon^{-1}p^{(1)}(\mathbf{x}) + \epsilon^0 p^{(2)}(\mathbf{x}) + \epsilon^1 p^{(3)}(\mathbf{x}) + \dots, \tag{14}$$

where $\mathbf{x} = (x,y)$ and the functions $\mathbf{v}^{(j)}$, $p^{(j)}$ and are of $\mathcal{O}(1)$. According to the assumptions $\mathrm{Re} = \mathcal{O}(\epsilon)$ and $\alpha_0 = \mathcal{O}(\epsilon^3)$, and thus on setting

$$\mathrm{Re} = \epsilon\,\mathcal{R}, \qquad \alpha_0 = \epsilon^3\,\alpha, \tag{15}$$

we obtain $\mathcal{R}$ and α are of order one.

Inserting the asymptotic expansions (13) and (14) into Equations (7) and (8), we obtain at $\mathcal{O}(\epsilon^{-3})$ in (1) that $p^{(0)} = p^{(0)}(x)$, i.e., independent of y; at $\mathcal{O}(\epsilon^{-2})$ in (1), at $\mathcal{O}(\epsilon^{-1})$ in (2) and at $\mathcal{O}(\epsilon^0)$ in (9) and (10)

$$\frac{\mathrm{d}p^{(0)}}{\mathrm{d}x} = \frac{\partial^2 u^{(1)}}{\partial y^2}, \tag{16}$$

$$\frac{\partial p^{(1)}}{\partial y} = 0, \tag{17}$$

$$\frac{\partial v^{(1)}}{\partial y} = 0, \tag{18}$$

$$u^{(1)} = U^0, \quad v^{(1)} = 0, \qquad \text{at } y = 0, \tag{19}$$

$$u^{(1)} = U^h, \quad v^{(1)} = 0, \qquad \text{at } y = h(x), \tag{20}$$

where U^0 and U^h denote the tangential velocities at $y = 0$ and at $y = h(x)$.

From (18)–(20) it follows that

$$v^{(1)}(x,y) \equiv 0. \tag{21}$$

We have moreover that $p^{(1)} = p^{(1)}(x)$ from (17). Using Equation (16) and the boundary conditions (19) and (20) for $u^{(1)}$ we conclude that

$$u^{(1)}(x,y) = U^0 + \frac{U^h - U^0}{h(x)}\,y + \frac{y(y-h(x))}{2}\frac{\mathrm{d}p^{(0)}}{\mathrm{d}x}. \tag{22}$$

The equations at the next order read as

$$\frac{\mathrm{d}p^{(1)}}{\mathrm{d}x} = \frac{1}{2}\left[\frac{1}{2}(n-4)\alpha\left(\frac{\partial u^{(1)}}{\partial y}\right)^2\frac{\partial^2 u^{(1)}}{\partial y^2} + (n-2)\alpha\left(\frac{\partial u^{(1)}}{\partial y}\right)^2\frac{\partial^2 u^{(1)}}{\partial y^2}\right.$$
$$\left. + \alpha\left(\frac{\partial u^{(1)}}{\partial y}\right)^2\frac{\partial^2 u^{(1)}}{\partial y^2} + 2\frac{\partial^2 u^{(2)}}{\partial y^2}\right], \tag{23}$$

$$\frac{\partial p^{(2)}}{\partial y} = -\frac{\partial^2 u^{(1)}}{\partial x \partial y}, \tag{24}$$

$$\frac{\partial v^{(2)}}{\partial y} = -\frac{\partial u^{(1)}}{\partial x}, \tag{25}$$

$$v^{(2)} = u^{(2)} = 0, \qquad \text{at } y = 0, \tag{26}$$

$$v^{(2)} = U^h h'(x), \quad u^{(2)} = 0 \qquad \text{at } y = h(x). \tag{27}$$

Equations (24)–(27) can be expressed as a one-dimensional Stokes system for $(v^{(2)}, p^{(2)})$. Assuming that this system is solvable, it follows from (23) and (25) that $u^{(2)}$ satisfies the second-order ODE (in y)

$$\frac{\partial^2 u^{(2)}}{\partial y^2} = \frac{\mathrm{d}p^{(1)}}{\mathrm{d}x} - \frac{3}{4}(n-2)\alpha\left(\frac{\partial u^{(1)}}{\partial y}\right)^2 \frac{\partial^2 u^{(1)}}{\partial y^2}, \tag{28}$$

$$u^{(2)} = 0, \qquad \text{at } y = 0, \tag{29}$$

$$u^{(2)} = 0, \qquad \text{at } y = h(x). \tag{30}$$

On the other hand, the Stokes system (24)–(27) is solvable if and only if the compatibility condition

$$\int_0^{h(x)} \frac{\partial u^{(1)}}{\partial x}\, \mathrm{d}y = -\, h'(x)\, U^h \tag{31}$$

is satisfied. Using the Leibniz rule, this amounts to

$$\frac{\mathrm{d}}{\mathrm{d}x}\int_0^{h(x)} u^{(1)}\, \mathrm{d}y = 0, \tag{32}$$

The classical Reynolds equation for $p^{(0)}$ is then a consequence of (22), namely

$$\frac{\mathrm{d}}{\mathrm{d}x}\left[\frac{h^3}{12}\frac{\mathrm{d}p^{(0)}}{\mathrm{d}x}\right] = \frac{1}{2}\frac{\mathrm{d}h}{\mathrm{d}x}(U^0 + U^h). \tag{33}$$

The equations at $\mathcal{O}(\epsilon^0)$ in (1), at $\mathcal{O}(\epsilon^1)$ in (2) and at $\mathcal{O}(\epsilon^2)$ in (9) and (10) read as

$$\frac{\partial^2 u^{(3)}}{\partial y^2} = \dots, \tag{34}$$

$$\frac{\partial p^{(3)}}{\partial y} + \frac{\partial^2 v^{(3)}}{\partial y^2} = \dots, \tag{35}$$

$$\frac{\partial v^{(3)}}{\partial y} = -\frac{\partial u^{(2)}}{\partial x}, \tag{36}$$

$$u^{(3)} = v^{(3)} = 0, \quad \text{at} \quad y = 0, \tag{37}$$

$$u^{(3)} = -\frac{1}{2}h'^2 U^h, \qquad v^{(3)} = 0, \qquad \text{at } y = h(x), \tag{38}$$

where by ... we have denoted terms which are known from the previous equations and are thus irrelevant for the following computations.

Equations (35)–(38) form a one-dimensional Stokes system for $(v^{(3)}, p^{(3)})$ and the velocity component $u^{(3)}$ satisfies the second-order ODE

$$\frac{\partial^2 u^{(3)}}{\partial y^2} = \dots, \tag{39}$$

$$u^{(3)} = 0, \qquad \text{at} \quad y = 0, \tag{40}$$

$$u^{(3)} = -\frac{1}{2}h'^2 U^h, \qquad \text{at} \quad y = h(x). \tag{41}$$

arising from Equation (34) and from the boundary conditions (37) and (38). Recall that terms indicated by ... have already been determined by the lower order equations.

The one-dimensional Stokes system for $(v^{(3)}, p^{(3)})$ is solvable if and only if the following compatibility condition is satisfied

$$\int_0^{h(x)} \frac{\partial u^{(2)}}{\partial x}\, \mathrm{d}y = 0\,. \tag{42}$$

Using the Leibniz integral rule, this condition becomes

$$\frac{\mathrm{d}}{\mathrm{d}x} \int_0^{h(x)} u^{(2)}\, \mathrm{d}y = 0\,. \tag{43}$$

Integrating by parts, one sees that

$$\int_0^{h(x)} u^{(2)}\, \mathrm{d}y = \frac{1}{2} \int_0^{h(x)} y\,(y - h(x))\, \frac{\partial^2 u^{(2)}}{\partial y^2}\, \mathrm{d}y\,. \tag{44}$$

Substituting (28) into the previous expression leads to a Reynolds type equation for the first-order pressure correction

$$\begin{aligned} \frac{\mathrm{d}}{\mathrm{d}x}\left[\frac{h^3}{12}\frac{\mathrm{d}p^{(1)}}{\mathrm{d}x}\right] = \frac{\alpha(n-2)}{320} &\left[5h'\frac{\mathrm{d}p^{(0)}}{\mathrm{d}x}\left(4(U^h - U^0)^2 + h^4\left(\frac{\mathrm{d}p^{(0)}}{\mathrm{d}x}\right)^2\right)\right. \\ &\left. + h\frac{\mathrm{d}^2p^{(0)}}{\mathrm{d}x^2}\left(20(U^h - U^0)^2 + 3h^4\left(\frac{\mathrm{d}p^{(0)}}{\mathrm{d}x}\right)^2\right)\right]\,. \end{aligned} \tag{45}$$

2.2.2. Case $\alpha_0 = \mathcal{O}(\epsilon^2)$

Using again the ansatz (13) and (14) and making the same assumptions as in the previous case, except for the material parameter α_0 which is written as $\alpha_0 = \alpha\,\epsilon^2$, we obtain at $\mathcal{O}(\epsilon^{-3})$ in (1) that $\frac{\partial p^{(0)}}{\partial y} = 0$, i.e., $p^{(0)}$ is independent of y.

At $\mathcal{O}(\epsilon^{-2})$ in (1), at $\mathcal{O}(\epsilon^{-1})$ in (2) and at $\mathcal{O}(\epsilon^{0})$ in (9) and (10), we obtain, respectively

$$\frac{\mathrm{d}p^{(0)}}{\mathrm{d}x} = \left(1 + \frac{\alpha}{2}\left(\frac{\partial u^{(1)}}{\partial y}\right)^2\right)^{\frac{n-4}{2}} \left[\frac{\partial^2 u^{(1)}}{\partial y^2} + \frac{\alpha}{2}(n-1)\left(\frac{\partial u^{(1)}}{\partial y}\right)^2 \frac{\partial^2 u^{(1)}}{\partial y^2}\right], \tag{46}$$

$$\frac{\partial p^{(1)}}{\partial y} = 0\,, \tag{47}$$

$$\frac{\partial v^{(1)}}{\partial y} = 0\,, \tag{48}$$

$$u^{(1)} = U^0\,, \quad v^{(1)} = 0\,, \qquad \text{at } y = 0\,, \tag{49}$$

$$u^{(1)} = U^h\,, \quad v^{(1)} = 0\,, \qquad \text{at } y = h(x)\,, \tag{50}$$

where U^0 and U^h denote the tangential velocities at $y = 0$ and at $y = h(x)$. From (48)–(50) it follows that $v^{(1)}(x,y) \equiv 0$, and from (47), we also see that $p^{(1)} = p^{(1)}(x)$.

At the next order in ϵ, the equations are

$$\frac{\mathrm{d}p^{(1)}}{\mathrm{d}x} = \frac{\alpha}{4}(n-4)\frac{\partial u^{(1)}}{\partial y}\frac{\partial^2 u^{(1)}}{\partial y^2}\frac{\partial u^{(2)}}{\partial y}\left(\frac{\alpha}{2}\left(\frac{\partial u^{(1)}}{\partial y}\right)^2+1\right)^{\frac{n-6}{2}}\left(\alpha(n-1)\left(\frac{\partial u^{(1)}}{\partial y}\right)^2+2\right)$$
$$+\left(\frac{\alpha}{2}\left(\frac{\partial u^{(1)}}{\partial y}\right)^2+1\right)^{\frac{n-4}{2}}\left[\frac{\partial^2 u^{(2)}}{\partial y^2}+\frac{\alpha}{2}(n-1)\frac{\partial u^{(1)}}{\partial y}\left(2\frac{\partial u^{(2)}}{\partial y}\frac{\partial^2 u^{(1)}}{\partial y^2}+\frac{\partial u^{(1)}}{\partial y}\frac{\partial^2 u^{(2)}}{\partial y^2}\right)\right], \quad (51)$$

$$\frac{\partial p^{(2)}}{\partial y} = \left(\frac{\alpha}{2}\left(\frac{\partial u^{(1)}}{\partial y}\right)^2+1\right)^{\frac{n-4}{2}}\left(\alpha(n-2)\frac{\partial u^{(1)}}{\partial x}\frac{\partial u^{(1)}}{\partial y}\frac{\partial^2 u^{(1)}}{\partial y^2}\right.$$
$$\left.+\frac{\partial^2 u^{(1)}}{\partial x\partial y}\left(1-\frac{\alpha}{2}(n-3)\left(\frac{\partial u^{(1)}}{\partial y}\right)^2\right)\right), \quad (52)$$

$$\frac{\partial v^{(2)}}{\partial y} = -\frac{\partial u^{(1)}}{\partial x}, \quad (53)$$

$$v^{(2)} = u^{(2)} = 0, \qquad \text{at } y = 0, \quad (54)$$

$$v^{(2)} = U^h h'(x), \quad u^{(2)} = 0 \qquad \text{at } y = h(x). \quad (55)$$

The Stokes system (52)–(55) is solvable if and only if the compatibility condition

$$\int_0^{h(x)} \frac{\partial u^{(1)}}{\partial x}\,\mathrm{d}y = -\,h'(x)\,U^h \quad (56)$$

is satisfied. Using the Leibniz rule, this amounts to

$$\frac{\mathrm{d}}{\mathrm{d}x}\int_0^{h(x)} u^{(1)}\,\mathrm{d}y = 0. \quad (57)$$

On the other hand, rewriting (46) as

$$\left(1+\frac{\alpha}{2}\left(\frac{\partial u^{(1)}}{\partial y}\right)^2\right)^{2-\frac{n}{2}}\frac{\mathrm{d}p^{(0)}}{\mathrm{d}x} = \frac{\partial^2 u^{(1)}}{\partial y^2}+\frac{\alpha}{2}(n-1)\left(\frac{\partial u^{(1)}}{\partial y}\right)^2\frac{\partial^2 u^{(1)}}{\partial y^2}, \quad (58)$$

multiplying both sides by $y(y-h(x))$ and integrating with respect to y from 0 to $h(x)$, yields

$$\frac{\mathrm{d}p^{(0)}}{\mathrm{d}x}\int_0^{h(x)} y(y-h(x))\left(1+\frac{\alpha}{2}\left(\frac{\partial u^{(1)}}{\partial y}\right)^2\right)^{2-\frac{n}{2}} dy =$$
$$= \int_0^{h(x)} y(y-h(x))\frac{\partial^2 u^{(1)}}{\partial y^2}\,dy + \frac{\alpha(n-1)}{2}\int_0^{h(x)} y(y-h(x))\left(\frac{\partial u^{(1)}}{\partial y}\right)^2\frac{\partial^2 u^{(1)}}{\partial y^2}\,dy \quad (59)$$
$$= -h(x)(U^0+U^h)+2\int_0^{h(x)} u^{(1)}\,dy - \frac{\alpha(n-1)}{6}\int_0^{h(x)}(2y-h(x))\left(\frac{\partial u^{(1)}}{\partial y}\right)^3 dy.$$

Differentiating both sides of the previous equation with respect to x, and taking into account the compatibility condition (57) we finally obtain the following equation for $p^{(0)}$:

$$\frac{\mathrm{d}}{\mathrm{d}x}\left[\frac{\mathrm{d}p^{(0)}}{\mathrm{d}x}\int_0^{h(x)} y(y-h(x))\left(1+\frac{\alpha}{2}\left(\frac{\partial u^{(1)}}{\partial y}\right)^2\right)^{2-\frac{n}{2}} dy\right] =$$

$$= -h'(x)(U^0+U^h) - \frac{\alpha(n-1)}{6}\frac{\mathrm{d}}{\mathrm{d}x}\left[\int_0^{h(x)} (2y-h(x))\left(\frac{\partial u^{(1)}}{\partial y}\right)^3 dy\right]. \tag{60}$$

Equations (46) and (60) form a dimensionally reduced nonlinear system of differential equations for the (main parts) of the velocity and the pressure fields. *Note that when $n = 2$ (or $\alpha = 0$) Equation (46) reduces to (16) and Equation (60) simplifies to the classical Reynolds Equation (33).*

Remark 1. *We have been concerned here with the power-law model (3). Other power-law type stress-velocity gradient relationships of course exist, see, e.g., expressions (2.10) and (2.14) in [24]. However, neither one of these latter models allows one to derive novel Reynolds type equations. Consider, for example, the deviatoric stress tensor*

$$\mathbf{S} = 2\left(\mu_0 + \mu_1\left(|\mathbf{D}|^2\right)^{\frac{n-2}{2}}\right)\mathbf{D} \tag{61}$$

(see (2.14) in [24]), and write it as

$$\mathbf{S} = 2\mu_0\left(1+\beta_1\,|\mathbf{D}|^{n-2}\right)\mathbf{D}, \tag{62}$$

with $\beta_1 = \mu_1/\mu_0$. Given that the modulus of $\mathbf{D}$, denoted by $|\mathbf{D}|$, is of order ϵ^{-1} since the derivatives in the vertical direction are of order ϵ^{-1} and the velocities are of order ϵ^0, β_1 needs to be of order ϵ^{n-2} for a Reynolds type equation to be possible. This choice is model-dependent and leads essentially to the classic Reynolds equation.

3. Examples

We will now use our corrections to the Reynolds equation to examine the influence of the power-law exponent on the lubrication characteristics. We will consider two classic examples: the flow between a rolling rigid cylinder and a rigid plane and the flow in an eccentric journal bearing, cf. [2]. We take that the pressures are not too high so that we can assume constant classical viscosity and ignore the possible deformation of the rolling cylinder and the dependence of the viscosity on the pressure.

3.1. Rolling Cylinder

Let h_0 be the minimum distance between the cylinder of radius R and the plane, cf. Figure 1, and let $\epsilon = h_0/R$. The non-dimensional film thickness $h = h(\theta)$ can be expressed in terms of the angular coordinate θ as

$$h(\theta) = \epsilon + 1 - \cos\theta\,. \tag{63}$$

We will consider the cases $\alpha_0 = \mathcal{O}(\epsilon^3)$ and $\alpha_0 = \mathcal{O}(\epsilon^2)$ separately and assume throughout that $U^0 = 0$ and $U^h = 1$.

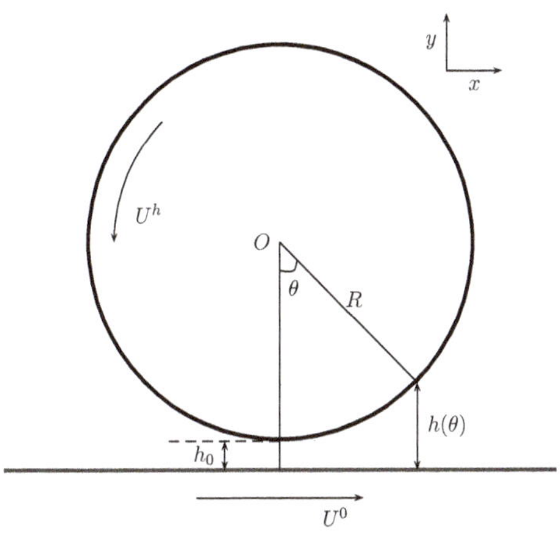

Figure 1. Cross-sectional geometry of an infinite cylinder rolling counter-clockwise on a plane. The fluid is between the cylinder and the flat surface. Reprinted from [14], with permission from Elsevier.

3.1.1. $\alpha_0 = \mathcal{O}(\epsilon^3)$

The Reynolds Equation (33) reduces to

$$\frac{\mathrm{d}}{\mathrm{d}x}\left[\frac{h^3}{12}\frac{\mathrm{d}p^{(0)}}{\mathrm{d}x}\right] = \frac{h'}{2}. \tag{64}$$

Letting θ_1 denote the unknown position of the liquid-cavity interface where $\frac{\mathrm{d}p^{(0)}}{\mathrm{d}x} = 0$ and making the transformation $\theta = \sin^{-1} x$, we obtain

$$\frac{\mathrm{d}p^{(0)}}{\mathrm{d}\theta} = 6\cos\theta\,\frac{h(\theta) - h(\theta_1)}{h^3(\theta)}. \tag{65}$$

Assuming that the continuous film starts at $\theta = -\frac{\pi}{2}$ and using the second Swift-Stieber boundary condition $p^{(0)}(\theta_1) = 0$, cf. [2], we may determine θ_1 from the condition

$$\int_{-\frac{\pi}{2}}^{\theta_1} \frac{\mathrm{d}p^{(0)}}{\mathrm{d}\theta}\,\mathrm{d}\theta = 0. \tag{66}$$

The Reynolds equation for the pressure correction reads as

$$\frac{\mathrm{d}}{\mathrm{d}x}\left[\frac{h^3}{12}\frac{\mathrm{d}p^{(1)}}{\mathrm{d}x}\right] = \frac{\alpha m}{160}\left[5h'\frac{\mathrm{d}p^{(0)}}{\mathrm{d}x}\left(4 + h^4\left(\frac{\mathrm{d}p^{(0)}}{\mathrm{d}x}\right)^2\right) + h\frac{\mathrm{d}^2p^{(0)}}{\mathrm{d}x^2}\left(20 + 3h^4\left(\frac{\mathrm{d}p^{(0)}}{\mathrm{d}x}\right)^2\right)\right], \tag{67}$$

where we have redefined the power-law exponent as

$$m = (n-2)/2$$

so that for $m \in [-0.5, 0)$ the fluid has the ability to shear thin and for $m > 0$ to shear thicken, see Málek et al. [25] for a general discussion of properties of fluids with shear-dependent viscosity. Making the change of variables $\theta = \sin^{-1} x$ and expressing $\mathfrak{p} = p^{(0)} + \epsilon\, p^{(1)}$, we thus obtain

$$\frac{\mathrm{d}\mathfrak{p}}{\mathrm{d}\theta} = \frac{\cos\theta}{h^3(\theta)}\left[6h(\theta) + \epsilon\frac{3\,\alpha m}{40}\int_{\theta_2}^{\theta}\frac{5h'(\theta)}{\cos\theta}\frac{\mathrm{d}p^{(0)}}{\mathrm{d}\theta}\left(4+\frac{h^4(\theta)}{\cos^2\theta}\left(\frac{\mathrm{d}p^{(0)}}{\mathrm{d}\theta}\right)^2\right)\mathrm{d}\theta\right.$$
$$\left. + \epsilon\frac{3\,\alpha\,m}{40}\int_{\theta_2}^{\theta}\frac{h(\theta)}{\cos\theta}\frac{\mathrm{d}^2p^{(0)}}{\mathrm{d}\theta^2}\left(20+\frac{3\,h^4(\theta)}{\cos^2\theta}\left(\frac{\mathrm{d}p^{(0)}}{\mathrm{d}\theta}\right)^2\right)\mathrm{d}\theta\right]. \tag{68}$$

where we have assumed that $\frac{\mathrm{d}\mathfrak{p}}{\mathrm{d}\theta} = 0$ at θ_2, the corrected position of the liquid-cavity interface. As above, θ_2 can be determined from the condition

$$\int_{-\frac{\pi}{2}}^{\theta_2}\frac{\mathrm{d}\mathfrak{p}}{\mathrm{d}\theta}\,\mathrm{d}\theta = 0\,, \tag{69}$$

assuming that $\mathfrak{p}(\theta_2) = 0$.

In Figure 2, we plot $\mathfrak{p}$ for $\epsilon = 0.1$, considering the power-law exponent values $m = -0.5$, -0.375, 1, and compare the pressure profiles with $p^{(0)}$.

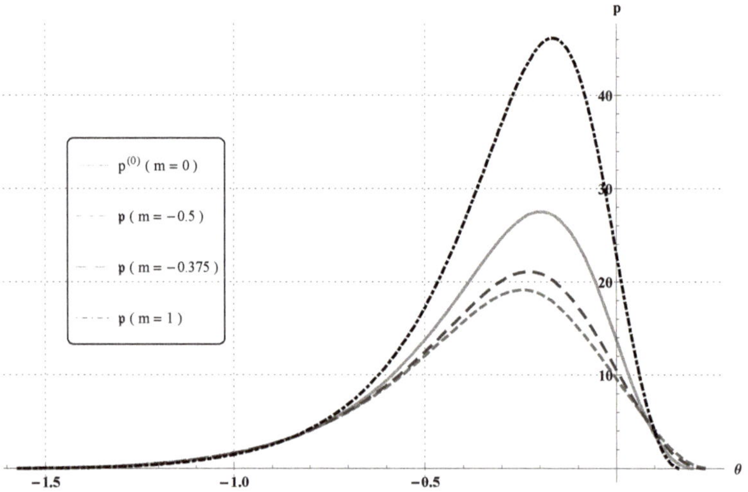

Figure 2. Pressure distributions computed from the classical Reynolds equation, $p^{(0)}(\theta)$ $(m = 0)$, and from the power-law system of Reynolds type equations, $\mathfrak{p}(\theta)$ $(m = -0.5, -0.375, 1)$, with $\epsilon = 0.1$ and $\alpha_0 = \mathcal{O}(\epsilon^3)$.

For $\epsilon = 0.1$, the value of the cavitation angle in the classical case $(m = 0)$ is $\theta_1 = 0.197408$ and the normal force is $\omega'_{\mathfrak{n}} = \int_{-\frac{\pi}{2}}^{\theta_1}\cos\theta\, p^{(0)}(\theta)\mathrm{d}\theta = 13.985$ in non-dimensional units. In Table 1, we present the values of the cavitation angle θ_2 and the corresponding normal forces for the power-law lubrication approximations.

Table 1. Cavitation angles θ_2 and the normal forces $\omega'_{\mathfrak{p}}$, computed for $m = -0.5, -0.375, 1$, when $\epsilon = 0.1$.

m	θ_2 **(rad)**	$\omega'_{\mathfrak{p}}$
−0.5	0.245119	10.7558
−0.375	0.229178	11.5351
1	0.167257	20.8097

In Figure 3 and Table 2, we document the same results when $\epsilon = 0.01$. The classical ($m = 0$) cavitation angle is now $\theta_1 = 0.0666263$ (radians) and the normal force $\omega'_{\mathfrak{p}} = 221.456$.

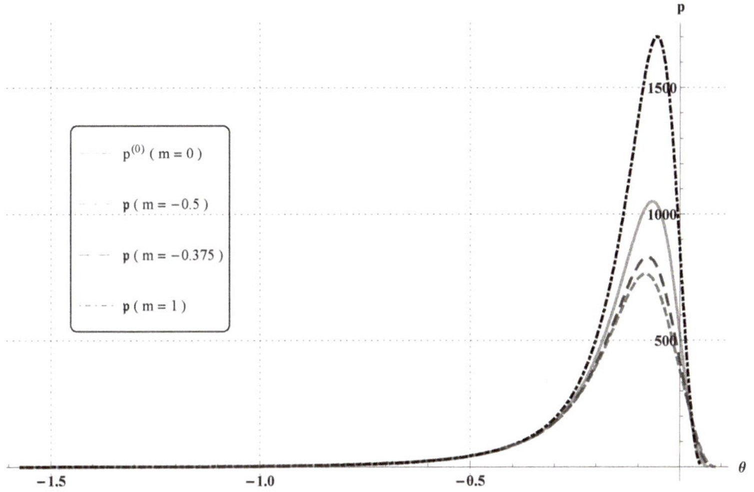

Figure 3. Pressure distributions computed from the classical Reynolds equation, $p^{(0)}(\theta)$ ($m = 0$), and from the power-law system of Reynolds type equations, $\mathfrak{p}(\theta)$ ($m = -0.5, -0.375, 1$), with $\epsilon = 0.01$ and $\alpha_0 = \mathcal{O}(\epsilon^3)$.

Table 2. Cavitation angles θ_2 and the normal forces $\omega'_{\mathfrak{p}}$, computed for $m = -0.5, -0.375, 1$, when $\epsilon = 0.01$.

m	θ_2 **(rad)**	$\omega'_{\mathfrak{p}}$
−0.5	0.0827747	181.54
−0.375	0.077616	191.168
1	0.0553576	306.272

3.1.2. $\alpha_0 = \mathcal{O}(\epsilon^2)$

In this case, we need to solve (numerically) the following nonlinear system for $p^{(0)}$ and $u^{(1)}$:

$$\frac{dp^{(0)}}{d\theta} = \cos\theta\left(1+\frac{\alpha}{2}\left(\frac{\partial u^{(1)}}{\partial y}\right)^2\right)^{m-1}\left[\frac{\partial^2 u^{(1)}}{\partial y^2}+\frac{\alpha}{2}(n-1)\left(\frac{\partial u^{(1)}}{\partial y}\right)^2\frac{\partial^2 u^{(1)}}{\partial y^2}\right],$$

$$\frac{dp^{(0)}}{d\theta}\int_0^{h(\theta)} y(y-h(\theta))\left(1+\frac{\alpha}{2}\left(\frac{\partial u^{(1)}}{\partial y}\right)^2\right)^{1-m} dy \tag{70}$$

$$= -\cos\theta\left[h(\theta)+\frac{\alpha(n-1)}{6}\int_0^{h(\theta)}(2y-h(\theta))\left(\frac{\partial u^{(1)}}{\partial y}\right)^3 dy\right]_{\theta_2}^{\theta}.$$

The boundary conditions for the pressure are the Swift-Stieber conditions, i.e.,

$$p^{(0)}(\theta_2) = 0, \qquad \frac{dp^{(0)}}{d\theta}(\theta_2) = 0, \tag{71}$$

and for the velocity those given in (49) and (50).

The computed pressure distributions have been plotted in Figure 4, for $\epsilon = 0.1$, and in Figure 5, for $\epsilon = 0.01$. In each case, we have considered the values of power-law exponent $-0.5, -0.375$ and 1 and compared the distributions with that of the classical Newtonian case ($m = 0$). The values of the corresponding cavitation angles and the normal forces are given in Tables 3 and 4.

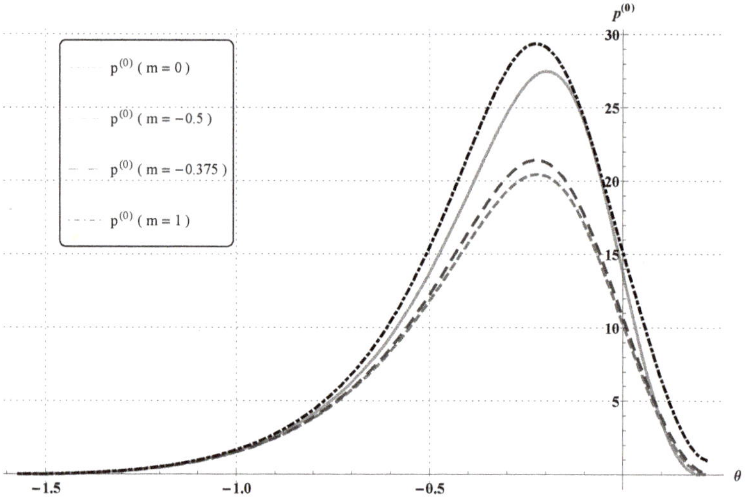

Figure 4. Pressure distributions computed from the classical Reynolds equation, $p^{(0)}(\theta)$ ($m = 0$), and from the power-law system of Reynolds type equations, $p^{(0)}(\theta)$ ($m = -0.5, -0.375, 1$), with $\epsilon = 0.1$ and $\alpha_0 = \mathcal{O}(\epsilon^2)$.

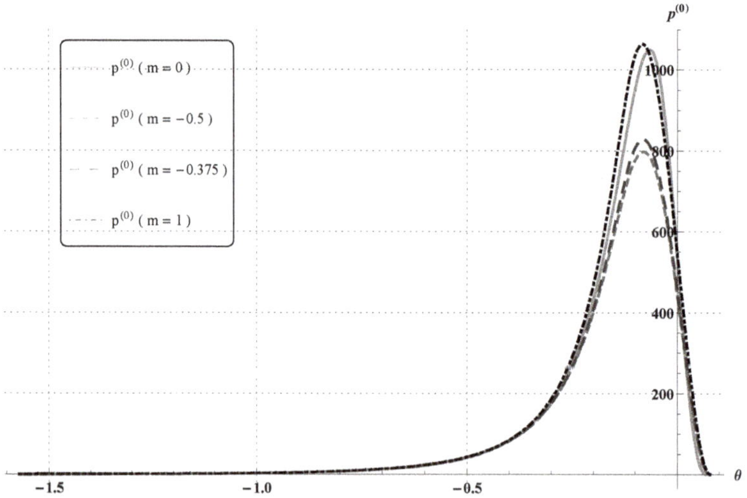

Figure 5. Pressure distributions computed from the classical Reynolds equation, $p^{(0)}(\theta)$ ($m = 0$), and from the power-law system of Reynolds type equations, $p^{(0)}(\theta)$ ($m = -0.5, -0.375, 1$), with $\epsilon = 0.01$ and $\alpha_0 = \mathcal{O}(\epsilon^2)$.

Table 3. Cavitation angles θ_2 and the normal forces ω'_p, computed for $m = -0.5, -0.375, 0, 1$ when $\epsilon = 0.1$.

m	θ_2 (rad)	ω'_p
−0.5	0.221825	11.105
−0.375	0.225363	11.5739
0	0.197408	13.985
1	0.227645	15.2337

Table 4. Cavitation angles θ_2 and the normal forces ω'_p, computed for $m = -0.5, -0.375, 0, 1$ when $\epsilon = 0.01$.

m	θ_2 (rad)	ω'_p
−0.5	0.0747584	189.875
−0.375	0.0760101	194.649
0	0.0666264	221.456
1	0.0774216	233.142

In general, the results are similar to the linear case. However, the maxima for the power-law pressures are now shifted to the left with respect to the classical Newtonian case.

3.2. *Journal Bearing*

Consider an (infinitely) long journal bearing consisting of a cylinder (journal) rotating eccentrically, at an angular velocity Ω, inside another cylinder (bearing) filled with an incompressible power-law fluid.

We denote the journal radius by R_0, the bearing radius by R ($R_0 < R$) and the distance between centers by a, see Figure 6. We define $\delta \equiv a/R_0$ and $\epsilon \equiv (R - R_0)/R_0$ and assume that the non-dimensional parameters δ and ϵ are small. The fluid film thickness H can be expressed as a function of the angle φ measured counter-clockwise from the bearing radius perpendicular to the journal. Measured along the bearing radius, the film thickness is, up to an error of the order of $O(\delta^3)$ (see [2]), given by

$$H(\varphi) = R - R_0 + a\,\cos(\varphi) + \frac{a^2}{2\,R_0}\,\sin^2(\varphi)\,. \tag{72}$$

The non-dimensional film thickness $h(\varphi)$ is defined through $H = R_0\epsilon h$ and, up to an error of the order of $\mathcal{O}(\epsilon^2)$, can be written as

$$h(\varphi) = 1 + \chi\,\cos(\varphi) + \epsilon\,\frac{\chi^2}{2}\,\sin^2(\varphi)\,, \tag{73}$$

where $\chi = a/(R - R_0)$ is the eccentricity ratio.

The equations we have derived do not take into account the curvature effects. In order to do so, we should express the equations in natural orthogonal coordinates (cf. Nazarov and Videman [22]). Since curvature brings a first-order correction term to the classical Reynolds equations, in geometries such as the journal bearing (see [22]), we will drop the last term in (73) and do not consider the case where the non-Newtonian effects appear at the first-order. The equations for the journal bearing become thus similar to those in Section 3.1, except for the absence of the cosine of the angle and for the angular coordinate being the non-dimensional $x = \varphi$. In other words, we make $\cos\theta \to 1$ and $\theta \to \varphi$ in the equations of Section 3.1. The boundary conditions for the pressure $p^{(0)}$ are $p^{(0)}(0) = p^{(0)}(2\pi)$ and $p^{(0)}(\pi) = 0$. Moreover, we choose $U^0 = 0$ and $U^h = 1$ (the velocity scale is $U = 2\pi\Omega R_0$).

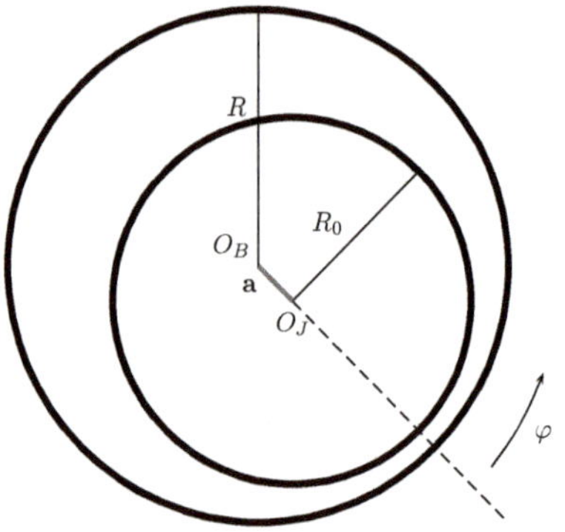

Figure 6. Cross-section of an infinite journal bearing. The journal is centered at O_J and rotates counter-clockwise within the fixed bearing centered at O_B. The fluid occupies the space between the two cylinders. Reprinted from [14], with permission from Elsevier.

3.2.1. $\alpha_0 = \mathcal{O}(\epsilon^3)$

We do not consider this case since, as explained above, the curvature effects are of the same order as the first order correction $p^{(1)}(\varphi)$.

3.2.2. $\alpha_0 = \mathcal{O}(\epsilon^2)$

The following pressure profiles were computed choosing (a) $\chi = 0.1$, (b) $\chi = 0.5$, (c) $\chi = 0.95$, see Figures 7–9, respectively.

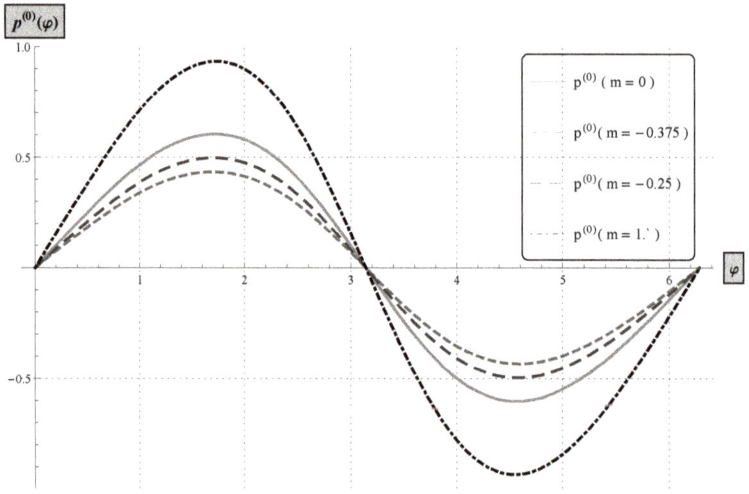

Figure 7. Pressure distributions computed from the classical Reynolds equation, $p^{(0)}(\theta)$ $(m = 0)$, and from the power-law system of Reynolds type equations, $p^{(0)}(\theta)$ $(m = -0.375, -0.25, 1)$, with $\chi = 0.1$.

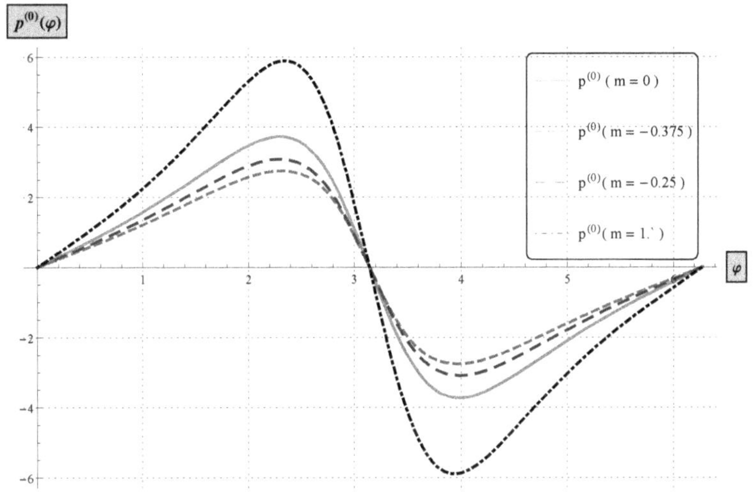

Figure 8. Pressure distributions computed from the classical Reynolds equation, $p^{(0)}(\theta)$ $(m = 0)$, and from the power-law system of Reynolds type equations, $p^{(0)}(\theta)$ $(m = -0.375, -0.25, 1)$, with $\chi = 0.5$.

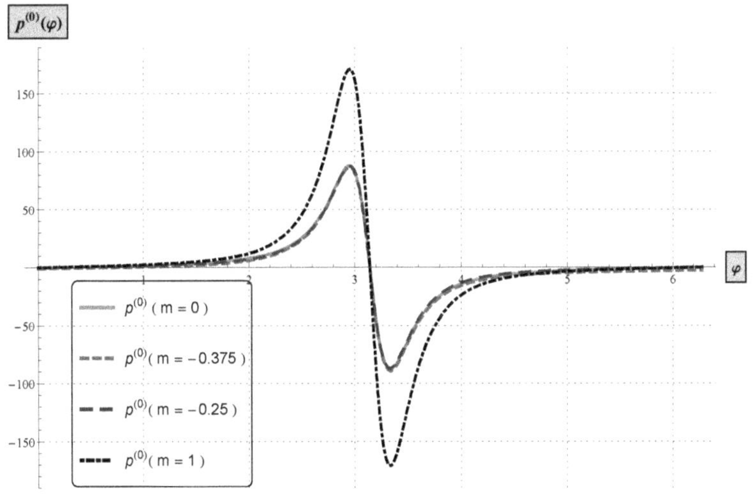

Figure 9. Pressure distributions computed from the classical Reynolds equation, $p^{(0)}(\theta)$ $(m = 0)$, and from the power-law system of Reynolds type equations, $p^{(0)}(\theta)$ $(m = -0.375, -0.25, 1)$, with $\chi = 0.95$.

We find that the increase of χ, namely from $\chi = 0.1$ to $\chi = 0.95$, results in a change in the pressure profile, regardless of the value of n. The maxima are higher for larger values of m, as in the previous example. No change is visible to the zeros of the pressure profile and no change of phase is obtained with regard to changes in n, χ, or ϵ.

4. Conclusions

We have given a rigorous derivation of Reynolds type lubrication approximations for a family of non-Newtonian power-law models. Based on a formal perturbation analysis, we have shown that, in the cases considered, lubricating film flows for fluids with shear-dependent viscosities can be approximated by simplified dimensionally reduced models. We have corroborated our theoretical results by presenting numerical computations which show that the pressure profiles behave as expected for shear-thinning fluids.

Author Contributions: Conceptualization, J.H.V.; Investigation, F.S.C.; Software, B.M.M.P.; Writing—original draft, G.A.S.D.; Writing—review and editing, K.R.R.

Funding: This research was funded by the Portuguese government through Fundação para a Ciência e a Tecnologia (FCT), Instituto Público, under the project UTAP-EXPL/MAT/0017/2017. G.A.S.D. was partially funded by the FCT fellowship SFRH/BPD/70578/2010.

Acknowledgments: The authors would like to thank Tom Gustafsson for valuable suggestions.

Conflicts of Interest: The authors declare no conflict of interest. The funders had no role in the design of the study; in the collection, analyses, or interpretation of data; in the writing of the manuscript, or in the decision to publish the results.

References

1. Reynolds, O. On the theory of lubrication and its application to Mr Tower's experiments. *Philos. Trans. R. Soc. Lond.* **1886**, *177*, 159–209.
2. Szeri, A.Z. *Fluid Film Lubrication*, 2nd ed.; Cambridge University Press: Cambridge, MA, USA, 2010.
3. Bair, S.; Khonsari, M.M. Reynolds equations for common generalized Newtonian models and an approximate Reynolds–Carreau equation. *Proc. Inst. Mech. Eng. Part J J. Eng.* **2006**, *220*, 365–374. [CrossRef]
4. Myers, T.G. Application of non-Newtonian models to thin film flow. *Phys. Rev. E* **2005**, *72*, 066302. [CrossRef] [PubMed]
5. Yang, Q.; Huang, P.; Fang, Y. A novel Reynolds equation of non-Newtonian fluid for lubrication simulation. *Tribol. Int.* **2016**, *94*, 458–463. [CrossRef]
6. Shukla, J.B.; Prasad, K.R.; Chandra, P. Effects of consistency variation of power law lubricants in squeeze film. *Wear* **1982**, *76*, 299–319. [CrossRef]
7. Park, O.O.; Kwon, M.H. Study on the lubrication approximation of power law fluid. *Korean J. Chem. Eng.* **1989**, *6*, 150–153. [CrossRef]
8. Johnson, M.W., Jr.; Mangkoesoebroto, S. Analysis of lubrication theory for the power law fluid. *J. Tribol.* **1993**, *115*, 71–77. [CrossRef]
9. Bourgin, P. Fluid-Film Flows of Differential Fluids of Complexity n Dimensional Approach—Applications to Lubrication Theory. *ASME J. Lubr. Technol.* **1979**, *101*, 140–144. [CrossRef]
10. Kacou, A.; Rajagopal, K.R.; Szeri, A.Z. A thermodynamic analysis of journal bearings lubricated by non-Newtonian fluids. *J. Tribol.* **1998**, *110*, 414–428. [CrossRef]
11. Harnoy, A. Three dimensional analysis of the elastico-viscous lubrication in short journal bearings. *Rheol. Acta* **1977**, *16*, 51–60. [CrossRef]
12. Ballal, B.Y.; Rivlin, R.S. Flow of a Newtonian fluid between eccentric rotating cylinders: Inertial effects. *Arch. Ration. Mech. Anal.* **1976**, *62*, 237–294. [CrossRef]
13. Andres, A.S.; Szeri, A.Z. Flow between Eccentric Rotating Cylinders. *J. Appl. Mech.* **1984**, *51*, 869–878. [CrossRef]
14. Cal, F.S.; Dias, G.A.S.; Pereira, B.M.M.; Pires, G.E.; Rajagopal, K.R.; Videman, J.H. On the lubrication approximation for a class of viscoelastic fluids. *Int. J. Nonlinear Mech.* **2016**, *87*, 30–37. [CrossRef]
15. Barus, C. Isothermals, isopiestics and isometrics relative to viscosity. *Am. J. Sci.* **1893**, *45*, 87–96. [CrossRef]
16. Bair, S. *High Pressure Rheology for Quantitative Elastohydrodynamics*; Elsevier: Amsterdam, The Netherlands, 2007.
17. Rajagopal, K.R.; Szeri, A. On an inconsistency in the derivation of the equations of elastohydrodynamic lubrication. *Proc. R. Soc. Lond. A Math. Phys. Sci.* **2003**, *459*, 2771–2786. [CrossRef]
18. Gustafsson, T.; Rajagopal, K.R.; Stenberg, R.; Videman, J. Nonlinear Reynolds equation for hydrodynamic lubrication. *Appl. Math. Model.* **2015**, *39*, 5299–5309. [CrossRef]
19. Fusi, L.; Farina, A.; Rosso, F.; Roscani, S. Pressure driven lubrication flow of a Bingham fluid in a channel: A novel approach. *J. Non-Newton. Fluid Mech.* **2015**, *221*, 66–75. [CrossRef]
20. Rajagopal, K.R. Implicit constitutive relations for fluids. *J. Fluid Mech.* **2006**, *550*, 243–249. [CrossRef]
21. Rajagopal, K.R.; Srinivasa, A.R. A Gibbs-potential-based formulation for obtaining the response functions for a class of viscoelastic materials. *Proc. R. Soc. Lond. A Math. Phys. Sci.* **2011**, *467*, 39–58. [CrossRef]

22. Nazarov, S.A.; Videman, J.H. A modified nonlinear Reynolds equation for thin viscous flows in lubrication. *Asymptot. Anal.* **2007**, *52*, 1–36.
23. Rajagopal, K.R. Remarks on the notion of "pressure". *Int. J. Nonlinear Mech.* **2015**, *71*, 165–172. [CrossRef]
24. Málek, J. Mathematical properties of flows of incompressible power-law-like fluids that are described by implicit constitutive relations. *Electron. Trans. Numer. Anal.* **2008**, *31*, 110–125.
25. Málek, J.; Rajagopal, K.R.; Ruzicka, M. Existence and regularity of solutions and stability of the rest state for fluids with shear dependent viscosity. *Math. Model. Methods Appl. Sci.* **1995**, *6*, 789–812. [CrossRef]

fluids

MDPI

Article

Application of a Projection Method for Simulating Flow of a Shear-Thinning Fluid

Masoud Jabbari [1,*], James McDonough [2], Evan Mitsoulis [3] and Jesper Henri Hattel [4]

1 School of Mechanical, Aerospace and Civil Engineering, University of Manchester, Manchester M13 9PL, UK
2 Departments of Mechanical Engineering and Mathematics, University of Kentucky, Lexington, KY 40506, USA
3 School of Mining Engineering and Metallurgy, National Technical University of Athens, 15780 Zografou, Greece
4 Departments of Mechanical Engineering, Technical University of Denmark, Nils Koppels Allé, 2800 Kgs. Lyngby, Denmark
* Correspondence: M.Jabbari@Manchester.ac.uk; Tel.: +44-161-306-4601

Received: 29 May 2019; Accepted: 3 July 2019; Published: 8 July 2019

Abstract: In this paper, a first-order projection method is used to solve the Navier–Stokes equations numerically for a time-dependent incompressible fluid inside a three-dimensional (3-D) lid-driven cavity. The flow structure in a cavity of aspect ratio $\delta = 1$ and Reynolds numbers (100, 400, 1000) is compared with existing results to validate the code. We then apply the developed code to flow of a generalised Newtonian fluid with the well-known Ostwald–de Waele power-law model. Results show that, by decreasing n (further deviation from Newtonian behaviour) from 1 to 0.9, the peak values of the velocity decrease while the centre of the main vortex moves towards the upper right corner of the cavity. However, for $n = 0.5$, the behaviour is reversed and the main vortex shifts back towards the centre of the cavity. We moreover demonstrate that, for the deeper cavities, $\delta = 2, 4$, as the shear-thinning parameter n decreased the top-main vortex expands towards the bottom surface, and correspondingly the secondary flow becomes less pronounced in the plane perpendicular to the cavity lid.

Keywords: lid-driven cavity; projection method; shear-thinning; aspect ratio; *Re* numbers

1. Introduction

Flow inside a cubic cavity with a top lid moving at a constant velocity has been the centre of many fundamental fluid flow studies [1–3] as well as practical applications [4–6]. Flow inside a lid-driven cavity (LDC)—see Figure 1—has been reportedly studied both experimentally and numerically [7–15], and used as a reliable benchmark for validating Navier–Stokes (NS) solvers. Recently, Shankar and Deshpande [1] have comprehensively reviewed the LDC flow and identified crucial aspects open for further investigations.

Steady two-dimensional (2-D) LDC flow has been studied in literature at different Reynolds number [12,16]—$Re = U_{lid}L/\nu$, where U_{lid} is the lid velocity, L the cavity length and ν the kinematic viscosity of the fluid. The 2-D LDC, however, deviates from actual experiments in 3-D mainly due to the no-slip conditions imposed by end walls in the cross-flow direction span [9–11,17]. Albensoeder et al. [17] reported that the fluid bifurcates from steady to periodic, and eventually to turbulent flow when it goes above the critical Reynolds number, $Re_{cr} \sim \mathcal{O}\left(10^3\right)$. This is, however, dependent on cavity aspect ratios, i.e., length-to-width, $\lambda = L/W$, and depth-to-width, $\delta = H/L$ [18,19].

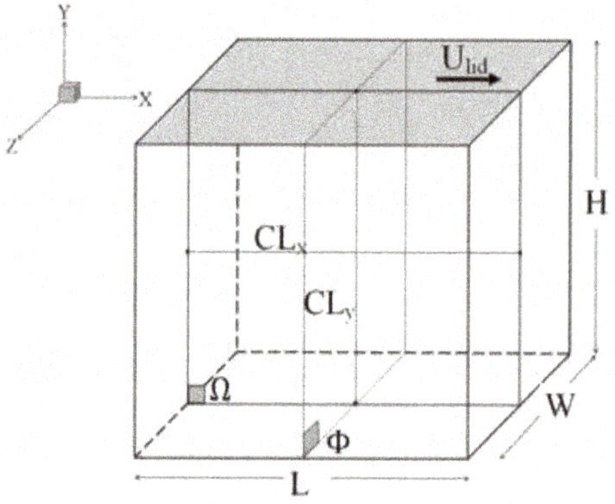

Figure 1. Schematic of a lid-driven cavity in three dimensions.

Most fluids in nature exhibit Newtonian behaviour, in which the dynamic viscosity, μ, at constant temperature and pressure is equal to $\tau/\dot{\gamma}$—τ is the shear stress, and $\dot{\gamma}$ is the shear rate [20,21]. For an incompressible fluid, the equation governing the tensorial shear stress (in Cartesian coordinates) is [22,23]

$$\underline{\underline{\tau}} = \mu \underline{\underline{\dot{\gamma}}}, \tag{1}$$

where μ is the dynamic viscosity, $\underline{\underline{\tau}}$ is the shear stress tensor, and $\underline{\underline{\dot{\gamma}}}$ the shear rate tensor. Here, $\underline{\underline{\dot{\gamma}}}$ is the rate of strain tensor, $\underline{\underline{\dot{\gamma}}} = \nabla \boldsymbol{u} + \nabla \boldsymbol{u}^\top$, which is given by [24–26]

$$|\dot{\gamma}| = \sqrt{\frac{1}{2} II_{\dot{\gamma}}} = \left[\frac{1}{2}\left\{\underline{\underline{\dot{\gamma}}} : \underline{\underline{\dot{\gamma}}}\right\}\right]^{1/2}, \tag{2}$$

where $II_{\dot{\gamma}}$ is the second invariant of $\underline{\underline{\dot{\gamma}}}$, $\nabla \boldsymbol{u}$ is the velocity gradient tensor and $\nabla \boldsymbol{u}^\top$ is its transpose. Similarly, for the stress tensor, $\underline{\underline{\tau}}$, we have [27–29]

$$|\tau| = \sqrt{\frac{1}{2} II_{\tau}} = \left[\frac{1}{2}\left\{\underline{\underline{\tau}} : \underline{\underline{\tau}}\right\}\right]^{1/2}, \tag{3}$$

where II_{τ} is the second invariant of $\underline{\underline{\tau}}$. The correlation between shear stress and shear rate defines the flow behaviour of a liquid. For a Newtonian fluid, there is a linear correlation between the shear stress and shear rate, and the slope of this line is the dynamic viscosity, μ. Recognition that stress in a fluid can have a nonlinear or temporal (or both) dependence on the rate of deformation developed in the late nineteenth and early twentieth centuries; we now refer to such materials as non-Newtonian fluids. For a non-Newtonian fluid, a.k.a nonlinear or complex fluid, the slope of shear stress versus shear rate curve is not constant, and hence is a function of τ or $\dot{\gamma}$. The most common category of the complex fluids is the pure viscous fluid in which $\underline{\underline{\dot{\gamma}}}$ is determined only by the current value of $\underline{\underline{\tau}}$ at any particular point in the flow domain.

Shear-thinning is perhaps the most widely encountered type of time-dependent non-Newtonian fluid behaviour in engineering practice for ceramics [30–33]. It is characterised by an apparent viscosity μ_a which gradually decreases with increasing shear rate:

$$\underline{\underline{\tau}} = \mu_a \underline{\underline{\dot{\gamma}}}, \tag{4}$$

or, in terms of the apparent viscosity,

$$\mu_a = k\underline{\underline{\dot{\gamma}}}^{n-1} \Rightarrow \underline{\underline{\tau}} = \left(k\underline{\underline{\dot{\gamma}}}^{n-1}\right)\underline{\underline{\dot{\gamma}}}. \tag{5}$$

Equation (5) is known as the Ostwald–de Waele power-law model. Varying n between $0 < n < 1$ will yield $d\mu/d\dot{\gamma} < 0$; i.e., shear-thinning fluids are characterised by a value of n (power-law index) smaller than unity. Many polymer melts and solutions exhibit a value of n in the range 0.3–0.7 depending upon the concentration and molecular weight, etc., of the polymer. Even smaller values of power-law index (n~0.1–0.15) are encountered with fine particle suspensions such as, e.g., kaolin-in-water and bentonite-in-water. Naturally, the smaller the value of n, the more shear-thinning the material is. The other constant, k, in Equation (5) (consistency index) is a measure of consistency of the substance, which is well described by Jabbari and Hattel [31].

A computational fluid dynamics (CFD) approach to problems showing non-Newtonian behaviour is generally more complicated than that for a corresponding Newtonian case due to the complexity introduced by the constitutive equation of the employed model in the diffusion terms of the Navier–Stokes equations. In particular, it is clear from Equation (5) that the proposed model for viscosity introduces nonlinearities in the diffusion terms—the highest-order differential operators. Different numerical methods have been employed in the past in order to simulate viscous and viscoplastic fluid flows; the most widely known simulations have employed power-law and Bingham models. However, for the LDC problem, most available studies have focused on modelling elastic and viscoelastic fluid flows: see, for example, [34–36]. There are also some studies related to the LDC problem for viscoplastic fluids (power-law and Bingham models) in which rather different approaches were used for solving the NS equations, viz., the smoothed particle hydrodynamics (SPH) method [37], and the PIM (point interpolation method) meshfree method [38]. Despite remarks by these authors, there seems to be no clear-cut overall advantage to non-standard methods in the present context.

In this work, we present results from simulation of flow inside a 3-D LDC. We begin with the mathematical formulation and brief discussion of the numerical analytic procedures employed, all of which are fairly well known. We validate a particular Fortran code in a range of Re numbers corresponding to steady flow by using existent, highly-accurate, direct numerical simulations for comparisons with our results. We then proceed to a problem involving a generalised Newtonian fluid with the shear-thinning power-law behaviour. The influence on the flow pattern in the 3-D cavity with several depth-to-width aspect ratios, δ, as well as Reynolds number, are investigated. These values of Reynolds number and aspect ratios are relevant to processing and manufacturing of ceramics and polymers.

2. Mathematical Model

Projection methods such as proposed by Chorin [39], Bell et al. [40] and Gresho [41], e.g., are used in problems involving time-dependent incompressible fluids; the diffusion–convection equations are first solved to predict intermediate velocities which are then projected onto the space of divergence-free vector fields [41]. The incompressible NS equations (the differential form of momentum equation combined with the continuity equation) may be written as

$$\frac{\partial \boldsymbol{u}}{\partial t} + \boldsymbol{u} \cdot \nabla \boldsymbol{u} = -\frac{1}{\rho} \nabla p + \nu \Delta \boldsymbol{u}\,, \tag{6a}$$

$$\nabla \cdot \boldsymbol{u} = 0\,, \tag{6b}$$

where ρ is density, and ν is kinematic viscosity. These equations are solved in the interior of a domain Ω consisting of the unit cube, $(0,1) \times (0,1) \times (0,1)$, shown in Figure 1, with initial conditions $\boldsymbol{u} \equiv 0$ and no-slip boundary conditions.

In the projection 1 method analyzed in [41], and employed here, an intermediate velocity, $\boldsymbol{u}^*$, is computed using the momentum equations with pressure gradient terms absent—given here in semi-discrete form:

$$\frac{\boldsymbol{u}^* - \boldsymbol{u}^n}{\Delta t} = -\boldsymbol{u}^n \cdot \nabla \boldsymbol{u}^n + \nu \Delta \boldsymbol{u}^n\,, \tag{7}$$

where $\boldsymbol{u}^n$ is the velocity at the nth time level ($\boldsymbol{u} = (u, v, w)^T$). Equation (7) comprises an easily-solved multi-dimensional system of Burgers' equations. In the present work, the system is coded using a Fortran 77/90 research code based on Douglas and Gunn time splitting [42] with quasilinearisation (a function-space Newton's method) of nonlinear advective terms.

In the next (projection) step, the velocity at the $(n+1)$th time level is computed as follows:

$$\frac{\boldsymbol{u}^{n+1} - \boldsymbol{u}^*}{\Delta t} = -\frac{1}{\rho}\nabla p^{n+1}, \tag{8a}$$

$$\Leftrightarrow \quad \boldsymbol{u}^{n+1} = \boldsymbol{u}^* - \frac{\Delta t}{\rho}\nabla p^{n+1}. \tag{8b}$$

The right-hand side of the above equation requires knowledge of the pressure, p, at $(n+1)$ level. This is obtained by taking the divergence and requiring that $\nabla \cdot \boldsymbol{u}^{n+1} = 0$, which is the divergence-free (continuity) condition, thereby leading to the following Poisson equation for p^{n+1}:

$$\Delta p^{n+1} = \frac{\rho}{\Delta t}\nabla \cdot \boldsymbol{u}^*. \tag{9}$$

It is important to emphasize, however, that this is not the true, physical pressure although it differs from physical pressure by $\sim \mathcal{O}(\Delta t/Re)$.

Since the boundary condition for $\boldsymbol{u}^*$ in Equation (7) is no slip, this leads to a boundary condition for the approximate pressure in the form (see, e.g., Foias et al. [43])

$$\frac{\partial p^{n+1}}{\partial n} = 0 \qquad \text{on } \partial\Omega. \tag{10}$$

It is important to note that boundary conditions must be imposed on physical boundaries, and, in the context of a staggered grid (see Figure 2), this implies that Dirichlet conditions on velocity components can be directly applied (on the boundary) only for the normal-direction component on each boundary. The tangential components must be averaged across the boundary by employing an "image" cell outside the solution domain. Neumann conditions used here for pressure are also implemented by means of image cells, as must also be done in the context of unstaggered gridding. This is done in all cases to maintain second-order accuracy of interior-point discretizations.

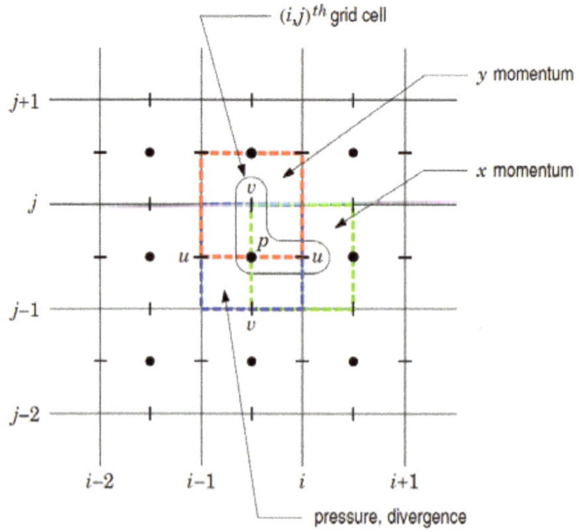

Figure 2. Illustration of a staggered grid in 2-D [44].

While using transient simulation, it is important to make sure that the Courant–Friedrichs–Lewy (CFL) condition is preserved. For a given resolution size, the CFL condition enforces an upper limit on the local time step [45]. In general, the CFL condition for a three-dimensional, transient problem is as follows:

$$\frac{u_x \Delta t}{\Delta x} + \frac{u_y \Delta t}{\Delta y} + \frac{u_z \Delta t}{\Delta z} \leq C_{\max}, \tag{11}$$

where $C_{\max}$ is normally 1 for explicit solvers. For implicit or semi-implicit, this value can increase up to a bigger value (e.g., 30), and, hence, choosing larger time step size will not harm the accuracy of the solution. However, in the presence of nonlinear phenomena, i.e., very strong shockwaves and non-Newtonian viscosity, an implicit scheme linearising the governing equations will exhibit a restraint on $C_{\max}$. This is due to a variation in time of the linearisation Jacobian from one iteration to the next. In the light of above a relatively small time step size, $\Delta t = 0.003$ s is used in this study.

Both Equations (6a) and (7) are valid in the case of constant viscosity. However, inserting a non-Newtonian viscosity will increase the nonlinearity of the system of equations. In order to account for the non-Newtonian behaviour of the fluid and reduce the nonlinearity of the system, one possibility is to replace the generalised apparent kinematic viscosity, Equation (6a), by the constant kinematic viscosity derived from Equation (5) [46]. The kinematic viscosity, however, is a function of $\Delta \boldsymbol{u}$, which makes the viscous term nonlinear. According to Taylor's series of expansion, the viscous terms become:

$$\begin{aligned} \nu\left(|\mathcal{D}(\Delta \boldsymbol{u})|\right)\mathcal{D}(\Delta \boldsymbol{u}) &= \nu\left(\left|\mathcal{D}\left(\Delta \boldsymbol{u}^{\text{old}}\right)\right|\right)\mathcal{D}\left(\Delta \boldsymbol{u}^{\text{old}}\right) \\ &+ \frac{d\left[\nu\left(|\mathcal{D}(\Delta \boldsymbol{u})|\right)\mathcal{D}(\Delta \boldsymbol{u})\right]}{d(\Delta \boldsymbol{u})}\left(\Delta \boldsymbol{u}^{\text{old}}\right)\left(\Delta u - \Delta \boldsymbol{u}^{\text{old}}\right) + \mathcal{O}\left(\left(\Delta u - \Delta \boldsymbol{u}^{\text{old}}\right)^2\right) \end{aligned} \tag{12}$$

in which $\mathcal{D}(\Delta \boldsymbol{u}) = \frac{1}{2}\left(\Delta \boldsymbol{u} + \Delta \boldsymbol{u}^{\top}\right)$, and

$$\begin{aligned} \frac{d\left[\nu\left(|\mathcal{D}(\Delta \boldsymbol{u})|\right)\mathcal{D}(\Delta \boldsymbol{u})\right]}{d(\Delta \boldsymbol{u})}\left(\Delta \boldsymbol{u}^{\text{old}}\right) &= \nu\left(\left|\mathcal{D}\left(\Delta \boldsymbol{u}^{\text{old}}\right)\right|\right)\frac{1}{2}\left(I + I^{\top}\right) \\ &+ \frac{d\left[\nu\left(|\mathcal{D}(\Delta \boldsymbol{u})|\right)\right]}{d(\Delta \boldsymbol{u})}\left(\Delta \boldsymbol{u}^{\text{old}}\right)\mathcal{D}\left(\Delta \boldsymbol{u}^{\text{old}}\right). \end{aligned} \tag{13}$$

Inserting Equation (13) into Equation (12) and neglecting the higher order term $\mathcal{O}\left(\left(\Delta u - \Delta \boldsymbol{u}^{\text{old}}\right)^2\right)$ will result in the Newton type iteration. Neglecting the second term on the right-hand side of Equation (13), the viscous term becomes linearised and fixed-point iterations will be achieved as follows:

$$\nu\left(|\mathcal{D}(\Delta \boldsymbol{u})|\right)\mathcal{D}(\Delta \boldsymbol{u}) \approx \nu\left(\left|\mathcal{D}\left(\Delta \boldsymbol{u}^{\text{old}}\right)\right|\right)\mathcal{D}(\Delta \boldsymbol{u}). \tag{14}$$

The rheological parameters (k and n) of the fluid are those previously studied by Jabbari and Hattel [31], as a common shear-thinning non-Newtonian fluid used in producing thin substrates for fuel cells and magnetic refrigeration applications. Results from the rheology experiment conducted by Jabbari et al. [30] showed that the $La_{0.85}Sr_{0.15}MnO_3$ (LSM) ceramic slurry follows the Ostwald–de Waele power-law behaviour with $k = 3.31$ and $n = 0.9$. In the present study, while keeping k constant, the n parameter is chosen to have different values of 0.9 and 0.5 to investigate the influence of greater deviations from Newtonian fluids (shown in Figure 3).

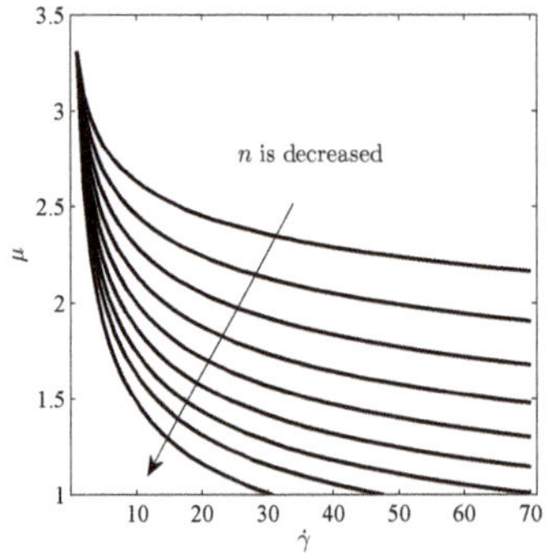

Figure 3. Representation of influence of power-law index, n, on non-Newtonian behaviour of a fluid.

3. Results and Discussion

We first present a fairly extensive set of results for a Newtonian fluid (applying $n = 1$ in Equation (5)) in a cubical cavity, for which much is available in the existing literature [7,11,17,47,48]. This provides thorough validation of the new code. Then, we conduct similar studies for a generalised shear-thinning case with different aspect ratios as well as the *Re* numbers.

3.1. Code Validation for Newtonian Fluid

Prior to proceeding with validation in the context of shear-thinning fluids, it is essential to ensure grid independence of the computations for a simpler Newtonian fluid. The aim is to achieve a maximum change of no more than 1% for a certain variable from one grid level to a more refined one. For this purpose, three different uniform grids with the same number of divisions $N_x = N_y = N_z$ in each coordinate direction were used: namely, 128^3, 256^3, and 384^3 grids.

Results are shown in Figure 4 for the normalised x-velocity component $\overline{u} = u/U_{lid}$ along the vertical centreline for a Newtonian fluid. Selecting as a criterion the value of $\overline{u}^{min}$, it can be seen that grid independence is achieved with the 256^3 grid (refining further to 384^3 showed a change of less than 1%), and, therefore, based on this grid refinement study, this grid is used for all later computations.

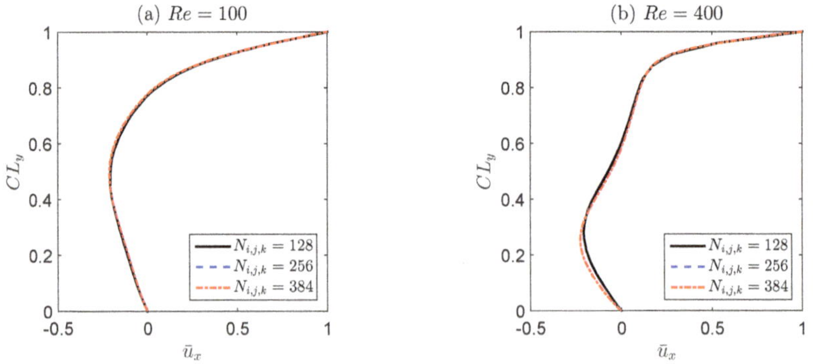

Figure 4. *Cont.*

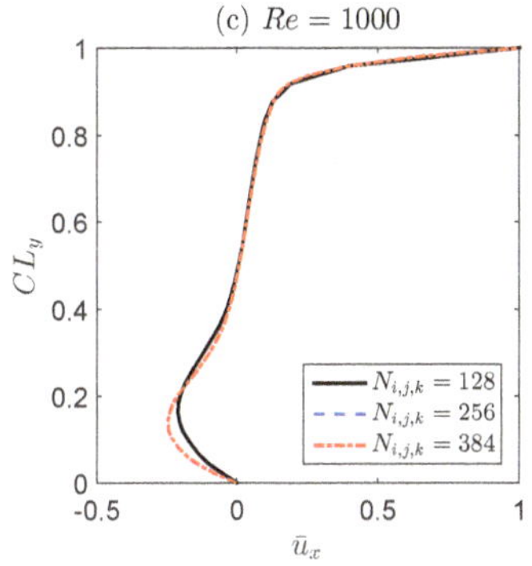

Figure 4. $\bar{u}$ component along the vertical centreline (CL_y) of the cavity for three different grid-resolutions at different Reynolds numbers of (**a**) $Re = 100$; (**b**) $Re = 400$; and (**c**) $Re = 1000$.

To further validate the developed code, we present a comparison of velocity profiles obtained from the present simulations with the Navier–Stokes solutions published in the literature by Albensoeder and Kuhlmann [47]. These cases correspond to flow in a lid-driven cubical cavity ($\delta = 1$) at different Reynolds numbers. Shown in Figure 5 are the normalised x- and y-components of velocity profiles along the vertical (CL_y) and horizontal (CL_x) centreline planes, respectively, in Ω (see Figure 1). Solid lines represent results obtained from the present simulations, whereas the discrete points denote the 3-D pseudo-spectral solution of Navier–Stokes equations in [47]. It is seen that the comparison is excellent at all Reynolds numbers considered here.

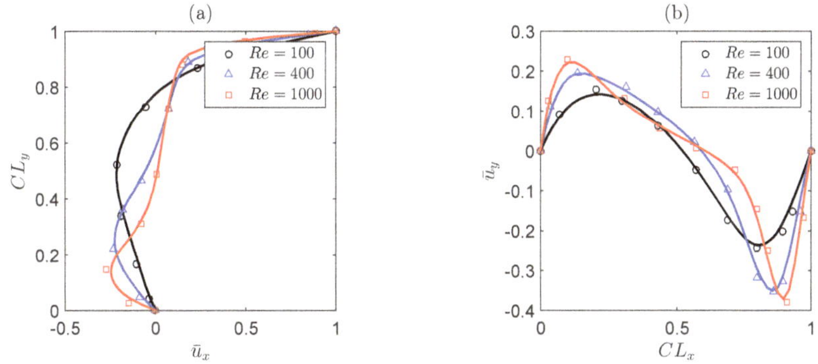

Figure 5. Normalised velocity profiles (**a**) $\bar{u}$ at CL_y, and (**b**) $\bar{v}$ at CL_x, discrete points from [47].

3.2. *Shear-Thinning Fluid*

Results of simulation for generalised Newtonian power-law behaviour in the LDC are presented here for three different Reynolds numbers ($Re = 100, 400, 1000$). For the power-law fluid, the Re is calculated by $Re = \rho L^n / k \cdot U_{lid}^{n-2}$. The shear rate of the flow is updated in every time step through the generalised kinematic viscosity. In order to validate the present code for generalised Newtonian power-law behaviour, we have compared the results for $\delta = 1$ and $n = 0.5$ at $Re = 100$ with those published by Neofytou [48]. Figure 6 depicts the normalised x-components of velocity profiles in the vertical (CL_y) centreline plane, showing a fairly good agreement. Apparent discrepancies are due to the fact that the referenced work was a 2-D calculation. The critical values for Reynolds number, Re_{cr}, for the transition from steady to periodic flow (a Hopf bifurcation), and for different values of k and n (shear-thinning parameters) are shown in Figure 7. It is seen that, for the current study, $k = 3.31$, for all n-values, the critical Reynolds number is above 1000.

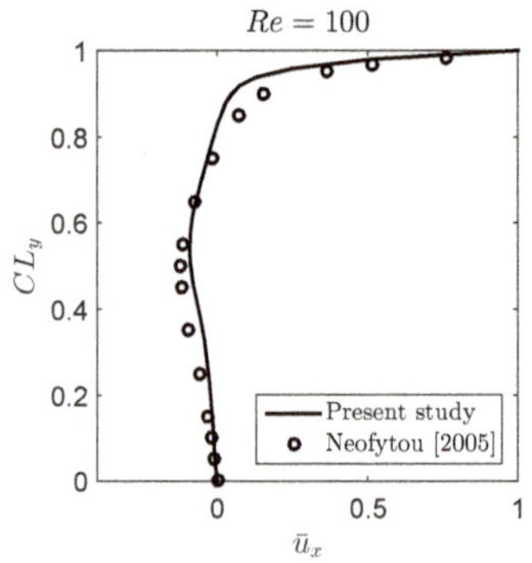

Figure 6. Normalised velocity profiles $\overline{u}$ at CL_y from the present study and its comparison to the work done by Neofytou [48].

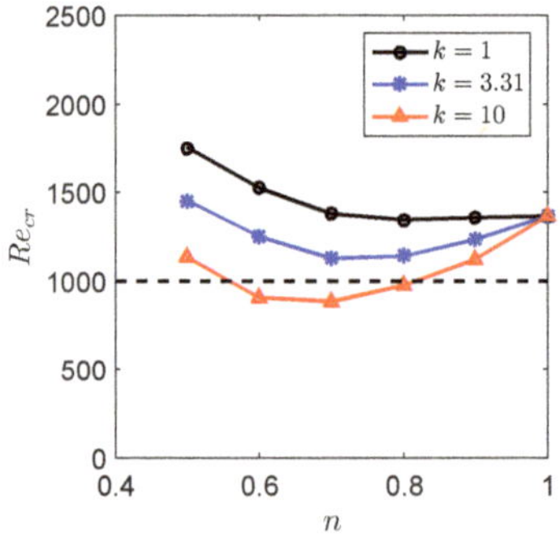

Figure 7. The critical Reynolds number for different k values with different n values.

Results for $\delta = 1$ together with three different n values (see Equation (5)) for the shear-thinning parameter are illustrated in Figure 8. It should be noted that the more n is decreased the greater the deviation from Newtonian behaviour. The results show that for all cases increasing Re causes higher gradients in addition to increasing peak values for u and v; this can be attributed to the gradual reduction of viscous effects that occurs with increasing Re.

The corresponding streamlines coloured by the velocity magnitude inside the cavity for the three different Reynolds numbers as well as three different n values are illustrated in Figure 9. At low Reynolds number, the top row, the region occupied by the primary vortex is relatively small due to the highly diffusive transport of vorticity, and considerable thickness of the boundary layers near the walls. On the other hand, at high Reynolds numbers, the middle and bottom rows, the primary vortex grows in size and occupies most of the central region of the cavity cross-section; the boundary layer is seen to be limited to very thin regions near the walls, as should be expected. In these cases, the vorticity transport is dominated by advection created due to motion of the top lid. The plot of streamlines revealed that a pair of small secondary vortices appear near the bottom corners of the cavity.

As Re is increased, the centre of the primary vortex (x_{vor}, y_{vor}) shifts towards the downstream wall, and simultaneously it moves down towards the bottom wall of the cavity (as seen in [47]). In 2-D simulations, the centre of the primary vortex has been observed to shift towards the downstream side of the cavity until the Reynolds number increases to about 100, but there onwards it asymptotically moves towards the geometric centre of the cavity [16] with further increase in the Reynolds number. It has been reported [14,16] that the results of 2-D and 3-D simulations agree very well for $Re = 0.01$ and $Re = 100$. However, for $Re = 400$ and $Re = 1000$, the predicted locations (x_{vor}, y_{vor}) from 2-D and 3-D flow models are significantly different [14,16]. We believe this difference is due to enhanced motion in the transverse direction. In particular, the entire flow field becomes unsteady at much lower *Re* in 3-D, and the vortex centre location, itself, does not remain in a fixed location. In particular, although results are still steady for $Re = 1000$, they transition to periodic behaviour at slightly higher *Re*, suggesting very long transients at $Re = 1000$. It is also noted that by increasing the shear-thinning effect, decreasing n, the regions with high velocity magnitudes become narrower (while shifting to the top surface), and hence showing boundary layer effect. Moreover, the secondary vortex tends to diminish with decreasing n at all three Reynolds numbers.

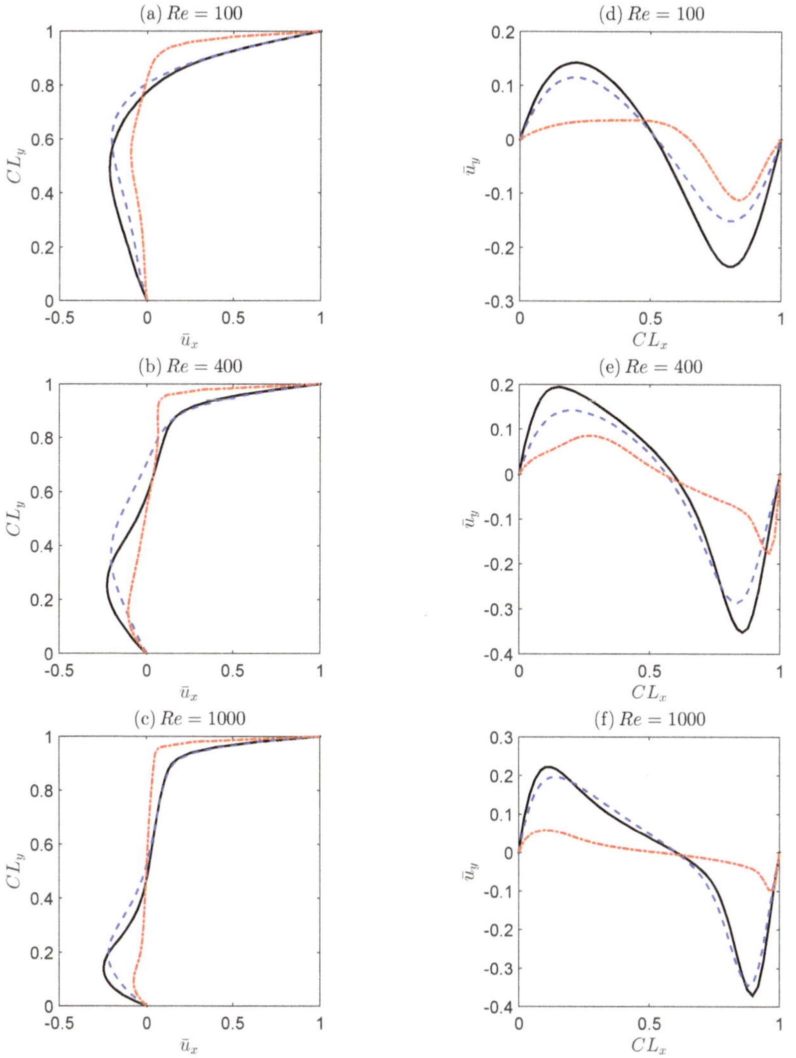

Figure 8. Normalised x-components of velocity profiles for aspect ratio $\delta = 1$ and Reynolds numbers (**a**) $Re = 100$; (**b**) $Re = 400$; (**c**) $Re = 1000$, and normalised y-components of velocity profiles for aspect ratio $\delta = 1$ and Reynolds numbers; (**d**) $Re = 100$; (**e**) $Re = 400$; and (**f**) $Re = 1000$;—$n = 1$ (in black),—$n = 0.9$ (in blue), and —$n = 0.5$ (in red).

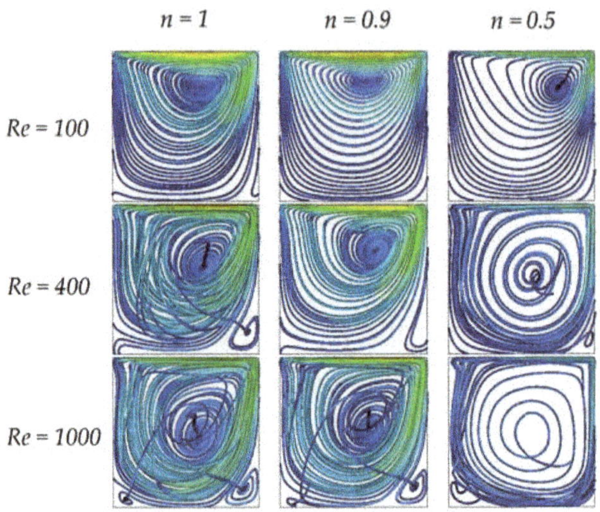

Figure 9. Flow streamlines coloured by velocity magnitude in central x–y plane of Ω for the three different Re numbers as well as three different n values.

The effect on the corresponding displacement of the primary vortex core (physical point in a 3-D domain) is summarized in Table 1 which presents centre-plane x–y coordinates. It can be seen that the centre of the main vortex shifts slightly towards the lid with decreasing Re in all cases, whereas, for shear-thinning fluids, the displacement is more marked. It is interesting to note that, by reducing the shear-thinning parameter, n, from 1 to 0.9, the main vortex moves towards the lid; however, by further decreasing n to 0.5, the main vortex shifts back to the centre of the cavity. This can be easily interpreted from the velocity profiles shown in Figure 8, where the velocity gradients show less variation along the CL_y and CL_x for $n = 0.5$, especially for higher Re.

Table 1. Location, (x, y), of the primary vortex core.

	$n = 1$	$n = 0.9$	$n = 0.5$
$Re = 100$	$(0.6161, 0.7634)$	$(0.5890, 0.7967)$	$(0.7508, 0.7540)$
$Re = 400$	$(0.6193, 0.5836)$	$(0.6564, 0.7004)$	$(0.5760, 0.5446)$
$Re = 1000$	$(0.5935, 0.4723)$	$(0.6084, 0.5130)$	$(0.5334, 0.4657)$

Results of simulations for deeper ($\delta > 1$) cavities are presented with a spanwise aspect ratio of $\lambda = 1$. Figure 10 presents the velocity streamlines coloured by the vorticity magnitude for $Re = 1000$ with $\delta = 2$ (top row) and $\delta = 4$ (bottom row), as well as for different n values ($n = 1, 0.9, 0.5$). It can be seen that for a Newtonian fluid (top-left) a deep cavity with $\delta = 2$ possesses another main vortex in the bottom region, which has an opposite span-direction compared to the main-top-vortex. The phenomenon occurs for a Newtonian fluid with $\delta = 4$ (bottom-left) leading to a third main vortex in the bottom region which has a different span-direction compared to the vortex above it, but the same as that of the main-top vortex.

For a shear-thinning fluid with a slight deviation from Newtonian (top-mid), $n = 0.9$, the secondary main vortex (in the bottom region) has expanded somewhat resulting in smaller area being occupied by the main-top vortex. This is much more pronounced for the same fluid with a deeper cavity (bottom-mid), $\delta = 4$, where the secondary main vortex has grown and diffused to the third main-vortex region. It is interesting to see that, for a fluid with even lower n (top-left), the main-top-vortex occupies most of the cavity volume. This phenomenon is due to more shear-thinning effects of the fluid with lower n close to the top surface.

Impact of Re on the main-top vortex for deeper ($\delta = 2, 4$) cavities with $n = 0.5$ are illustrated in Figure 11. As it can be seen, by increasing the Re (from left to right column), the main-top vortex expands towards the bottom surface of the cavity.

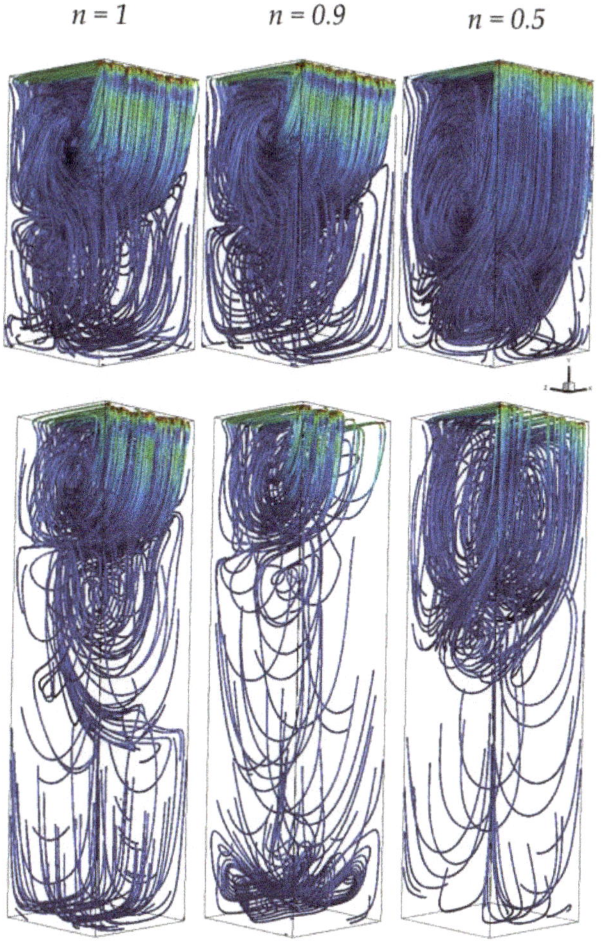

Figure 10. Flow streamlines coloured by vorticity magnitude in deeper cavities (δ = 2, 4) for Re = 1000 as well as three different n values.

It should be highlighted that, for $\delta > 1$, there exists a critical Reynolds number above that the flow near the bottom wall will lose symmetry about the line CL_y. The flow symmetry will be regained below the critical Reynolds number or by increasing the aspect ratio [19]. Velocity vector in the $x = 0.5$ plane (Φ in Figure 1) inside the cavity for $\delta = 2$, together with different Reynolds numbers ($Re = 100, 400, 1000$) and different values of $n = 1, 0.9, 0.5$, are given in Figure 12. It is noticed that, by increasing the shear-thinning effect, going from left column to the right, the secondary flow becomes weaker. This is due to the shear-thinning effect, for which the fluid tends to flow more in the in-plane direction (x) rather than in the y–z plane. It can be also seen that, in the plane Φ, a pair of secondary-flow vortices is found at small Reynolds number ($Re = 100$), and these gradually shift towards the lower corner of the cavity side-walls with increasing Reynolds numbers (going from the top row to the bottom row of the figure). In addition, two pairs of longitudinal vortices begin to appear at the upper corner of the side-walls (in fact, a closer examination revealed the presence of a few additional weak vortices, thus making the flow field quite complex). These secondary vortices gain in strength with increasing Reynolds number. The above mentioned phenomena are somewhat less pronounced for fluid with a high shear-thinning effect. Our examination also showed that the same trend exists for the deeper cavity ($\delta = 4$).

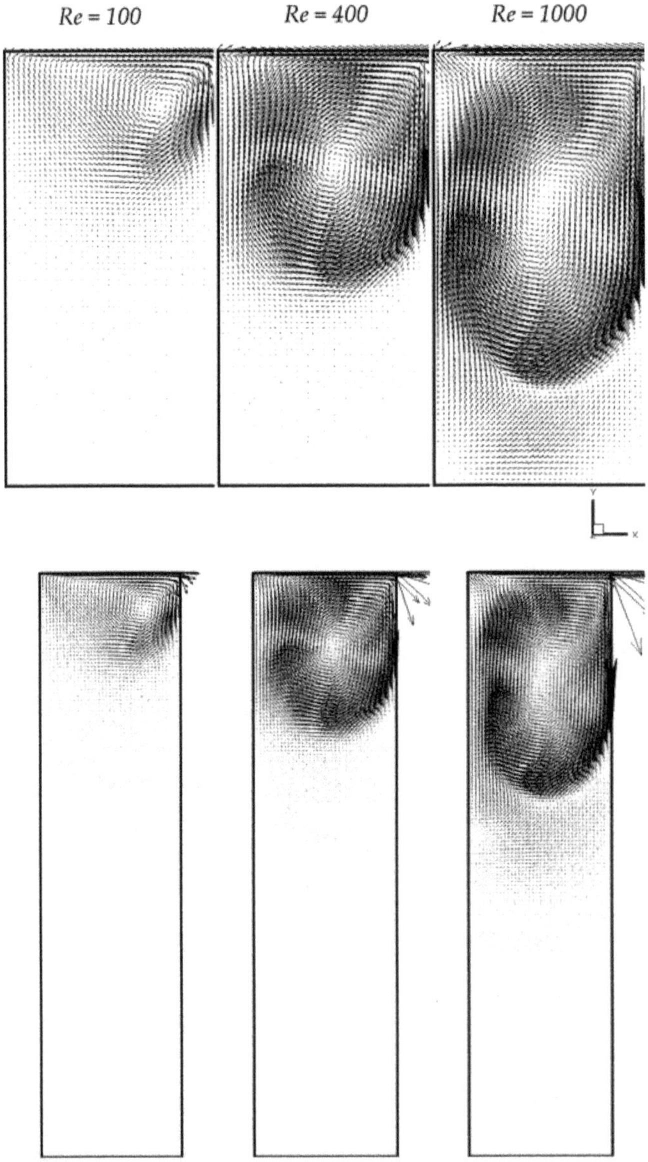

Figure 11. Velocity vector in central x–y plane of Ω for deeper cavities ($\delta = 2$, 4) as well as different Reynolds number. The non-Newtonian parameter n is equal to 0.5.

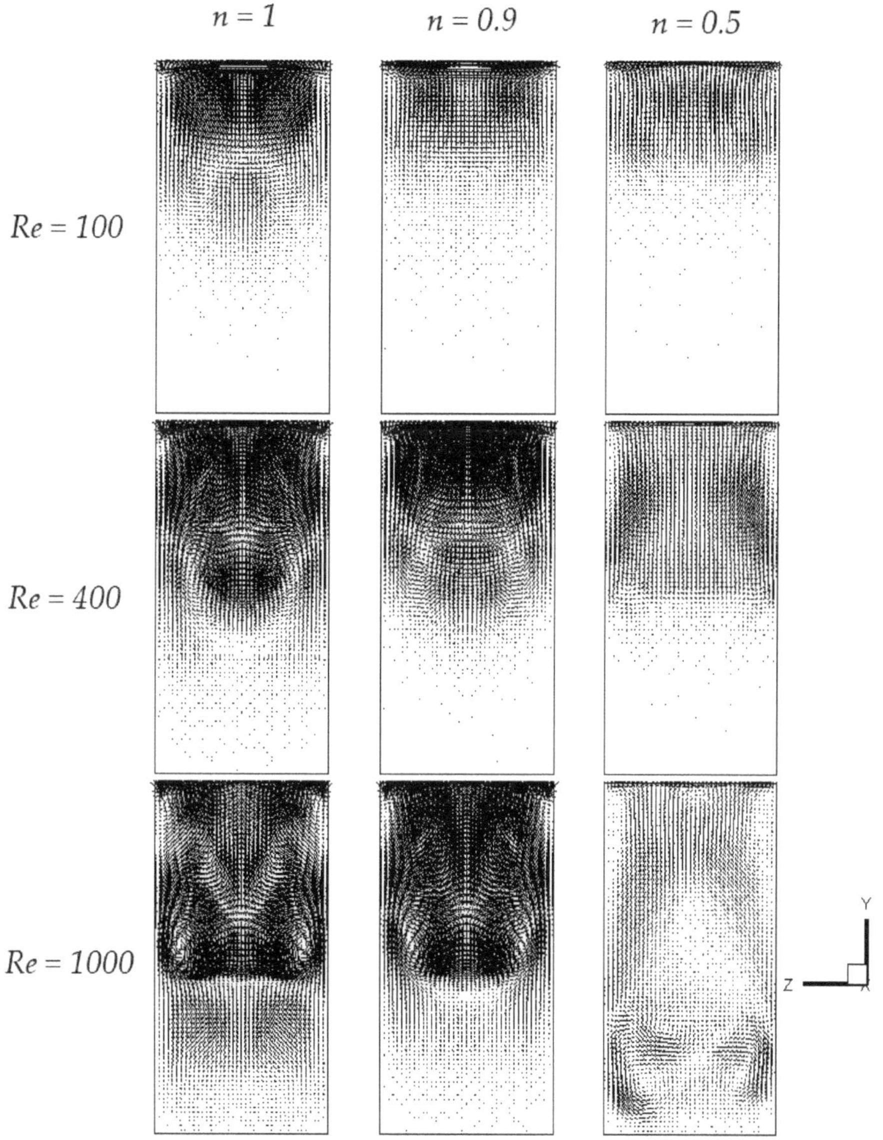

Figure 12. Velocity vector in plane of Φ for a cavity of $\delta = 2$ together with different Reynolds numbers as well as different values of n.

4. Conclusions

In this paper, the 3-D LDC flow has been studied in rectangular cavities using a first-order projection method. The developed model has been validated against already published literature, and is further extended to simulate flows with generalised Newtonian power-law behaviour as a shear-thinning fluid.

The results show that the vortex structure in the cubed cavity ($\delta = 1$) changes considerably with *Re*, where increasing *Re* results in a second primary vortex appearing beside the first primary vortex. For deep cavities ($\delta > 1$), the results show the existence of a series of counter-rotating primary vortices that are placed vertically along the cavity-height. The size and number of these vortices depend on the cavity aspect ratio (δ) and the Reynolds number.

The present study, moreover, shows that the centre of the main vortex shifts towards the lid with decreasing Re in all cases, whereas, for a shear-thinning fluid, the displacement is more marked. It is, moreover, noted that by reducing the shear-thinning parameter, n, from 1 to 0.9, the main vortex shifts towards the lid, however, by further decreasing n to 0.5, the main vortex shifts back to the centre of the box.

Complex flow problems, such as transition to turbulence, and turbulent flow of non-Newtonian power-law fluids in the cavities with different aspect ratios may be of interest for future studies. Migration of a secondary phase, i.e., particles, in the cavity flow of non-Newtonian fluids could be another interesting phenomenon to investigate. This will be beneficial to applications in processing and manufacturing of ceramics and polymers where particles are suspended in non-Newtonian fluid.

Author Contributions: Conceptualization, M.J.; Investigation, M.J.; Software, M.J. and J.M.; Writing—original draft, M.J. and J.M.; Writing—review and editing, J.M., E.M. and J.H.H.

Funding: This research received no external funding.

Conflicts of Interest: The authors declare no conflict of interest.

References

1. Shankar, P.; Deshpande, M. Fluid Mechanics in the Driven Cavity. *Annu. Rev. Fluid Mech.* **2000**, *32*, 93–136. [CrossRef]
2. Shuja, S.; Yilbas, B.; Iqbal, M. Mixed Convection in a Square Cavity due to Heat Generating Rectangular Body: Effect of Cavity Exit Port Locations. *Int. J. Numer. Methods Heat Fluid Flow* **2000**, *10*, 824–841. [CrossRef]
3. Wong, J.C.F. Numerical Simulation of Two-dimensional Laminar Mixed-convection in a Lid-driven Cavity Using the Mixed Finite Element Consistent Splitting Scheme. *Int. J. Numer. Methods Heat Fluid Flow* **2007**, *17*, 46–93. [CrossRef]
4. Aidun, C.K.; Triantafillopoulos, N.; Benson, J. Global Stability of a Lid-driven Cavity with Throughflow: Flow Visualization Studies. *Phys. Fluids A* **1991**, *3*, 2081–2091. [CrossRef]
5. Jana, S.C.; Metcalfe, G.; Ottino, J. Experimental and Computational Studies of Mixing in Complex Stokes Flows: The Vortex Mixing Flow and Multicellular Cavity Flows. *J. Fluid Mech.* **1994**, *269*, 199–246. [CrossRef]
6. Alleborn, N.; Raszillier, H.; Durst, F. Lid-driven Cavity with Heat and Mass Transport. *Int. J. Heat Mass Trans.* **1999**, *42*, 833–853. [CrossRef]
7. Mills, R.D. On the Closed Motion of a Fluid in a Square Cavity. *J. R. Aeronaut. Soc.* **1965**, *69*, 116–120. [CrossRef]
8. Pan, F.; Acrivos, A. Steady flows in rectangular cavities. *J. Fluid Mech.* **1967**, *28*, 643–655. [CrossRef]
9. Koseff, J.; Street, R. Visualization Studies of a Shear Driven Three-dimensional Recirculating Flow. *J. Fluid Eng.* **1984**, *106*, 21–27. [CrossRef]
10. Koseff, J.; Street, R. The lid-driven cavity flow: A synthesis of qualitative and quantitative observations. *J. Fluid Eng.* **1984**, *106*, 390–398. [CrossRef]
11. de Vahl Davis, G.; Mallinson, G. An Evaluation of Upwind and Central Difference Approximations by a Study of Recirculating Flow. *Comput. Fluids* **1976**, *4*, 29–43. [CrossRef]
12. Schreiber, R.; Keller, H.B. Driven Cavity Flows by Efficient Numerical Techniques. *J. Comput. Phys.* **1983**, *49*, 310–333. [CrossRef]
13. Ku, H.C.; Hirsh, R.S.; Taylor, T.D. A Pseudospectral Method for Solution of the Three-dimensional Incompressible Navier–Stokes Equations. *J. Comput. Phys.* **1987**, *70*, 439–462. [CrossRef]
14. Iwatsu, R.; Ishii, K.; Kawamura, T.; Kuwahara, K.; Hyun, J.M. Numerical Simulation of Three-dimensional Flow structure in a Driven Cavity. *Fluid Dyn. Res.* **1989**, *5*, 173. [CrossRef]
15. Chiang, T.; Sheu, W.; Hwang, R.R. Effect of Reynolds Number on the Eddy Structure in a Lid-driven Cavity. *Int. J. Numer Methods Fluids* **1998**, *26*, 557–579. [CrossRef]
16. Ghia, U.; Ghia, K.N.; Shin, C. High-Re Solutions for Incompressible Flow Using the Navier–Stokes Equations and a Multigrid Method. *J. Comput. Phys.* **1982**, *48*, 387–411. [CrossRef]

17. Albensoeder, S.; Kuhlmann, H.; Rath, H. Three-dimensional Centrifugal-flow Instabilities in the Lid-driven-cavity Problem. *Phys. Fluids* **2001**, *13*, 121–135. [CrossRef]
18. Leriche, E.; Gavrilakis, S. Direct Numerical Simulation of the Flow in a Lid-driven Cubical Cavity. *Phys. Fluids* **2000**, *12*, 1363–1376. [CrossRef]
19. De, S.; Nagendra, K.; Lakshmisha, K. Simulation of Laminar Flow in a Three-dimensional Lid-driven Cavity by Lattice Boltzmann Method. *Int. J. Numer. Method Heat Fluid Flow* **2009**, *19*, 790–815. [CrossRef]
20. Tanner, R.I.; Walters, K. *Rheology: An Historical Perspective*; Elsevier: Amsterdam, The Netherlands, 1998; Volume 7.
21. Reid, R.C.; Prausnitz, J.M.; Poling, B.E. *The Properties of Gases and Liquids*; McGraw Hill Book Co.: New York, NY, USA, 1987.
22. Blackery, J.; Mitsoulis, E. Creeping Motion of a Sphere in Tubes Filled with a Bingham Plastic Material. *J. Non-Newton. Fluid Mech.* **1997**, *70*, 59–77. [CrossRef]
23. Beaulne, M.; Mitsoulis, E. Creeping Motion of a Sphere in Tubes Filled with Herschel-Bulkley Fluids. *J. Non-Newton. Fluid Mech.* **1997**, *72*, 55–71. [CrossRef]
24. Mitsoulis, E. Numerical Simulation of Confined Flow of Polyethylene Melts Around a Cylinder in a Planar Channel. *J. Non-Newton. Fluid Mech.* **1998**, *76*, 327–350. [CrossRef]
25. Missirlis, K.; Assimacopoulos, D.; Mitsoulis, E.; Chhabra, R. Wall Effects for Motion of Spheres in Power-law Fluids. *J. Non-Newton. Fluid Mech.* **2001**, *96*, 459–471. [CrossRef]
26. Mitsoulis, E.; Zisis, T. Flow of Bingham Plastics in a Lid-driven Square Cavity. *J. Non-Newton. Fluid Mech.* **2001**, *101*, 173–180. [CrossRef]
27. Chatzimina, M.; Xenophontos, C.; Georgiou, G.; Argyropaidas, I.; Mitsoulis, E. Cessation of Annular Poiseuille Flows of Bingham Plastics. *J. Non-Newton. Fluid Mech.* **2007**, *142*, 135–142. [CrossRef]
28. Mitsoulis, E.; Hatzikiriakos, S. Steady Flow Simulations of Compressible PTFE Paste Extrusion Under Severe Wall Slip. *J. Non-Newton. Fluid Mech.* **2009**, *157*, 26–33. [CrossRef]
29. Muravleva, L.; Muravleva, E.; Georgiou, G.; Mitsoulis, E. Numerical Simulations of Cessation Flows of a Bingham Plastic with the Augmented Lagrangian Method. *J. Non-Newton. Fluid Mech.* **2010**, *165*, 544–550. [CrossRef]
30. Jabbari, M.; Bulatova, R.; Hattel, J.; Bahl, C. Quasi–steady State Power Law Model for the Flow of $(La_{0.85}Sr_{0.15})_{0.9}MnO_3$ Ceramic Slurry in Tape Casting. *Mater. Sci. Technol.* **2013**, *29*, 1080–1087. [CrossRef]
31. Jabbari, M.; Hattel, J. Numerical Modeling of the Side Flow in Tape Casting of a Non-Newtonian Fluid. *J. Am. Ceram. Soc.* **2013**, *96*, 1414–1420. [CrossRef]
32. Jabbari, M.; Bulatova, R.; Hattel, J.; Bahl, C. An Evaluation of Interface Capturing Methods in a VOF Based Model for Multiphase Flow of a Non-Newtonian Ceramic in Tape Casting. *App. Math. Model.* **2014**, *38*, 3222–3232. [CrossRef]
33. Jabbari, M.; Spangenberg, J.; Hattel, J. Particle Migration Using Local Variation of the Viscosity (LVOV) Model in Flow of a Non-Newtonian Fluid for Ceramic Tape Casting. *Chem. Eng. Res. Des.* **2016**, *109*, 226–233. [CrossRef]
34. Pakdel, P.; Spiegelberg, S.; McKinley, G. Cavity Flows of Elastic Liquids: Two-dimensional Flows. *Phys. Fluids* **1997**, *9*, 3123–3140. [CrossRef]
35. Pakdel, P.; McKinley, G. Cavity Flows of Elastic Liquids: Purely Elastic Instabilities. *Phys. Fluids* **1998**, *10*, 1058–1070. [CrossRef]
36. Yapici, K.; Karasozen, B.; Uludag, Y. Finite Volume Simulation of Viscoelastic Laminar Flow in a Lid-driven Cavity. *J. Non-Newton. Fluid Mech.* **2009**, *164*, 51–65. [CrossRef]
37. Rafiee, A. Modelling of Generalized Newtonian Lid-Driven Cavity Flow Using An SPH Method. *ANZIAM J.* **2008**, *49*, 411–422. [CrossRef]
38. Shamekhi, A.; Sadeghy, K. Cavity Flow Simulation of Carreau-Yasuda Non-Newtonian Fluids Using PIM Meshfree Method. *Appl. Math. Model.* **2009**, *33*, 4131–4145. [CrossRef]
39. Shamekhi, A.; Sadeghy, K. A Numerical Method for Solving Incompressible Viscous Flow Problems. *Appl. Math. Model.* **1967**, *2*, 12–26.
40. Bell, J.; Colella, P.; Glaz, H. A Second-Order Projection Method for the Incompressible Navier–Stokes Equations. *J. Comput. Phys.* **1989**, *85*, 257–283. [CrossRef]

41. Gresho, P. On the Theory of Semi-implicit Projection Methods for Viscous Incompressible Flow and its Implementation via a Finite Element Method that also Introduces a Nearly Consistent Mass Matrix. Part 1: Theory. *Int. J. Numer. Methods Fluids* **1990**, *11*, 587–620. [CrossRef]
42. Douglas, J.; Gunn, J. A General Formulation of Alternating Direction Method Part I. Parabolic and Hyperbolic Problems. *Numer. Math.* **1964**, *6*, 428–453. [CrossRef]
43. Foias, C.; Manley, O.; Rosa, R.; Temam, R. *Navier–Stokes Equations and Turbulence*; Cambridge University Press: Cambridge, UK, 2001; Volume 83.
44. McDonough, J.M. *Lectures in Computational Fluid Dynamics of Incompressible Flow: Mathematics, Algorithms and Implementations*; Lecture Notes for ME 691; Department of Mechanical Engineering, University of Kentucky: Lexington, KY, USA, 2013.
45. Gnedin, N.; Semenov, V.; Kravtsov, A. Enforcing the Courant–Friedrichs–Lewy Condition in Explicitly Conservative Local Time Stepping Schemes. *J. Comput. Phys.* **2018**, *359*, 93–105. [CrossRef]
46. Mitsoulis, E. Flows of Viscoplastic Materials: Models and Computations. *Rheol. Rev.* **2007**, *2007*, 135–178.
47. Albensoeder, S.; Kuhlmann, H. Accurate Three-dimensional Lid-driven Cavity Flow. *J. Comput. Phys.* **2005**, *206*, 536–558. [CrossRef]
48. Neofytou, P. A 3rd Order Upwind Finite Volume Method for Generalised Newtonian Fluid Flows. *Adv. Eng. Softw.* **2005**, *36*, 664–680. [CrossRef]

fluids

MDPI

Article

A New Exact Solution for the Flow of a Fluid through Porous Media for a Variety of Boundary Conditions

U. S. Mahabaleshwar [1], P. N. Vinay Kumar [2,*], K. R. Nagaraju [3], Gabriella Bognár [4] and S. N. Ravichandra Nayakar [5]

1 Department of Mathematics, Davangere University, Shivagangothri, Davangere 577 007, India
2 Department of Mathematics, SHDD Government First Grade College, Paduvalahippe, Hassan 573 211, India
3 Department of Mathematics, Government Engineering College, Hassan 573 201, India
4 Faculty of Mechanical Engineering and Informatics, Institute of Machine and Product Design, University of Miskolc, 3515 Miskolc-Egyetemvaros, Hungary
5 Department of Mathematics, University BDT College of Engineering, Davangere 577 004, India
* Correspondence: vinaykumarpn1981@gmail.com

Received: 23 May 2019; Accepted: 29 June 2019; Published: 8 July 2019

Abstract: The viscous fluid flow past a semi-infinite porous solid, which is proportionally sheared at one boundary with the possibility of the fluid slipping according to Navier's slip or second order slip, is considered here. Such an assumption takes into consideration several of the boundary conditions used in the literature, and is a generalization of them. Upon introducing a similarity transformation, the governing equations for the problem under consideration reduces to a system of nonlinear partial differential equations. Interestingly, we were able to obtain an exact analytical solution for the velocity, though the equation is nonlinear. The flow through the porous solid is assumed to obey the Brinkman equation, and is considered relevant to several applications.

Keywords: Brinkman equation; viscosity ratio; first- and second-order slip; similarity transformation; porous medium

1. Introduction

The flow of a fluid through a porous medium has numerous applications in industries dealing with polymer extrusion process, glass blowing, metallurgical processes, and geophysical and allied areas (see [1]). A variety of equations have been used to describe the flow of a fluid through a porous medium as it is one of the important key factors in maintaining the temperature in the medium. These equations due to [2–5] and others, are merely approximations to the appropriate balance laws. A variety of ideas have been suggested to model the flow of mixtures, and one such approach is that which follows from the seminal works of Darcy and Brinkman and has been given a formal structure by [6,7]. Several specific problems have been solved using such an approach (see [8–19]). Here, we study the flow of a fluid through a porous media that is governed by the Brinkman equation (see [20–30]) for a discussion of the status of the Brinkman equation within the context of mixture theory). The fact that we are able to obtain an analytical solution to the problem makes the study all the more interesting. Despite the fact that advanced computing facilities are available to obtain the numerical solution, investigators around the world are much more interested in providing the analytical solution due to their accuracy, relevance, and convenient to analyze physical process, in comparison to numerical solutions. The analytical solution can provide a better assessment of consistency and parameter estimates. Many authors have investigated the fluid flow through porous media and provided analytical solution (see [31–37]). The novelty of this study is our use of a variety of boundary conditions that subsumes those that have been considered earlier, in addition to new conditions concerning slip and proportional shearing at the boundary.

In this study, we consider the flow of a fluid through a semi-infinite porous media with one boundary subject to the slipping or adherence of the fluid, the solid being proportionally sheared, and the fluid being injected at the boundary (see Figure 1). We are able to obtain an analytical solution by introducing a similarity variable that greatly simplifies the governing equation. The effects of the boundary conditions on the flow through the porous media are determined.

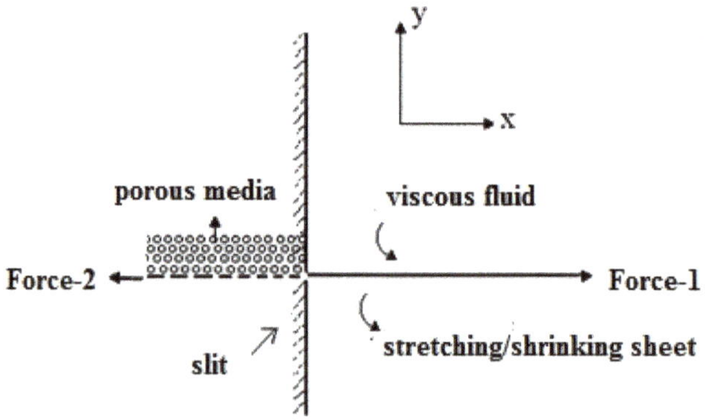

Figure 1. Schematic diagram showing stretching or contraction at the boundary.

2. Theoretical Model

Two dimensional laminar, steady, incompressible fluid flow through a porous media is considered. The x-axis is taken along the stretching of the sheet in the direction of the motion, and y-axis is perpendicular to the slit. In order to confine the fluid flow in the region $y > 0$, two forces of equal strength are applied along the x-axis. u and v denote the axial as well as transverse velocities in the flow field. Figure 1 depicts the physical flow problem subjected to the boundary conditions.

We considered the flow of the classical incompressible Navier-Stokes [38] fluid through a porous half-space. We assumed that the equations governing the flow are those given by the Brinkman equation for flow through porous media which assumes that the fluid is incompressible and, hence, the conservation of mass reduces to

$$\nabla \cdot \vec{q} = 0, \tag{1}$$

and the conservation of linear momentum that takes the form

$$\rho\left[\frac{1}{\phi}\vec{q}_t + \frac{1}{\phi}\left(\vec{q}\cdot\nabla\right)\vec{q}\right] = -\nabla p + \mu_{eff}\nabla^2\vec{q} - \frac{\mu}{K}\vec{q}, \tag{2}$$

where μ_{eff} represents the effective viscosity of the fluid(see [39,40]) provides the definition of the other parameters. The Brinkman equation can be shown to be obtained as a systematic approximation using mixture theory by assuming special structures for the interaction forces between the porous solid and fluid, and assuming the porous solid is rigid (see [41] for a detailed derivation).The transformed governing equations for the conservation of mass and the balance of linear momentum are given as

$$\frac{\partial u}{\partial x} + \frac{\partial v}{\partial y} = 0, \tag{3}$$

$$u\frac{\partial u}{\partial x} + v\frac{\partial u}{\partial y} = -\frac{\phi^2}{\rho_f}\frac{\partial p}{\partial x} + \phi^2 \nu_{eff}\frac{\partial^2 u}{\partial y^2} - \frac{\nu\phi^2}{K}u, \tag{4}$$

where $\nu = \frac{\mu}{\rho_f}$ and $\nu_{eff} = \frac{\mu_{eff}}{\rho_f}$. The Forchheimer term in the interaction is neglected as it produces little impact on the fluid flow in a porous medium governed by the Brinkman equation (see [42]). Also, the pressure gradient is neglected, and the time factor is zero for the steady case.

The governing boundary conditions are (see [43,44])

$$u(x,y) = d\alpha x + A\frac{\partial u}{\partial y} + B\frac{\partial^2 u}{\partial y^2}, \qquad \mathrm{v} = \mathrm{v}_c, \qquad \text{at} \qquad y = 0, \tag{5a}$$

$$u(x,y) \to 0, \qquad \text{as} \qquad y \to \infty. \tag{5b}$$

Here, d is the parameter of proportional shearing at the boundary, with $d \neq 0$ and $d = 0$ corresponding to the boundary at $y = 0$ and being either proportionally sheared or being fixed. The constants A and B represent the first- and second-order slip coefficients, respectively. Also, the mass transpiration parameter, v_c, represents suction or injection depending on $\mathrm{v}_c > 0$ or $\mathrm{v}_c < 0$, respectively.

In order to carry out the analysis, the physical stream functions in terms of similarity variables f and η are introduced as follows:

$$\psi = \sqrt{\alpha\, \nu_{eff}}\quad x\, f(\eta), \tag{6}$$

where

$$\eta = \frac{1}{\phi}\sqrt{\frac{\alpha}{\nu_{eff}}}\, y. \tag{7}$$

In terms of physical stream function ψ, the axial and transverse velocities can be rewritten as follows:

$$u = \frac{\partial \psi}{\partial y}, \quad \mathrm{v} = -\frac{\partial \psi}{\partial x}. \tag{8}$$

The Equation (8) satisfies the continuity equation. Upon substitution of Equations (6) and (7) into Equation (4), we obtain

$$\Lambda\frac{\partial^3 \psi}{\partial y^3} + \frac{\partial(\psi, \frac{\partial \psi}{\partial y})}{\partial(x,y)} - K_1\frac{\partial \psi}{\partial y} = 0. \tag{9}$$

Here, the second term indicates the Jacobian and subject. The appropriate boundary conditions are (see [43,44]) as follows:

$$\frac{\partial \psi}{\partial y} = d\alpha x + A\frac{\partial u}{\partial y} + B\frac{\partial^2 u}{\partial y^2}, \quad \frac{\partial \psi}{\partial x} = v_c \text{ at } \mathrm{y} = 0, \tag{10a}$$

$$\frac{\partial \psi}{\partial y} = 0, \qquad \text{as} \qquad \mathrm{y} \to \infty. \tag{10b}$$

Here, $\Lambda = \frac{\mu_{eff}}{\mu}$ is the Brinkman number or viscosity ratio. Using Equations (9) and (10) with Equation (6), the following transformed equation with constant coefficient is derived:

$$\Lambda f_{\eta\eta\eta} + f f_{\eta\eta} - f_\eta^2 - K_1 f_\eta = 0. \tag{11}$$

The governing boundary conditions for Equation (11) are given as

$$f(0) = V_c, \quad f_\eta(0) = d + \Gamma_1 f_{\eta\eta}(0) + \Gamma_2 f_{\eta\eta\eta}(0), \quad \text{at } \eta = 0, \tag{12}$$

$$f_\eta(\infty) \to 0 \quad \text{as } \eta \to \infty, \tag{13}$$

where $\Gamma_1 = A\sqrt{\frac{\alpha}{\nu}} > 0$ and $\Gamma_2 = B\frac{\alpha}{\nu} < 0$ are the first- and second-order slip parameters, and $K_1 = \frac{\nu\phi^2}{\alpha K}$ is the reciprocal of Darcy number $Da = \frac{l^2}{K}$, with $l = \phi\sqrt{\frac{\nu}{\alpha}}$ and $V_c = \frac{\mathrm{v}_c}{\sqrt{\alpha\,\rho}}$ as the mass suction/injection parameters. The subscript denotes the derivative with respect to η.

3. The Analytical Solution

The flow problem considered is the generalization of the classical works of [43–48]. In our problem, the viscous flow with first- and second-order velocity slips over a porous half-space that is stretched or contracted at the boundary (see Figure 1), and the flow being governed by the Darcy–Brinkman model is considered. One obtains nonlinear partial differentiation from Equations (3) and (4), which are mapped into systems of the nonlinear ordinary differential Equation (11), with a constant coefficient by means of similarity transformation subjected to the imposed boundary (12)–(13).The analytical solution for the velocity distribution is determined.

The exact analytical solution of Equation (11) subjected to the governing boundary conditions Equations (12) and (13) is derived. The condition in Equation (13) suggests choosing the equation of the form

$$f(\eta) = A_1 + B_1 \exp(-\beta\eta), \tag{14}$$

where $\beta > 0$ is to be determined later. Also, A_1 and B_1 are constants that are to be determined by using Equation (12):

$$A_1 = V_c + d\left(\frac{1}{\beta + \Gamma_1\beta^2 - \Gamma_2\beta^3}\right) \text{ and } B_1 = -d\left(\frac{1}{\beta + \Gamma_1\beta^2 - \Gamma_2\beta^3}\right). \tag{15}$$

It follows from Equation (11), (14), and (15) that

$$\Lambda\Gamma_2\beta^4 - (\Lambda\Gamma_1 + V_c\Gamma_2)\beta^3 + (V_c\Gamma_1 - \Lambda - K_1\Gamma_2)\beta^2 + (V_c + K_1\Gamma_1)\beta - (d + K_1) = 0. \tag{16}$$

Here, $\beta > 0$ is one of the real roots (see [49,50]).

By using the transformation variable $\xi = \beta + \frac{a_3}{4}$, Equation (16) transforms into

$$\xi^4 + p\,\xi^2 + q\,\xi + r = 0, \tag{17}$$

where $p = a_2 - \frac{3}{8}a_3^2$, $q = \left(A_1 - \frac{1}{2}A_2A_3 + \frac{1}{8}A_3^3\right)$, and $r = a_0 - \frac{1}{4}a_1a_3 + \frac{1}{16}a_2a_3^2 - \frac{3}{256}a_3^4$, and $a_3 = -\frac{\Lambda\Gamma_1 + V_c\Gamma_2}{\Lambda\Gamma_2}$, $a_2 = \frac{V_c\Gamma_1 - \Lambda - K_1\Gamma_2}{\Lambda\Gamma_2}$, $a_1 = \frac{V_c + K_1\Gamma_1}{\Lambda\Gamma_2}$, $a_0 = \frac{d + K_1}{\Lambda\Gamma_2}$.

The four corresponding roots of the algebraic Equation (17) are

$$\beta_1 = \frac{\sqrt{C}}{2} + \frac{1}{2}\sqrt{D_1} - \frac{a_3}{4}, \tag{18a}$$

$$\beta_2 = \frac{\sqrt{C}}{2} - \frac{1}{2}\sqrt{D_1} - \frac{a_3}{4}, \tag{18b}$$

$$\beta_3 = -\frac{\sqrt{C}}{2} + \frac{1}{2}\sqrt{D_1} - \frac{a_3}{4}, \tag{18c}$$

$$\beta_4 = -\frac{\sqrt{C}}{2} - \frac{1}{2}\sqrt{D_1} - \frac{a_3}{4}, \tag{18d}$$

where

$$D_1 = D - \frac{2q}{C},$$

$$C = -\tfrac{2p}{3} + 2^{1/3}\left(p^2 + 12r\right)\left[3\left(2p^3 + 27q^2 - 72pr + \sqrt{-4(p^2+12r)^3 + (2p^3 + 27q^2 - 72pr)^2}\right)^{1/3}\right]^{-1}$$
$$+\left(2^{1/3}\,3\right)^{-1}\left(2p^3 + 27q^2 - 72pr + \sqrt{-4(p^2+12r)^3 + (2p^3 + 27q^2 - 72pr)^2}\right)^{1/3}$$

and

$$D = -\frac{4p}{3} - 2^{1/3}\left(p^2 + 12r\right)\left[3\left(2p^3 + 27q^2 - 72pr - \sqrt{-4\left(p^2 + 12r\right)^3 + \left(2p^3 + 27q^2 - 72pr\right)^2}\right)^{1/3}\right]^{-1}$$
$$-\left(2^{1/3}3\right)^{-1}\left(2p^3 + 27q^2 - 72pr + \sqrt{-4\left(p^2 + 12r\right)^3 + \left(2p^3 + 27q^2 - 72pr\right)^2}\right)^{1/3}$$

Equation (18) gives the complete solution of Equation (17). However, it should be noted that there is only one feasible solution for the Equation (17) when $\Gamma_2 < 0$, and based on the flow field Equation (16) has feasible solutions for $\beta > 0$.

4. Results and Discussion

In this paper, we were able to establish the impact of various physical parameters on the velocity distribution. The solutions obtained are in good agreement with that of the classical works when suitably restricted to specific conditions. The main emphasis of this study is the effect of boundary conditions on the flow through porous media. There are several possibilities at the boundaries, namely, the fluid meeting the no-slip adherence condition, the Navier slip condition, the second-order slip condition, as well as the possibility of blowing of the fluid.

The effects of physical parameters such as the mass transpiration parameter (V_C), first-order Navier slip (Γ_1), second-order slip (Γ_2),Brinkman ratio (Λ), and proportional shearing parameter (K_1)are discussed graphically. As the velocity distribution is an exponential function with a negative argument, it decreases with the increase in η. Since β is a function of mass transpiration parameter (V_C), first-order Navier slip (Γ_1), second-order slip (Γ_2), Brinkman ratio (Λ), and proportional shearing parameter (K_1)both axial as well as transverse velocities are forced to decrease exponentially.

The solution domain of β in Equation (16)has only complex roots when $D_1 < 0$, only real roots when $D_1 > 0$,and real repeated roots when $D_1 = 0$. Figure 2a–d depicts the solution behavior of β verses V_C for various values of D_1 and Γ_1. Figure 3a–d depicts the solution domain of β versus V_C for different values of K_1. In fact, by choosing $\Gamma_2 = 0$, Equation (16) reduces to a cubic equation and with the proper choice of Γ_1, Λ, and K_1, and the results are reduced to those obtained by [43,51,52].

Figure 4a–c demonstrates the impact of first-order velocity slip with proportional shearing on the solution domain of β versus V_c. The presence of a larger slip drags the separation curve towards the slit. The viscous fluid flow in a permeable medium with slip in a stretching boundary is quite different from that of a contracting boundary.

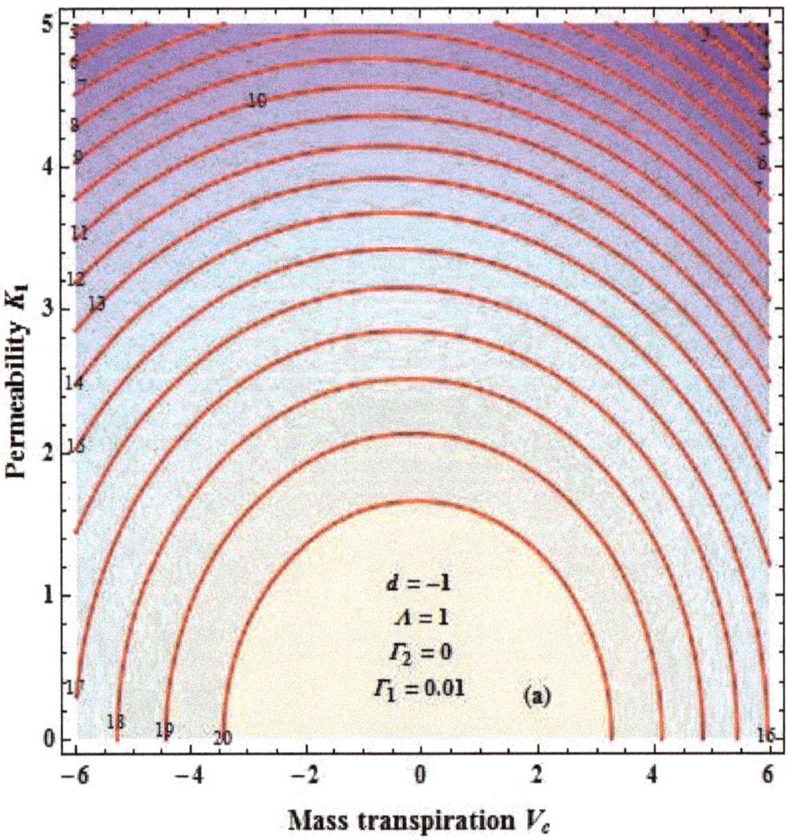

Figure 2. *Cont.*

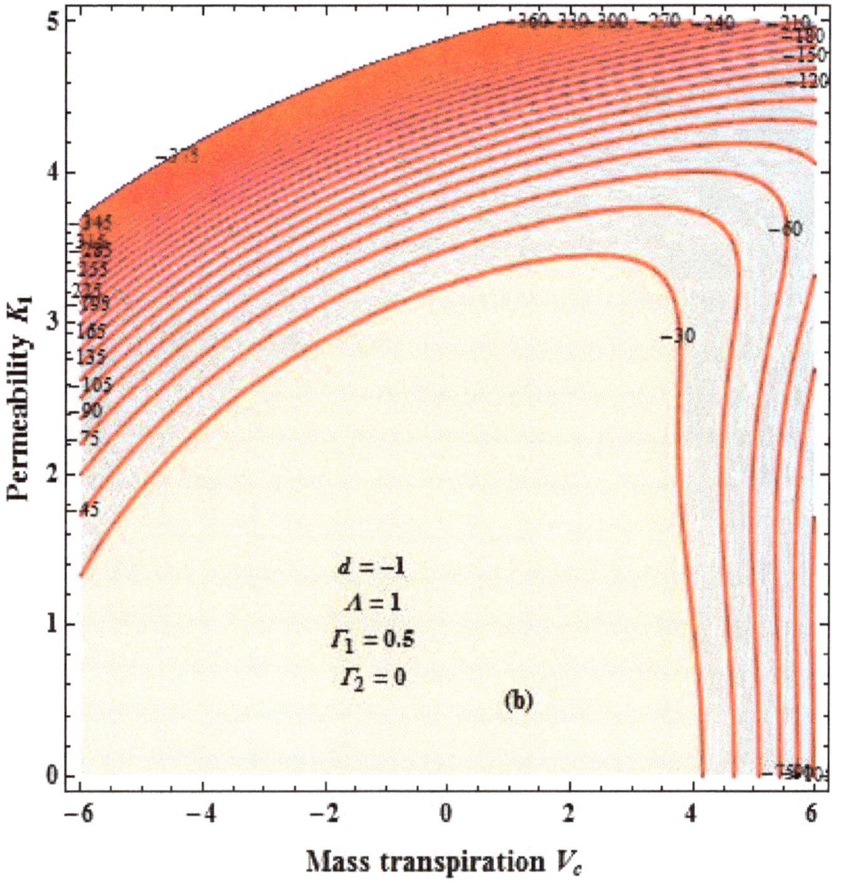

Figure 2. *Cont.*

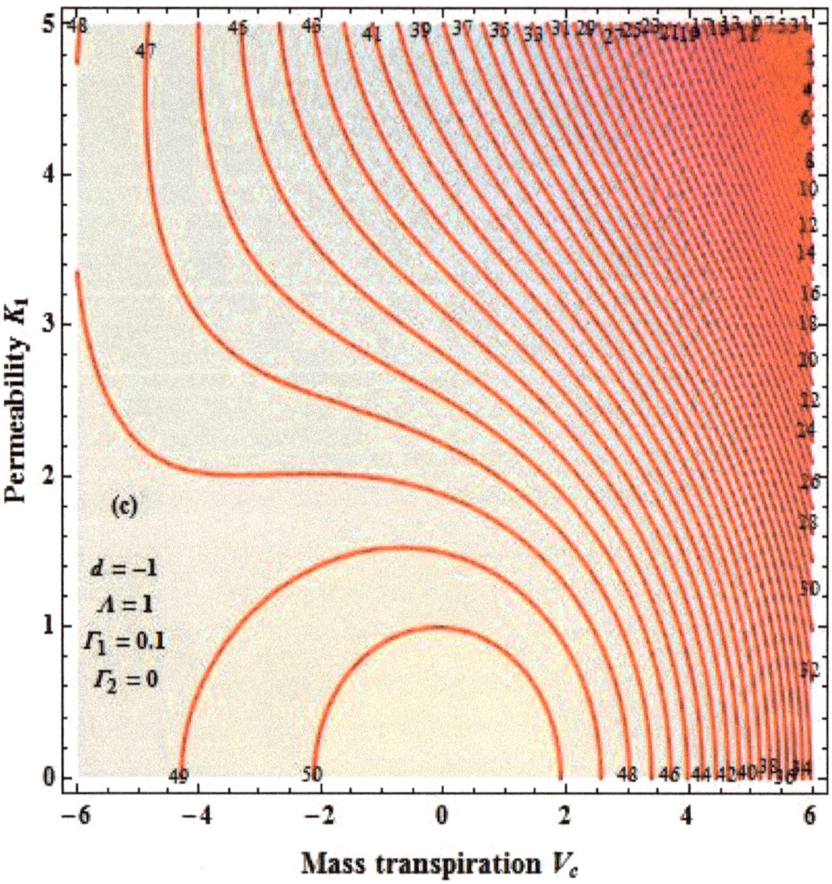

Figure 2. *Cont.*

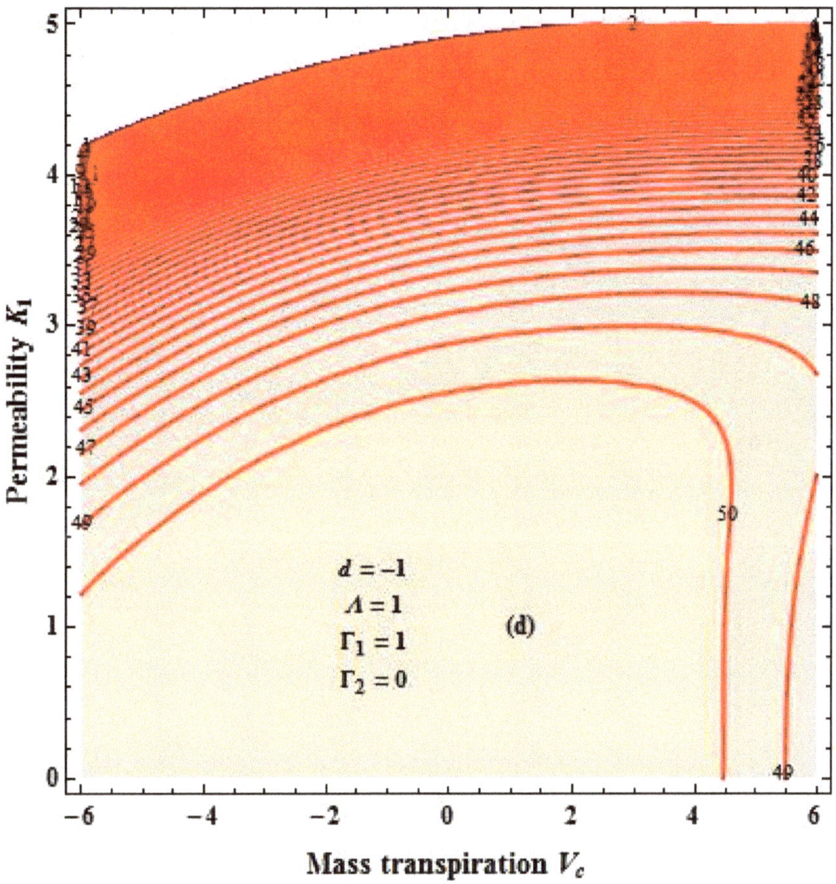

Figure 2. (**a–d**) The solution domain of D_1 for K_1 verses mass transpiration V_c with Brinkman ratio $\Lambda = 1$ for the case of a shrinking boundary with different choices of Γ_1 and Γ_2.

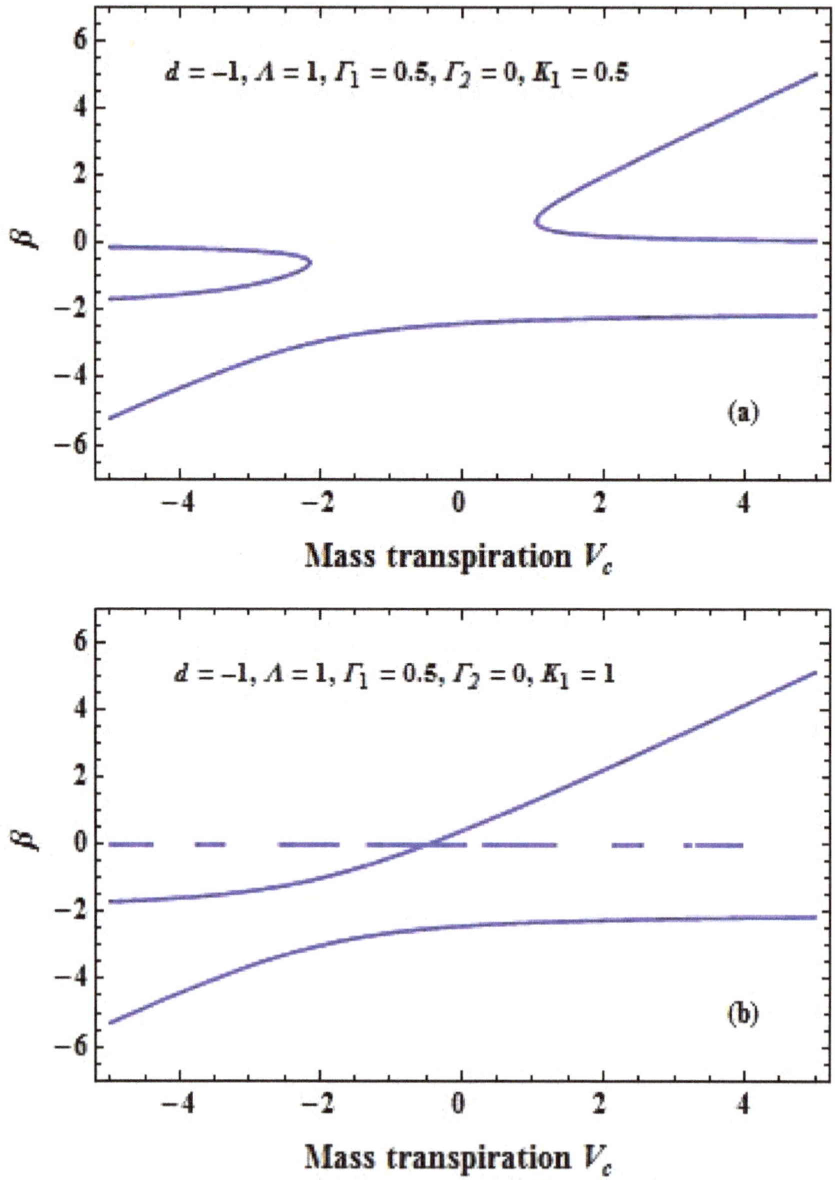

Figure 3. *Cont.*

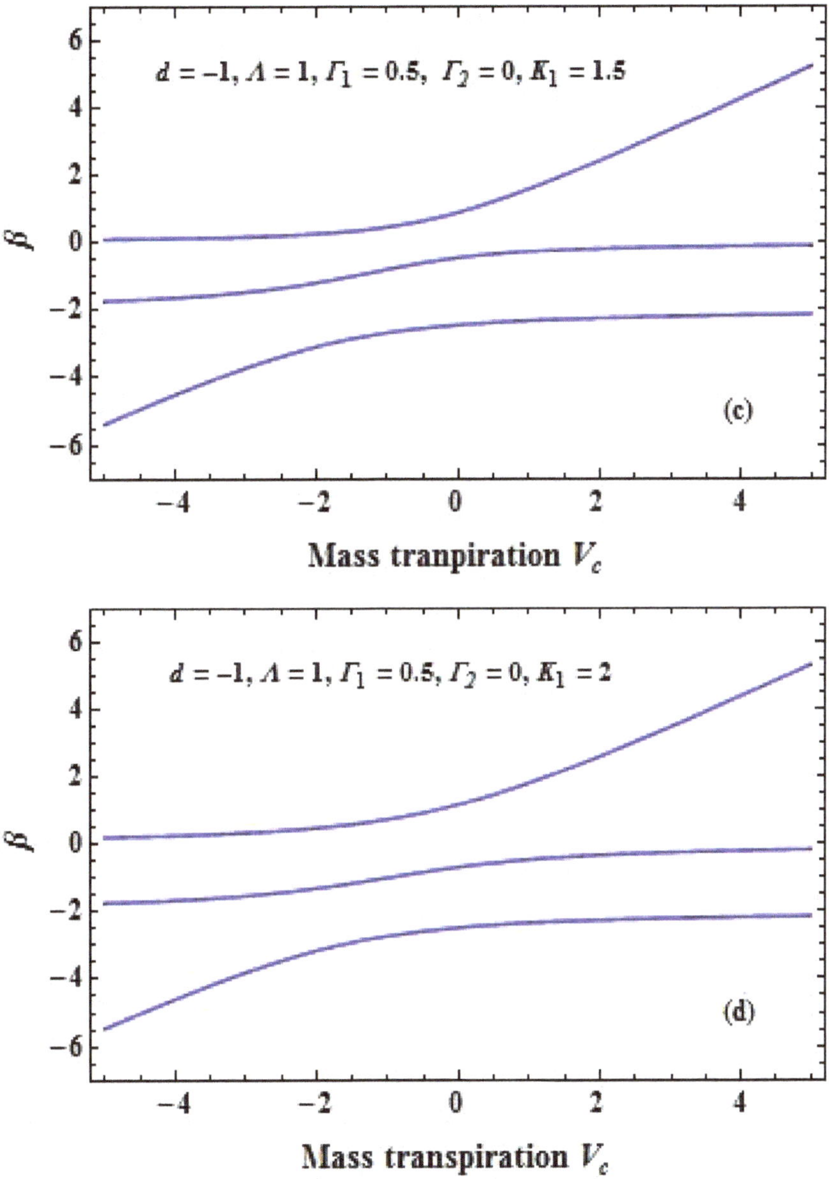

Figure 3. (**a–d**) The solution domain of β versus V_c for different values of K_1 for the case of a shrinking boundary.

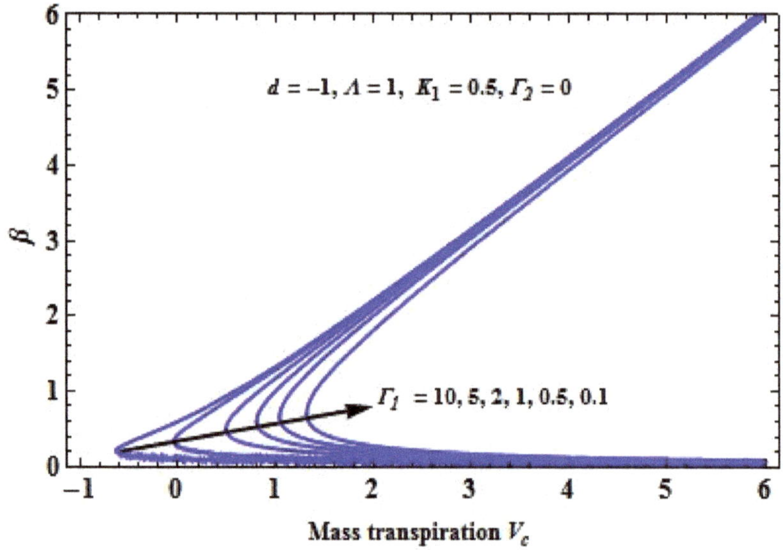

(a)

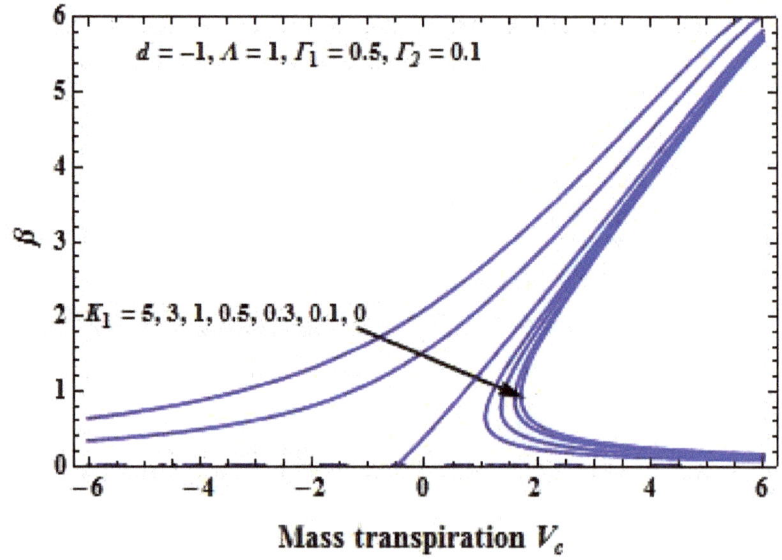

(b)

Figure 4. *Cont.*

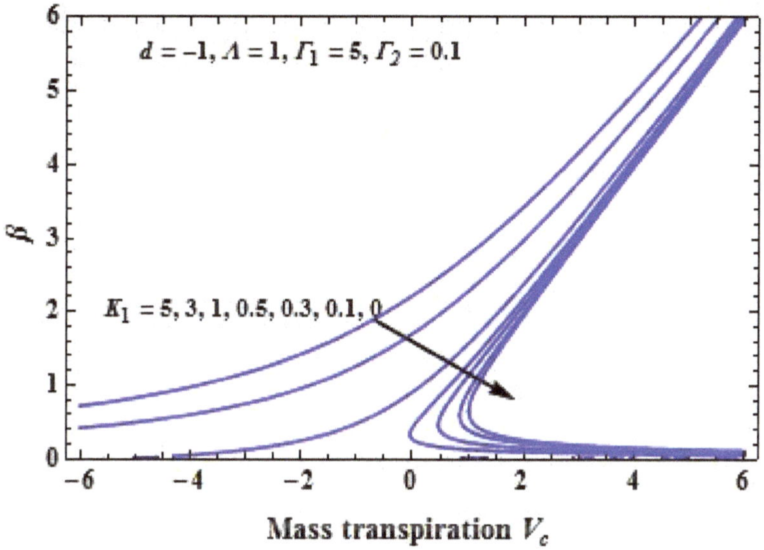

(c)

Figure 4. (**a**) The solution domain for β versus V_c for different values of Γ_1 for the case of a shrinking boundary in the absence of Γ_2. (**b**) The solution domain for β versus V_c for different values of K_1 for the case of a shrinking boundary with $\Gamma_1 = 0.5$. (**c**) Solution domain for β versus V_c for different values of K_1 for the case of a shrinking boundary with $\Gamma_1 = 5$.

Figure 5a,b depicts the axial velocity profiles for various values of first-order slip parameters, and for the fixed values of other physical parameters. This plot clearly demonstrates that the increasing Navier's slip results in the reduction of the velocity boundary. In comparison to the lower branch solution, the boundary layer thickness decreases in the case of upper branch solution. Furthermore, under the given slip parameter and mass suction parameter, one can see the increasing velocity boundary thickness with the increase in proportional shearing parameter. Also, the reduction in mass suction leads to the decrease in velocity boundary for other physical parameters fixed. However, for the case of mass injection, the velocity boundary increases with increasing values of slip parameter. Thus, the flow geometry and the rate of change of velocity boundary layer thickness are significantly influenced by the slip parameter.

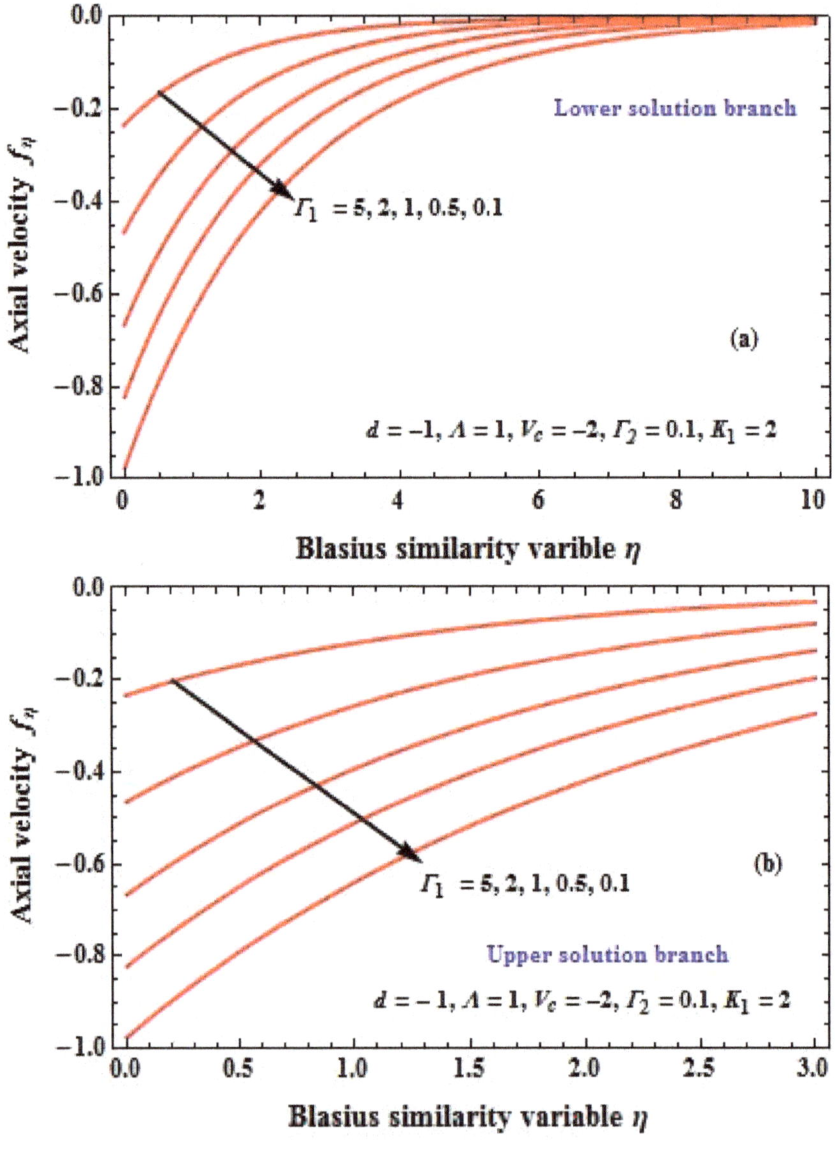

Figure 5. (**a**,**b**) Upper and lower solution branches of axial velocity profile, $f_\eta(\eta)$ verses η, for different values of Navier slip parameter Γ_1 when $K_1 = 0.5$ and $K_1 = 2$ for the case of a shrinking boundary.

Figure 6a–c demonstrates the effect of first- and second-order slip parameters on the axial, as well as transverse velocity profiles respectively, in the accelerating boundary. From the plots, it is clear that the increasing values of first-order slip for fixed values of various physical parameters results in the velocity boundary profiles decreasing, whereas the decreasing value of second-order slip for fixed values of other physical parameters results in the decreasing velocity boundary profiles.

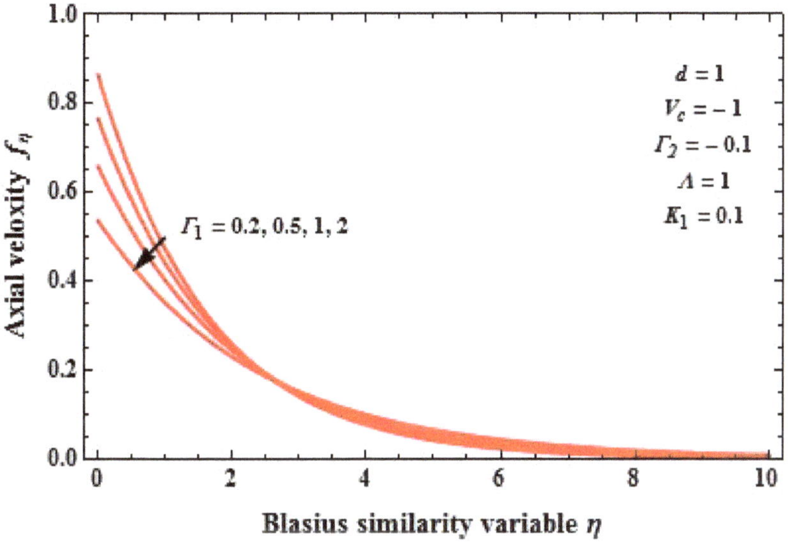

(a)

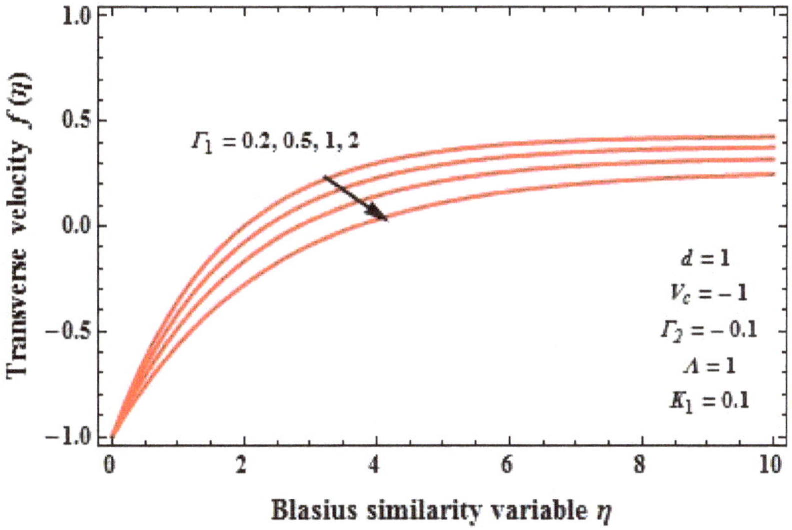

(b)

Figure 6. *Cont.*

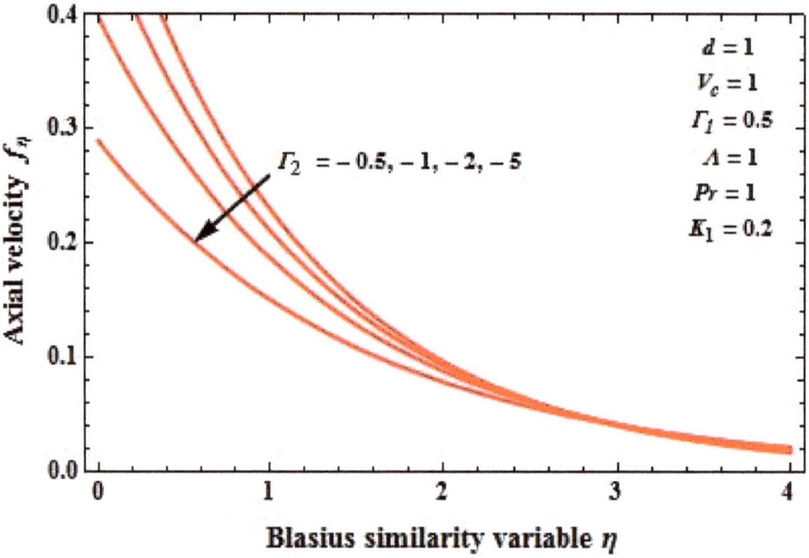

(**c**)

Figure 6. (**a**) Axial velocity $f_\eta(\eta)$ verses η for different values of Γ_1 for the case of a stretching boundary. (**b**) Transverse velocity $f(\eta)$ verses η for different values of Γ_1 with $\Gamma_2 = -0.1$ in the presence of K_1 for the case of a stretching boundary. (**c**) Axial velocity $f_\eta(\eta)$ verses η for different values of Γ_2 with $\Gamma_1 = 0.5$ in the presence of K_1 for the case of a stretching boundary.

Figure 7a,b demonstrates the effect of Brinkman ratio on the axial and transverse velocity profiles. From the plots, it can be seen that increasing values of Λ while keeping other physical parameters fixed results in enhanced boundary layer thickness, whereas in Figure 7c,d, the different values of V_C with fixed Brinkman ratio result in exactly the opposite.

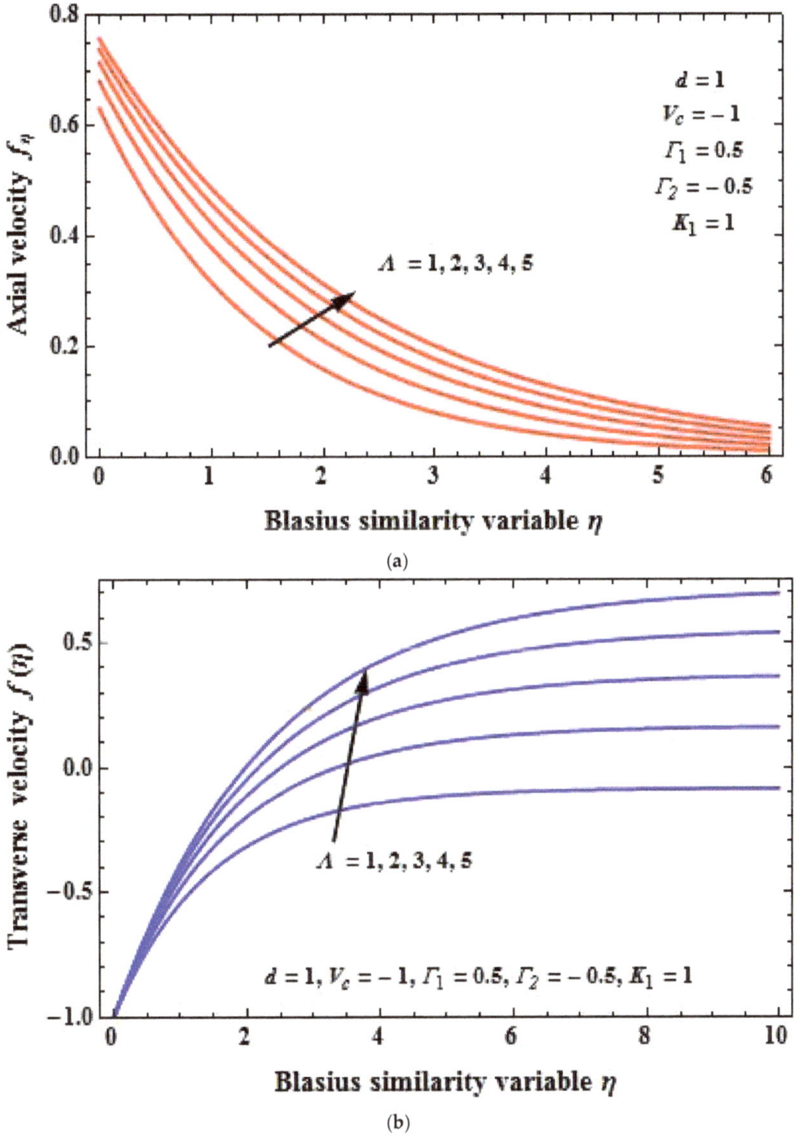

Figure 7. *Cont.*

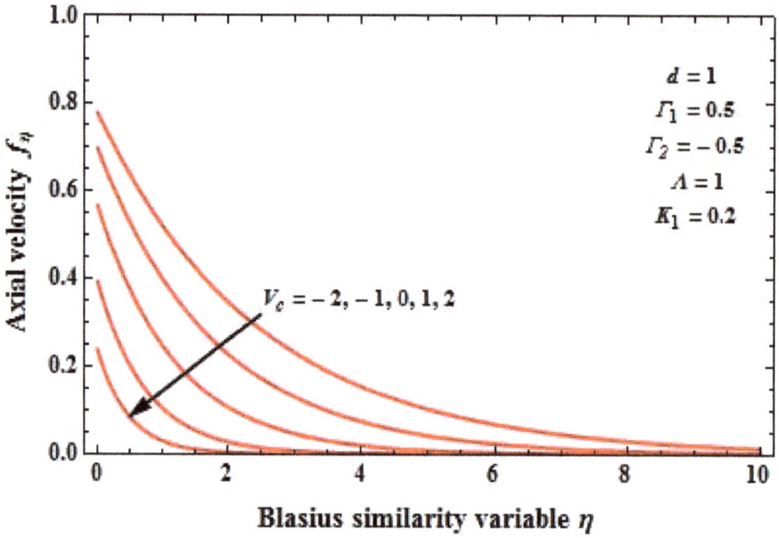

(c)

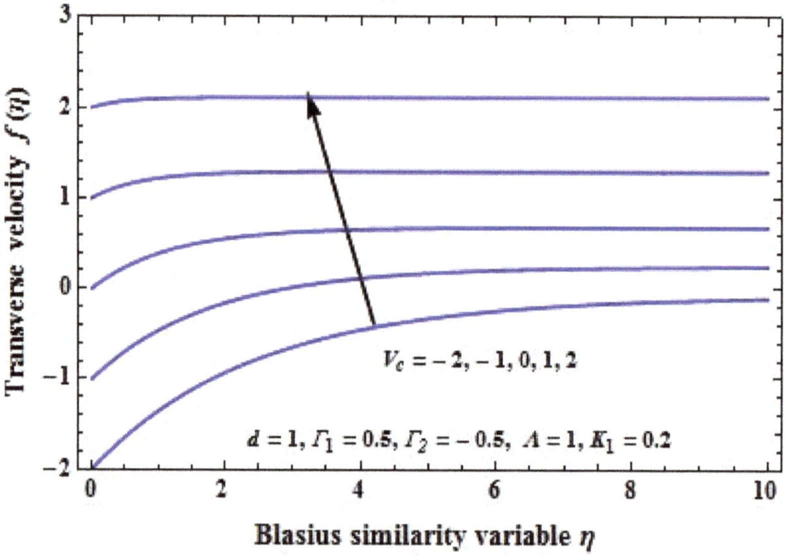

(d)

Figure 7. (**a**) Axial velocity $f_\eta(\eta)$ verses η for different values of Λ in the presence of K_1 for the case of a stretching boundary.(**b**) Transverse velocity $f(\eta)$ verses η for different values of Λ in the presence of K_1 for the case of a stretching boundary. (**c**) Axial velocity $f_\eta(\eta)$ verses η for different values of V_c in the presence of K_1 for the case of a stretching boundary. (**d**) Transverse velocity $f(\eta)$ verses η for different values of V_C in the presence of K_1 for the case of the stretching boundary.

Figure 8a,b depicts the effect of various physical parameters on the shear stress profile. In all of these plots, there are crossover points for the shear stress profiles, and the combined effects on the porous solid can be observed. The increase in the values of Γ_1, Γ_2, and K_1 results in increasing shear at the wall boundary. In the case of mass injection, the shear wall boundary decreases faster for a smaller value of second-order slip parameter. Interestingly, however, the increasing value of Brinkman ratio leads to decreasing shear wall boundary as seen in Figure 8a.

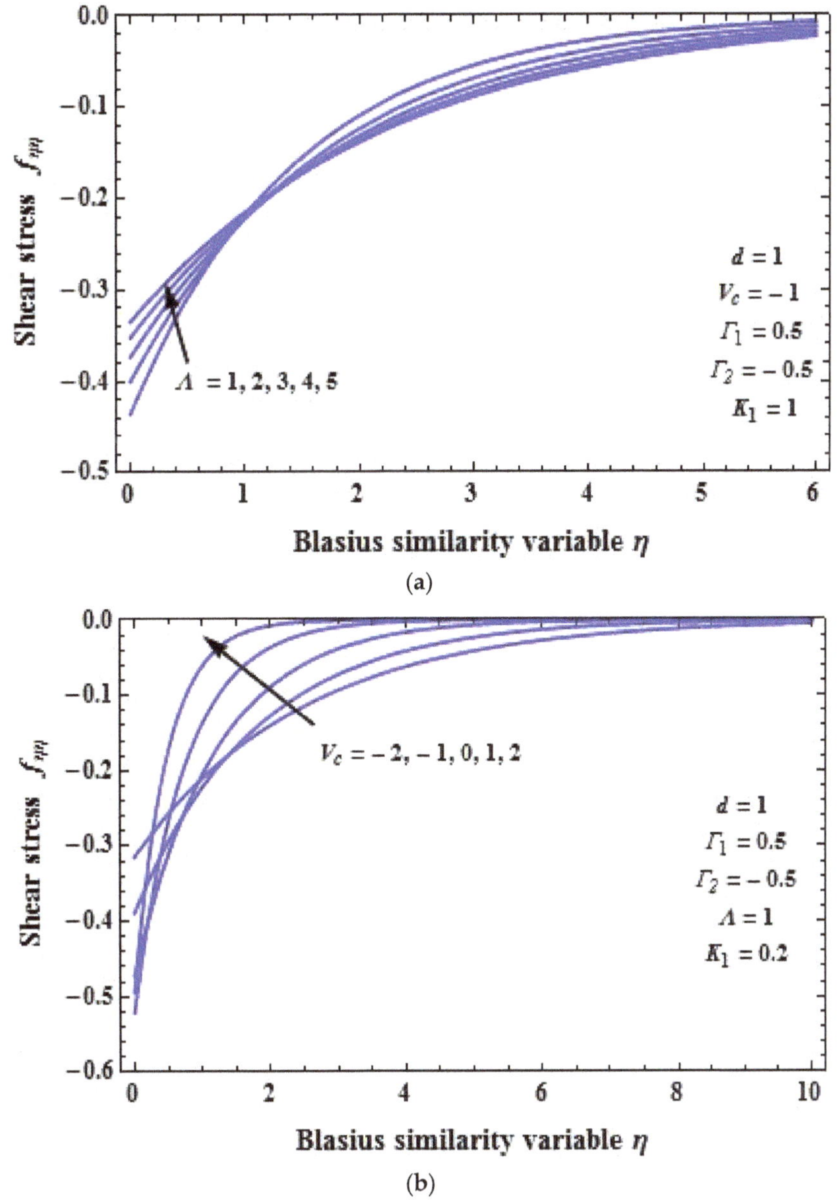

Figure 8. (**a**) Shear stress $f_{\eta\eta}(\eta)$ verses η for different values of Λ in the presence of K_1 for the case of the stretching boundary. (**b**) Shear stress $f_{\eta\eta}(\eta)$ verses η for different values V_c in the presence of K_1 for the case of a stretching boundary.

5. Concluding Remarks

In conclusion, the viscous fluid flow past a porous solid wherein the flow is governed by the Brinkman equation with first- and second-order slip in the presence of mass transpiration was solved, and the exact analytical solution for the governing nonlinear partial differential equation was obtained. The solution was analyzed for the effect of slip parameters, the mass transpiration parameter, the Brinkman ratio, and the extent of shearing or contraction. In the case of the boundary contracting, the solution branches (Figure 5a,b), whereas in the case of the boundary stretching there is only one branch of the solution, and depending on the mass transpiration parameter, the solution is branched (Figure 5a,b).

Author Contributions: Conceptualization, U.S.M., P.N.V.K. and K.R.N.; methodology, U.S.M. and G.B.; software, P.N.V.K., K.R.N., G.B. and S.N.R.N.; validation, G.B; formal analysis, U.S.M., P.N.V.K. and K.R.N.; investigation, K.R.N.; resources, S.N.R.N.; data curation, K.R.N.; writing—original draft preparation, U.S.M., P.N.V.K., K.R.N. and S.N.R.N.; writing—review and editing, P.N.V.K. and G.B.; visualization, G.B. and S.N.R.N.; supervision, U.S.M.; project administration, U.S.M.; funding acquisition, U.S.M.

Funding: This research received no external funding.

Acknowledgments: Author K.R. Nagaraju would like to thank Sri Karisiddappa, Hon'ble Vice-chancellor, VTU, Belagavi, India (Former Principal of Government Engineering College, Hassan-573 201, India) for his valuable support in carrying out his research work.

Conflicts of Interest: The authors declare no conflict of interest.

References

1. Fisher, B.G. *Extrusion of Plastics*, 3rd ed.; Newnes-Butterworld: London, UK, 1976.
2. Brinkman, H.C. On the permeability of the media consisting of closely packed porousparticles. *Appl. Sci. Res.* **1947**, *1*, 81–86. [CrossRef]
3. Darcy, H. *Les Fontaines Publiques De La Ville De Dijon*; Victor Dalmont: Paris, France, 1856.
4. Forchheimer, P. *Wasserbewegung Durch Boden*; Zeitschrift des Vereines Deutscher In-Geneieure: Düsseldorf, Germany, 1901; Volume 45, pp. 1782–1788.
5. Biot, M.A. Mechanics of Deformation and Acoustic Propagation in Porous Media. *J. Appl. Phys.* **1962**, *33*, 1482–1498. [CrossRef]
6. Truesdell, C. Sulla basi della thermomechanical. *Rend. Lincei* **1957**, *22a*, 158–166.
7. Truesdell, C. Sulla basi della thermomechanical. *Rend. Lincei* **1957**, *22b*, 33–38.
8. Chakrabarti, A.; Gupta, A.S. Hydromagnetic flow and heat transfer over a stretching sheet. *Q. Appl. Math.* **1979**, *37*, 73–78. [CrossRef]
9. Gupta, P.S.; Gupta, A.S. Heat and mass transfer on a stretching sheet with suction or blowing. *Can. J. Chem. Eng.* **1977**, *55*, 744–746. [CrossRef]
10. Fang, T. Boundary layer flow over a shrinking sheet with power-law velocity. *Int. J. Heat Mass Transf.* **2008**, *51*, 5838–5843. [CrossRef]
11. Milavcic, M.; Wang, C.Y. Viscous flow due to a shrinking sheet. *Q. Appl. Math.* **2006**, *64*, 283–290. [CrossRef]
12. Nakayama, A. *PC-Aided Numerical Heat Transfer and Convective Flow*; CRC Press: Boca Raton, FL, USA, 1995.
13. Sakiadis, B.C. Boundary-layer behavior on continuous solid surface. *AIChE J.* **1961**, *7*, 26–28. [CrossRef]
14. Sakiadis, B.C. Boundary-layer behavior on continuous solid surfaces. II. The boundary layer on a continuous flat surface. *AIChE J.* **1961**, *7*, 221–225. [CrossRef]
15. Rajagopal, K.R.; Tao, L. *Mechanics of Mixtures*; World Scientific: Singapore, 1995.
16. Givler, R.C.; Altobelli, S.A. A Determination of the Effective Viscosity for the Brinkman-Forchheimer Flow Model. *J. Fluid Mech.* **1994**, *258*, 355–370. [CrossRef]
17. Shao, Q.; Fahs, M.; Hoteit, H.; Carrera, J.; Ackerer, P.; Younes, A. A 3D semi-analytical solution for density-driven flow in porous media. *Water Res. Res.* **2018**, *54*, 10094–10116. [CrossRef]
18. Lesinigo, M.; D'Angelo, C.; Quarteroni, A. A multiscale Darcy-Brinkman model for fluid flow in fractured porous media. *Numer. Math.* **2011**, *117*, 717–752. [CrossRef]
19. Murali, K.; Naidu, V.K.; Venkatesh, B. Solution of Darcy-Brinkman-Forchheimer Equation for Irregular Flow Channel by Finite Elements Approach. *J. Phys. Conf. Ser.* **2019**, *1172*, 012033. [CrossRef]

20. Kumaran, V.; Tamizharasi, R. Pressure in MHD/Brinkman flow past a stretching sheet. *Commun. Nonlinear Sci. Numer. Simul.* **2011**, *16*, 4671–4681.
21. Nield, D.A.; Bejan, A. *Convection in Porous Media*; Springer Verlag Inc.: New York, NY, USA, 1998.
22. Nield, D.A.; Kuznetsov, A.V. Forced convection in porous media: Transverse heterogeneity effects and thermal development. In *Handbook of Porous Media*, 2nd ed.; Vafai, K., Ed.; Taylor and Francis: New York, NY, USA, 2005; pp. 143–193.
23. Nield, D.A. The modeling of viscous dissipation in a saturated porous medium. *J. Heat Transf.* **2007**, *129*, 1459–1463. [CrossRef]
24. Nield, D.A.; Bejan, A. *Convection in Porous Media*, 4th ed.; Springer: New York, NY, USA, 2013.
25. Pantokratoras, A. Flow adjacent to a stretching permeable sheet in a Darcy-Brinkman porous medium. *Transp. Porous Med.* **2009**, *80*, 223–227. [CrossRef]
26. Pop, I.; Ingham, D.B. Flow past a sphere embedded in a porous medium based on the Brinkman model. *Int. Commun. Heat Mass Transf.* **1996**, *23*, 865–874. [CrossRef]
27. Pop, I.; Na, T.Y. A note on MHD flow over a stretching permeable surface. *Mech. Res. Commun.* **1998**, *25*, 263–269. [CrossRef]
28. Pop, I.; Cheng, P. Flow past a circular cylinder embedded in a porous medium based on the Brinkman model. *Int. J. Eng. Sci.* **1992**, *30*, 257–262. [CrossRef]
29. Wang, C.Y. Darcy-Brinkman Flow with Solid Inclusions. *Chem. Eng. Commun.* **2010**, *197*, 261–274. [CrossRef]
30. Srinivasan, S.; Rajagopal, K.R. A thermodynamic basis for the derivation of the Darcy, Forchheimer and Brinkman models for flows through porous media and their generalizations. *Int. J. Non-Linear Mech.* **2014**, *58*, 162–166. [CrossRef]
31. Ingham, D.B.; Pop, I. *Transport in Porous Media*; Pergamon: Oxford, UK, 2002.
32. Magyari, E.; Keller, B. Exact solutions for self-similar boundary-layer flows induced by permeable stretching surfaces. *Eur. J. Mech. B* **2000**, *19*, 109–122. [CrossRef]
33. Mastroberardino, A.; Mahabaleshwar, U.S. Mixed convection in viscoelastic flow due to a stretching sheet in a porous medium. *J. Porous Media* **2013**, *16*, 483–500. [CrossRef]
34. Siddheshwar, P.G.; Mahabaleshwar, U.S. Effects of radiation and heat source on MHD flow of a viscoelastic liquid and heat transfer over a stretching sheet. *Int. J. Non-Linear Mech.* **2005**, *40*, 807–820. [CrossRef]
35. Siddheshwar, P.G.; Chan, A.; Mahabaleshwar, U.S. Suction-induced magnetohydrodynamics of a viscoelastic fluid over a stretching surface within a porous medium. *IMA J. Appl. Math.* **2014**, *79*, 445–458. [CrossRef]
36. Tamayol, A.; Hooman, K.; Bahrami, M. Thermal analysis of flow in a porous medium over a permeable stretching wall. *Transp. Porous Media* **2010**, *85*, 661–676. [CrossRef]
37. Shao, Q.; Fahs, M.; Younes, A.; Makradi, A. A High Accurate Solution for Darcy-Brinkman Double-Diffusive Convection in Saturated Porous Media. *J. Numer. Heat Transf. Part B Fundam.* **2015**, *69*, 26–47. [CrossRef]
38. Navier, C.L.M.H. Mémoire sur les lois du mouvement des fluids. *Mém. Acad. R. Sci. Inst. Fr.* **1827**, *6*, 389–440.
39. Brinkman, H.C. A calculation of the viscous force exerted by a flowing fluid on a dense swarm of particles. *Appl. Sci. Res.* **1949**, *1*, 27–34. [CrossRef]
40. Mahabaleshwar, U.S.; Nagaraju, K.R.; Kumar, P.N.V.; Baleanu, D.; Lorenzini, G. An exact analytical solution of the unsteady magnetohydrodynamics nonlinear dynamics of laminar boundary layer due to an impulsively linear stretching sheet. *Contin. Mech. Thermodyn.* **2017**, *29*, 559–567. [CrossRef]
41. Rajagopal, K.R. On a hierarchy of approximate models for flows of incompressible fluids through porous solids. *Math. Model. Methods Appl. Sci.* **2007**, *17*, 215–252. [CrossRef]
42. Vafai, K.; Tien, C.L. Boundary and inertia effects on flow and heat transfer in porous media. *Int. J. Heat Mass Transf.* **1981**, *24*, 195–203. [CrossRef]
43. Fang, T.; Aziz, A. Viscous flow with second-order slip velocity over a stretching sheet. *Zeitschrift Für Naturforschung A* **2010**, *65*, 1087–1092. [CrossRef]
44. Lin, W. Mass transfer induced slip effect on viscous gas flows above a shrinking/stretching sheet. *Int. J. Heat Mass Transf.* **2016**, *93*, 17–22.
45. Andersson, H.I. Slip flow past a stretching surface. *Acta Mech.* **2002**, *158*, 121–125. [CrossRef]
46. Crane, L.J. Flow past a stretching plate. *Z. Angew. Math. Phys.* **1970**, *21*, 645–647. [CrossRef]
47. Pavlov, K.B. Magnetohydrodynamic flow of an incompressible viscous liquid caused by deformation of plane surface. *Magn. Gidrodin.* **1974**, *4*, 146–147.

48. Wang, C.Y. Flow due to a stretching boundary with partial slip—An exact solution of the Navier-Stokes equations. *Chem. Eng. Sci.* **2002**, *57*, 3745–3747. [CrossRef]
49. Abramowitz, M.; Stegun, I.A. *Handbook of Mathematical Functions with Formulas, Graphs, and Mathematical Tables*, 9th ed.; Dover: New York, NY, USA, 1972; pp. 17–18.
50. Birkhoff, G.; MacLane, S. *A Survey of Modern Algebra*; Macmillan: New York, NY, USA, 1996; pp. 107–108.
51. Fang, T.-G.; Zhang, J.; Yao, S.-S. Slip Magnetohydrodynamic Viscous Flow over a Permeable Shrinking Sheet. *Chin. Phys. Lett.* **2010**, *27*, 124702. [CrossRef]
52. Fang, T.; Yao, S.; Zhang, J.; Aziz, A. Viscous flow over a shrinking sheet with a second order slip flow model. *Commun. Nonlinear Sci. Numer. Simul.* **2010**, *15*, 1831–1842. [CrossRef]

fluids

MDPI

Article

Semi-Analytical Solutions for the Poiseuille–Couette Flow of a Generalised Phan-Thien–Tanner Fluid

Ângela M. Ribau [1,*,†], Luís L. Ferrás [2,†], Maria L. Morgado [3,†], Magda Rebelo [4,†] and Alexandre M. Afonso [1,†]

[1] Centro de Estudos de Fenómenos de Transporte, Departamento de Engenharia Mecânica, Faculdade de Engenharia da Universidade do Porto, Rua Dr. Roberto Frias, s/n, 4200-465 Porto, Portugal

[2] Centro de Matemática & Departamento de matemática da Universidade do Minho, Campus de Azurém, 4800-058 Guimarães, Portugal

[3] Center for Computational and Stochastic Mathematics, Instituto Superior Técnico, Universidade de Lisboa & Department of Mathematics, University of Trás-os-Montes e Alto Douro, UTAD, 5001-801 Vila Real, Portugal

[4] Centro de Matemática e Aplicações (CMA) and Departamento de Matemática, Faculdade de Ciências e Tecnologia, Universidade NOVA de Lisboa, Quinta da Torre, 2829-516 Caparica, Portugal

* Correspondence: angelaribau@fe.up.pt

† These authors contributed equally to this work.

Received: 3 June 2019; Accepted: 9 July 2019; Published: 12 July 2019

Abstract: This work presents new analytical and semi-analytical solutions for the pure Couette and Poiseuille–Couette flows, described by the recently proposed (Ferrás et al., A Generalised Phan-Thien–Tanner Model, JNNFM 2019) viscoelastic model, known as the generalised Phan-Thien–Tanner constitutive equation. This generalised version considers the Mittag–Leffler function instead of the classical linear or exponential functions of the trace of the stress tensor, and provides one or two new fitting constants in order to achieve additional fitting flexibility. The analytical solutions derived in this work allow a better understanding of the model, and therefore contribute to improve the modelling of complex materials, and will provide an interesting challenge to computational rheologists, to benchmarking and to code verification.

Keywords: generalised simplified PTT; Phan-Thien–Tanner (PTT) model; Mittag–Leffler; Couette flow; Poiseuille–Couette flow

---

1. Introduction

It is well known that much can be learned about a physical phenomenon if a mathematical model exists that can mimic and predict its behavior. The world of complex fluids is no different, and, therefore, several models have been proposed over the years for that purpose. These models can be more or less complex, depending on the properties of the fluids that are taken into account.

In this work, we are interested in viscoelastic materials [1], for which several models have been proposed in the past. One can classify these models as: differential (that make use of the local deformation field only) and integral (that take into account all the past deformation at each instant). Differential models usually allow a faster numerical solution of the differential equations involved, while integral models are computationally expensive and may lead to error propagation. On the other hand, integral models allow a better modelling, since they incorporate the real world fluid memory (the present state is influenced by all past weighted deformations). It is therefore of major importance to improve the fitting capabilities of differential models and reduce the computational effort needed to compute integral models.

In a recent work, Ferrás et al. [2] proposed an improved differential model based on the model by Nhan Phan-Thien and Roger Tanner (PTT [3]), derived from the Lodge–Yamamoto type of network theory for polymeric fluids. The constitutive equation proposed by Nhan Phan-Thien and Roger Tanner, for the case of an isothermal flow, is given by:

$$f(\tau_{kk})\,\boldsymbol{\tau} + \lambda\overset{\square}{\boldsymbol{\tau}} = 2\eta_p\mathbf{D} \tag{1}$$

with

$$f(\tau_{kk}) = 1 + \frac{\varepsilon\lambda}{\eta_p}\tau_{kk}, \tag{2}$$

where $\mathbf{D}$ is the rate of deformation tensor, $\boldsymbol{\tau}$ is the stress tensor, λ is a relaxation time, η_p is the polymeric viscosity, τ_{kk} is the trace of the stress tensor, ε represents the extensibility parameter and $\overset{\square}{\boldsymbol{\tau}}$ represents the Gordon–Schowalter derivative defined as

$$\overset{\square}{\boldsymbol{\tau}} = \frac{\partial\boldsymbol{\tau}}{\partial t} + \mathbf{u}\cdot\nabla\boldsymbol{\tau} - (\nabla\mathbf{u})^T\cdot\boldsymbol{\tau} - \boldsymbol{\tau}\cdot(\nabla\mathbf{u}) + \xi\,(\boldsymbol{\tau}\cdot\mathbf{D} + \mathbf{D}\cdot\boldsymbol{\tau}). \tag{3}$$

Here, $\mathbf{u}$ is the velocity vector, $\nabla\mathbf{u}$ is the velocity gradient and the parameter ξ accounts for the slip between the molecular network and the continuous medium (it should be remarked that for the derivation of the analytical solutions we will consider $\xi = 0$). Later, Phan-Thien proposed a new model, based on an exponential function form [4] and showed that this new function would be quite adequate to represent the rate of destruction of junctions, but the parameter ε should be of the order 0.01. The function $f(\tau_{kk})$ is given by:

$$f(\tau_{kk}) = \exp\left(\frac{\varepsilon\lambda}{\eta_p}\tau_{kk}\right). \tag{4}$$

Ferrás et al. [2] considered a more general function for the rate of destruction of junctions, the Mittag–Leffler function where one or two fitting constants are included, in order to achieve additional fitting flexibility [2]. The Mittag–Leffler function is defined by,

$$E_{\alpha,\beta}(z) = \sum_{k=0}^{\infty}\frac{z^k}{\Gamma(\alpha k + \beta)}, \tag{5}$$

with α, β real and positive. When $\alpha = \beta = 1$, the Mittag–Leffler [5] function reduces to the exponential function. When $\beta = 1$, the original one-parameter Mittag–Leffler function, E_α, is obtained. Thus, the new function of the trace of stress tensor (now denoted by $K(.)$ instead of $f(.)$, to distinguish from the classical cases) describing the network destruction of junctions is written as:

$$K(\tau_{kk}) = \Gamma(\beta)\,E_{\alpha,\beta}\left(\frac{\varepsilon\lambda}{\eta_p}\tau_{kk}\right), \tag{6}$$

where Γ is the Gamma function and the normalisation $\Gamma(\beta)$ is used to ensure that $K(0) = 1$, for all choices of β.

The linear and the exponential model of the Phan-Thien–Tanner has been frequently used in the literature, and in fact Ferrás et al. [6] considered a new quadratic version of the PTT model, i.e., a second-order expansion of the exponential model given by:

$$f(\tau_{kk}) = 1 + \frac{\varepsilon\lambda}{\eta_p}\tau_{kk} + \frac{1}{2}\left(\frac{\varepsilon\lambda}{\eta_p}\tau_{kk}\right)^2. \tag{7}$$

Here, we compare the generalised Phan-Thien–Tanner (gPTT), given by Equation (6), with the linear, the exponential and the quadratic versions of the PTT (Equations (2), (4) and (7), respectively).

To compare these models, we study the dimensionless material properties in steady shear flow of the three versions of the PTT model and compare them with the new gPTT model, considering different values of α and β.

The material functions can be obtained considering a steady-state Couette flow in the x-direction, $\mathbf{u} = (\dot{\gamma}y, 0, 0)$, where $\dot{\gamma}$ is the shear rate. For this flow, considering the parameter $\xi = 0$, the constitutive Equation (1) reduces to:

$$\begin{cases} K(\tau_{kk})\tau_{xx} = 2\lambda\dot{\gamma}\tau_{xy} \\ K(\tau_{kk})\tau_{xy} = \eta_p\dot{\gamma}. \\ \tau_{yy} = \tau_{zz} = \tau_{xz} = \tau_{yz} = 0 \end{cases} \tag{8}$$

From the system of Equation (8), $\tau_{kk} = \tau_{xx}$ and applying some algebra in the first two equations, a relationship between the shear stress and the normal stress is found,

$$\tau_{xx} = 2\frac{\lambda}{\eta_p}\tau_{xy}^2. \tag{9}$$

We can also obtain the viscometric material functions: the steady shear viscosity, $\mu(\dot{\gamma})$, the first normal stress difference coefficient, $\Psi_1(\dot{\gamma})$, and the second normal stress difference coefficient, $\Psi_2(\dot{\gamma})$, which are given by:

$$\mu(\dot{\gamma}) = \frac{\tau_{xy}}{\dot{\gamma}}, \tag{10}$$

$$\Psi_1(\dot{\gamma}) = \frac{\tau_{xx} - \tau_{xy}}{\dot{\gamma}^2}, \tag{11}$$

$$\Psi_2(\dot{\gamma}) = \frac{\tau_{yy} - \tau_{zz}}{\dot{\gamma}^2}. \tag{12}$$

As for other versions of the simplified PTT models for which $\xi = 0$, the second normal stress coefficient is null, $\Psi_2(\dot{\gamma}) = 0$, so, we only need to find $\mu(\dot{\gamma})$ and $\Psi_1(\dot{\gamma})$. Therefore, manipulating the second equation of the system of Equations (8) we get,

$$\tau_{xy} = \frac{\eta_p\dot{\gamma}}{K(\tau_{xx})}. \tag{13}$$

The dimensionless expression for the steady shear viscosity becomes,

$$\frac{\mu(\dot{\gamma})}{\eta_p} = \frac{\tau_{xy}}{\eta_p\dot{\gamma}} = \frac{1}{K(\tau_{xx})} \tag{14}$$

and the dimensionless first normal stress coefficient is given by,

$$\frac{\Psi_1(\dot{\gamma})}{2\eta_p\lambda} = \frac{\tau_{xx}}{2\eta_p\lambda\dot{\gamma}^2} = \frac{1}{[K(\tau_{xx})]^2}. \tag{15}$$

In [6], it was shown that, for the linear PTT, the quadratic PTT and the exponential PTT, the dimensionless material functions depend on the generalised Deborah number, $\sqrt{\varepsilon}(\lambda\dot{\gamma})$. We show that the same happens for the gPTT model. To obtain the material function for the gPTT model, we need to solve the non-linear system of equations (Equation (8)), which can be written in terms of τ_{xx} in the non-linear form:

$$\frac{1}{2}K(\tau_{xx})^2\frac{\varepsilon\lambda}{\eta_p}\tau_{xx} = \varepsilon(\lambda\dot{\gamma})^2. \tag{16}$$

Giving values to $\frac{\varepsilon\lambda}{\eta_p}\tau_{xx}$, we can find $\sqrt{\varepsilon}(\lambda\dot{\gamma})$ using Equation (16). Then, the function $K(\tau_{xx})$ is directly calculated, allowing us to obtain the material functions given by Equations (14) and (15).

Figure 1 presents the dimensionless material properties for the steady-state Couette flow using three versions of the PTT (linear, quadratic, and exponential) and also the gPTT model. In Figure 1a, we set $\beta = 1$ and use different values of α, and, in Figure 1b, we set $\alpha = 1$ and use different values to β.

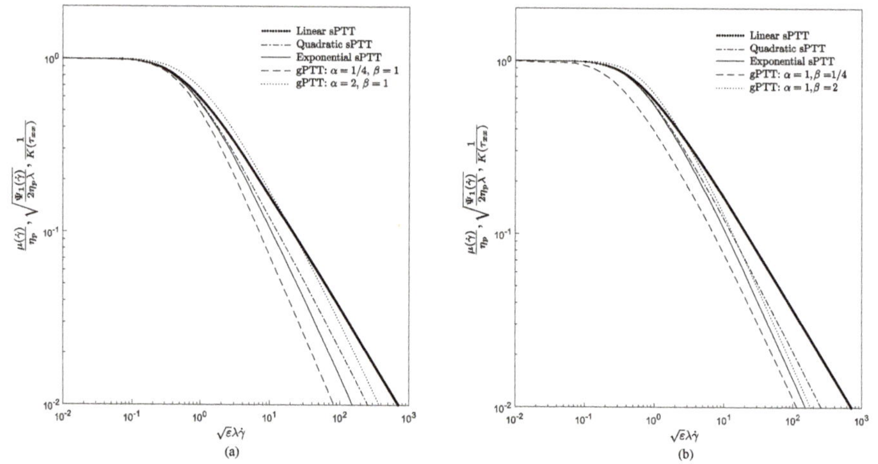

Figure 1. Dimensionless material properties in steady-state Couette flow using the three versions of the sPTT and for the gPTT model: (**a**) $\beta = 1$; and (**b**) $\alpha = 1$.

We observe that the new generalised function allows a broader description of the thinning properties of the fluid. Both the thinning rate and the onset of the thinning behavior can be controlled by the new model parameters. Therefore, this new model must be further explored for weak flows, such as Couette flows.

This model was extensively studied for strong flows in [2], where an explanation on the influence of the new model parameters was provided.

Note that the exponential version of the model was developed to take into account the strong destruction of network junctions, which occurs, for example, in strong flows (e.g., extensional flows). Although the exponential model was derived for such strong flows, it was shown in [2] that the gPTT model could slightly improve the fitting for shear (weak) flows, considering polymer solutions. Here, we consider polymer melts.

Figure 2 shows that the gPTT model provides a much better fitting to weak flows of polymer melts (low density polyethylene melt [7]), even when using only one extra parameter (α).

To quantify the error incurred during the fitting process, we used a mean square error given by

$$error = \sum_{i}^{N_\mu} \left[\log \mu(\dot{\gamma})_i - \log \mu(\dot{\gamma})_{fit_i}\right]^2 + \sum_{j}^{N_{\Psi_1}} \left[\log \Psi_1(\dot{\gamma})_j - \log \Psi_1(\dot{\gamma})_{fit_j}\right]^2, \quad (17)$$

$$error_\mu = \sum_{i}^{N_\mu} \left[\log \mu(\dot{\gamma})_i - \log \mu(\dot{\gamma})_{fit_i}\right]^2, \quad (18)$$

$$error_{\Psi_1} = \sum_{j}^{N_{\Psi_1}} \left[\log \Psi_1(\dot{\gamma})_j - \log \Psi_1(\dot{\gamma})_{fit_j}\right]^2, \quad (19)$$

with N_μ and N_{Ψ_1} the number of experimental points obtained for $\mu(\dot{\gamma})$ and $\Psi_1(\dot{\gamma})$, respectively.

A better fit was obtained for the new generalised model when compared to the original exponential PTT model. The total mean square error obtained for the exponential PTT model was 29.7, being five times the error obtained for its generalised version (for which a value of 6.0 was obtained). The new model allows a better fit for low and high shear rates for the first normal stress difference (where the $error_{\Psi_1}$ obtained for the exponential PTT model is 20 times higher than the error obtained for the gPTT). For the shear viscosity, the gPTT model predicts a lower value (when compared to experimental data) for high shear rates (although it should be remarked that the $error_\mu$ is four times smaller when compared to the exponential model).

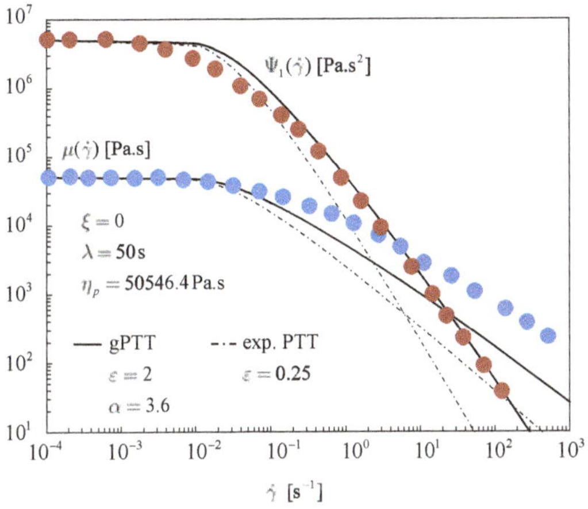

Figure 2. Fitting of the shear viscosity and the first normal stress difference coefficient to rheological data from Laun [7]. The generalised PTT model only considers the one-parameter Mittag–Leffler function, E_α. By adding only one parameter, we obtain a fitting error (Equation (17)) of 29.7 and 6 for the exponential and gPTT models, respectively. The symbols represent the experimental data from Laun [7] for a low density polyethylene melt.

Based on what is described above, this work presents analytical and semi-analytical solutions for pure Couette and Poiseuille–Couette flows, described by the generalised Phan-Thien–Tanner constitutive equation. It is well known that the rate of destruction of junctions increases for strong flows (e.g., extensional flows), but, in this case, we consider weak flows, and study the capability of this new model to describe them. This is done by performing a parametric study for the influence of the gPTT parameters.

2. Analytical Solution for the gPTT Model in Couette flow

In this section, we derive the analytical solution for the fully developed flow of the gPTT model considering both Couette and Poiseuille–Couette flows (cf. Figure 3). To obtain closed form analytical solutions, the slip parameter in the Gordon–Schowalter derivative is set to $\xi = 0$.

The equations governing the flow of an isothermal incompressible fluid are the continuity,

$$\nabla \cdot \mathbf{u} = 0 \tag{20}$$

and the momentum equation,

$$\rho \frac{D\mathbf{u}}{Dt} = -\nabla p + \nabla \cdot \boldsymbol{\tau} \tag{21}$$

together with the constitutive equation,

$$\Gamma\left(\beta\right) E_{\alpha,\beta}\left(\frac{\varepsilon\lambda}{\eta_p}\tau_{kk}\right)\boldsymbol{\tau} + \lambda \overset{\circ}{\boldsymbol{\tau}} = 2\eta_p \mathbf{D}, \tag{22}$$

where $\frac{D}{Dt}$ is the material derivative, p is the pressure, t is the time and ρ is the fluid density.

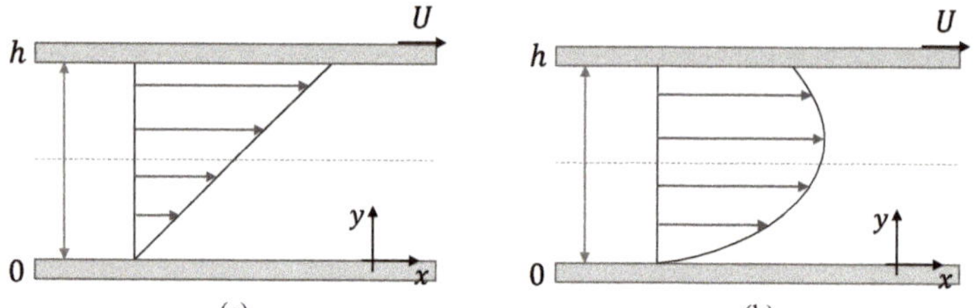

Figure 3. Geometry of: (**a**) the pure Couette flow; and (**b**) the Couette flow with an imposed pressure gradient (Poiseuille–Couette flow).

We consider a Cartesian coordinate system with x, y, and z being the streamwise, transverse and spanwise directions, respectively. The flow is assumed to be fully-developed and therefore the governing equations can be further simplified since

$$\frac{\partial}{\partial x} = 0(\text{except for pressure}), \frac{\partial v}{\partial y} = 0, \frac{\partial p}{\partial y} = 0. \tag{23}$$

Therefore, Equation (21) can be integrated, leading to the following general equation for the shear stress:

$$\tau_{xy} = P_x y + c_1, \tag{24}$$

where P_x is the pressure gradient in the x direction, τ_{xy} is the shear stress and c_1 is a stress constant. This equation is valid regardless of the rheological constitutive equation. The constitutive equations for the generalised PTT model describing this flow can be further simplified leading to:

$$K(\tau_{kk})\tau_{xx} = (2-\xi)(\lambda\dot{\gamma})\tau_{xy}, \tag{25}$$

$$K(\tau_{kk})\tau_{yy} = -\xi(\lambda\dot{\gamma})\tau_{xy}, \tag{26}$$

$$K(\tau_{kk})\tau_{xy} = \eta_p\dot{\gamma} + (1-\frac{\xi}{2})(\lambda\dot{\gamma})\tau_{yy} - \frac{\xi}{2}(\lambda\dot{\gamma})\tau_{xx}, \tag{27}$$

where the shear rate $\dot{\gamma}$ is a function of y ($\dot{\gamma}(y) \equiv \frac{du}{dy}$) and $\tau_{kk} = \tau_{xx} + \tau_{yy}$ is the trace of the stress tensor. Assuming $\xi = 0$, Equation (26) implies that $\tau_{yy} = 0$, and the trace of the stress tensor becomes $\tau_{kk} = \tau_{xx}$. Dividing Equation (25) by Equation (27), $K(\tau_{xx})$ cancels out, and we get the explicit relationship between the streamwise normal stress and the shear stress given by Equation (9).

Now, combining Equations (9), (24) and (27), the following shear rate profile is obtained,

$$\dot{\gamma}(y) = \Gamma(\beta)E_{\alpha,\beta}\left(\frac{2\varepsilon\lambda^2}{\eta_p^2}(P_x y + c_1)^2\right)\frac{(P_x y + c_1)}{\eta_p}. \tag{28}$$

The velocity profile can be obtained integrating the shear rate subject to the Couette boundary conditions (null velocity at the immobile wall),

$$u(0) = 0 \tag{29}$$

and an imposed constant velocity, U, at the moving wall,

$$u(h) = \text{U}. \tag{30}$$

This leads to the following velocity profile:

$$u(y) = \mathrm{U} - \frac{\Gamma(\beta)}{\eta_p P_x} \sum_{k=0}^{\infty} \left(\left(\frac{2\varepsilon\lambda^2}{\eta_p^2} \right)^k \frac{(P_x h + c_1)^{2k+2} - (P_x y + c_1)^{2k+2}}{\Gamma(\alpha k + \beta)\,(2k+2)} \right), \tag{31}$$

where c_1 can be obtained by solving numerically the following equation,

$$\mathrm{U} = \frac{\Gamma(\beta)}{\eta_p P_x} \sum_{k=0}^{\infty} \left(\left(\frac{2\varepsilon\lambda^2}{\eta_p^2} \right)^k \frac{1}{\Gamma(\alpha k + \beta)} \frac{(P_x h + c_1)^{2k+2} - c_1{}^{2k+2}}{2k+2} \right). \tag{32}$$

Combining Equations (31) and (32) leads to the following dimensionless velocity profile:

$$\bar{u}(\bar{y}) = \frac{\Gamma(\beta)}{\bar{P_x}} \sum_{k=0}^{\infty} \left(\left(2\varepsilon Wi^2 \right)^k \frac{1}{\Gamma(\alpha k + \beta)} \frac{(\bar{P_x}\bar{y} + \bar{c_1})^{2k+2} - \bar{c_1}{}^{2k+2}}{2k+2} \right) \tag{33}$$

with $\bar{y} = \frac{y}{h}$, $\bar{u}\,(y) = \frac{u(\bar{y})}{\mathrm{U}}$, $\bar{c_1} = \frac{c_1 h}{\eta_p \mathrm{U}}$, $\bar{P_x} = \frac{P_x h^2}{\eta_p \mathrm{U}}$ and $Wi = \frac{\lambda \mathrm{U}}{h}$ the Weissenberg number.

Remark 1. *Note that, if* $c_1 = -P_x \frac{h}{2}$, *Equation (31) becomes,*

$$u(y) = \mathrm{U} - \frac{\Gamma(\beta)}{\eta_p P_x} \sum_{k=0}^{\infty} \left(\left(\frac{2\varepsilon\lambda^2}{\eta_p^2} \right)^k \frac{1}{\Gamma(\alpha k + \beta)} \frac{\left(P_x \frac{h}{2} \right)^{2k+2} - \left(P_x \left(y - \frac{h}{2} \right) \right)^{2k+2}}{2k+2} \right), \tag{34}$$

and Equation (32) leads to $u(h) = 0$, *corresponding to Poiseuille flow with no slip boundary conditions.*

The velocity profile can be written in dimensionless form as:

$$\bar{u}(\bar{y}) = \frac{\Gamma(\beta)}{\bar{P_x}} \sum_{k=0}^{\infty} \left(\left(2\varepsilon Wi^2 \right)^k \frac{1}{\Gamma(\alpha k + \beta)} \frac{\left(\bar{P_x} \left(\bar{y} - \frac{1}{2} \right) \right)^{2k+2} - \left(\frac{\bar{P_x}}{2} \right)^{2k+2}}{2k+2} \right). \tag{35}$$

When we consider $\alpha = \beta = 1$, *this equation reduces to the one presented by Oliveira and Pinho [8] for the planar channel flow of an exponential PTT fluid:*

$$\bar{u}(\bar{y}) = \frac{1}{4\varepsilon Wi^2 \bar{P_x}} \left(\exp\left(2\varepsilon Wi^2 \bar{P_x}^2 \left(\bar{y} - \frac{1}{2} \right)^2 \right) - \exp\left(\frac{2\varepsilon Wi^2 \bar{P_x}^2}{4} \right) \right). \tag{36}$$

Figure 4 shows a comparison between the gPTT model and exponential PTT (Equations (35) and (36)) for different values of εWi^2. *As expected, the results are identical, confirming the solution limit for* $\alpha = \beta = 1$ *on the Mittag–Leffler function.*

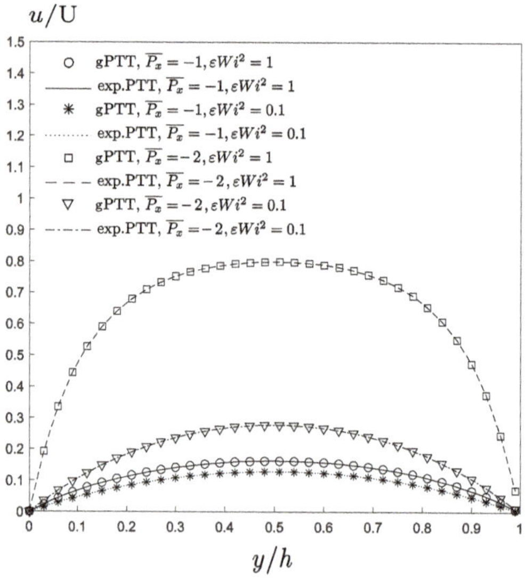

Figure 4. Comparison between the gPTT model and exponential PTT considering a Poiseuille flow with different values of εWi^2 and different values of imposed $\bar{P}_x$.

3. Analytical Solution for the gPTT Model in Pure Couette flow.

For the pure Couette flow, $P_x = 0$, therefore Equation (24) becomes,

$$\tau_{xy} = c_1. \tag{37}$$

The shear rate is then given by Equation (38),

$$\dot{\gamma}(y) = \Gamma(\beta) E_{\alpha,\beta}\left(\frac{2\varepsilon\lambda^2}{\eta_p^2} {c_1}^2\right)\frac{c_1}{\eta_p}. \tag{38}$$

Integrating Equation (38) and taking into account Equation (29), the velocity field for the pure Couette flow is obtained,

$$u(y) = \frac{\Gamma(\beta)}{\eta_p}\sum_{k=0}^{\infty}\left(\left(\frac{2\varepsilon\lambda^2}{\eta_p^2}\right)^k \frac{c_1^{2k+1}}{\Gamma(\alpha k+\beta)}\, y\right). \tag{39}$$

Making use of the boundary condition given in Equation (30), we obtain the following nonlinear equation on c_1, which must be solved numerically,

$$\frac{U}{h} = \frac{\Gamma(\beta)}{\eta_p}\sum_{k=0}^{\infty}\left(\left(\frac{2\varepsilon\lambda^2}{\eta_p^2}\right)^k \frac{c_1^{2k+1}}{\Gamma(\alpha k+\beta)}\right). \tag{40}$$

Equations (39) and (40) can be written in dimensionless form as:

$$\bar{u}(\bar{y}) = \Gamma(\beta)\bar{c_1}\bar{y}\sum_{k=0}^{\infty}\left(\left(2\varepsilon Wi^2 \bar{c_1}^2\right)^k \frac{1}{\Gamma(\alpha k+\beta)}\right) \tag{41}$$

and

$$1 = \Gamma(\beta)\bar{c_1} \sum_{k=0}^{\infty} \left(\left(2\varepsilon Wi^2 \bar{c_1}^2\right)^k \frac{1}{\Gamma(\alpha k + \beta)} \right), \tag{42}$$

respectively.

4. Discussion of Results

In the previous section, semi-analytical equations were derived for the gPTT model in Poiseuille–Couette flow. In this section, we investigate the influence of the Mittag–Leffler function parameters α and β on the velocity profile of the Poiseuille–Couette flow. We consider different εWi^2 values, and also different values of α and β, and we compare the results with the ones obtained for the exponential PTT model. Figure 5 shows the velocity profiles obtained for the Poiseuille–Couette flow considering two different εWi^2 values and different values of α ($\beta = 1$).

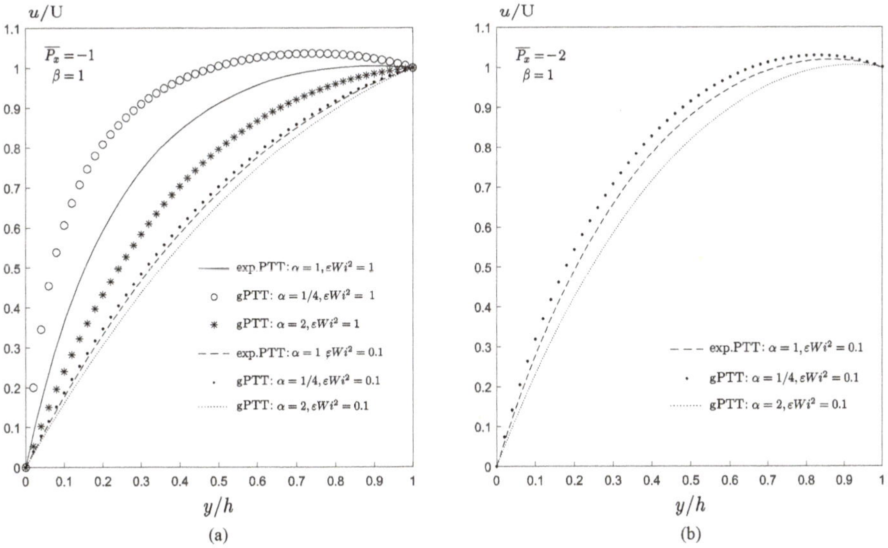

Figure 5. Velocity profiles obtained for the Poiseuille–Couette flow considering different values of εWi^2 and different values of α ($\beta = 1$): (**a**) $\bar{P}_x = -1$; and (**b**) $\bar{P}_x = -2$.

Figure 6 shows the velocity profiles obtained for the Poiseuille–Couette flow considering two different εWi^2 and different values of β ($\alpha = 1$).

We observe in Figure 5a that for $\alpha > 1$ the flow rate decreases while for $\alpha < 1$ it increases. As expected, for a constant pressure drop, the flow rate increases with εWi^2. In Figure 5b, we can observe that with the increase of the absolute value of the pressure drop, the velocity profile becomes more sensitive to small changes in α.

For the case of constant $\alpha = 1$ and varying β (Figure 6), the trends are similar to the ones obtained in Figure 5 (varying α), but now the velocity profile is less sensitive to large vales of β (with $\beta > 1$). In Figure 6b, we observe that the combined effects of pressure drop and large values of εWi^2 lead to a substantial increase of the flow rate for $\beta < 1$.

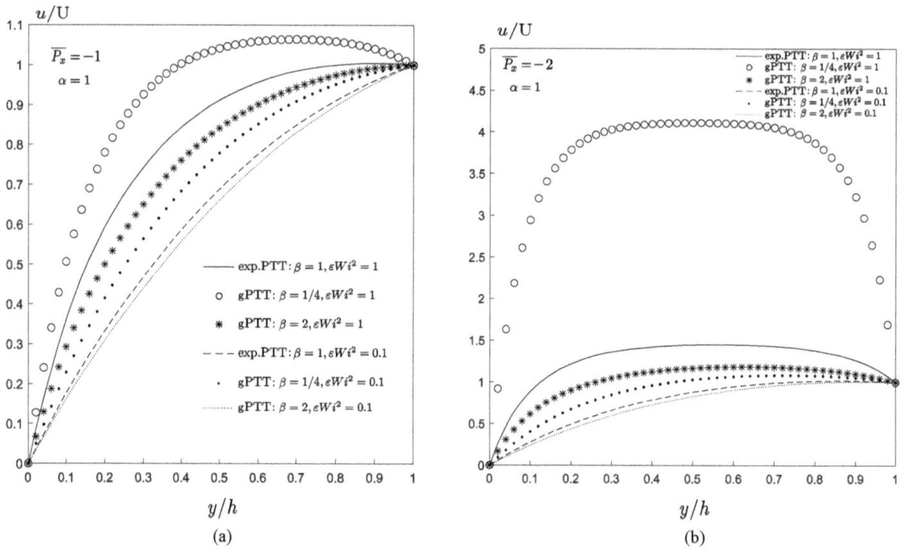

Figure 6. Velocity profiles obtained for the Poiseuille–Couette flow considering different values of εWi^2 and different values of β ($\alpha = 1$): (**a**) $\bar{P}_x = -1$; and (**b**) $\bar{P}_x = -2$.

5. Conclusions

In this work, we develop new analytical solutions for the Poiseuille–Couette flow of a viscoelastic fluid modelled by the generalised PTT model. We study the influence of the model's new parameters on the velocity profile and we discuss the role of the new function of the stress tensor on weak flows. The new model allows a broader description of flow behavior, and therefore it should be considered in the modelling of complex viscoelastic flows. The analytical solutions developed in this work are helpful for validating CFD codes, and also allow a further understanding of the model behavior in weak flows.

Author Contributions: Conceptualisation, Â.M.R., L.L.F., M.L.M., M.R. and A.M.A.; methodology, Â.M.R., L.L.F., M.L.M., M.R. and A.M.A.; software, Â.M.R.; validation, Â.M.R., L.L.F., M.L.M., M.R. and A.M.A.; formal analysis, Â.M.R., L.L.F., M.L.M., M.R. and A.M.A.; investigation, Â.M.R., L.L.F., M.L.M., M.R. and A.M.A.; writing—original draft preparation, Â.M.R., L.L.F., M.L.M., M.R. and A.M.A.; writing—review and editing, Â.M.R., L.L.F., M.L.M., M.R. and A.M.A.; and funding acquisition, A.M.A.

Funding: This research was funded by FEDER through COMPETE2020—Programa Operacional Competitividade e Internacionalização (POCI) and by national funds through FCT—Fundação para a Ciência e a Tecnologia, I.P. through Projects PTDC/EMS-ENE/3362/2014, POCI-01-0145-FEDER-016665, UID-MAT-00013/2013, and UID/MAT/00297/2013 as well as grant number SFRH/BPD/100353/2014. This work was partially supported by the Fundação para a Ciência e a Tecnologia (Portuguese Foundation for Science and Technology) through the project UID/MAT/00297/2019 (Centro de Matemática e Aplicações).

Acknowledgments: A.M. Afonso acknowledges the support by CEFT (Centro de Estudos de Fenómenos de Transporte). M.L. Morgado acknowledges the financial support of the Portuguese FCT - Fundação para a Ciência e a Tecnologia, through the project UID/Multi/04621/2019 of CEMAT/IST-ID, Center for Computational and Stochastic Mathematics, Instituto Superior Técnico, University of Lisbon. M. Rebelo acknowledges the support by Centro de Matemática e Aplicações (CMA). The authors acknowledge Gareth H. McKinley for his insightful comments that contributed to the creation of this work.

Conflicts of Interest: The authors declare no conflict of interest.

References

1. Bird, R.B.; Armstrong, R.C.; Hassager, O. *Dynamics of Polymeric Liquids. Volume 1: Fluid Mechanics*; John Wiley & Sons: Hoboken, NJ, USA, 1987.

2. Ferrás, L.L.; Morgado, M.L.; Rebelo, M.; McKinley Gareth, H.; Afonso, A.M. A Generalised Phan-Thien-Tanner Model. *J. Non-Newton. Fluid Mech.* **2019**, *269*, 88–99. [CrossRef]
3. Phan-Thien, N.; Tanner, R. A new constitutive equation derived from network theory. *J. Non-Newton. Fluid Mech.* **1977**, *2*, 353–365. [CrossRef]
4. Phan-Thien, N. A Nonlinear Network Viscoelastic Model. *J. Rheol.* **1978**, *22*, 259–283. [CrossRef]
5. Podlubny, I. *Fractional Differential Equations: An Introduction to Fractional Derivatives, Fractional Differential Equations, to Methods of Their Solution and Some of Their Applications*; Elsevier: Amsterdam, The Netherlands, 1998; Volume 198.
6. Ferrás, L.L.; Afonso, A.M.; Alves, M.A.; Nóbrega, J.M.; Pinho, F.T. Annular flow of viscoelastic fluids: Analytical and numerical solutions. *J. Non-Newton. Fluid Mech.* **2014**, *212*, 80–91. [CrossRef]
7. Laun, H.M. Description of the non-linear shear behaviour of a low density polyethylene melt. *Rheol. Acta* **1978**, *17*, 1–15. [CrossRef]
8. Oliveira, P.J.; Pinho, F.T. Analytical solution for fully developed channel and pipe flow of Phan-Thien-Tanner fluids. *J. Fluid Mech.* **1999**, *387*, 271–280. [CrossRef]

fluids

Article

Optimal Boundary Control of Non-Isothermal Viscous Fluid Flow

Evgenii S. Baranovskii *, Anastasia A. Domnich and Mikhail A. Artemov

Department of Applied Mathematics, Informatics and Mechanics, Voronezh State University, 394018 Voronezh, Russia

* Correspondence: esbaranovskii@gmail.com

Received: 29 June 2019; Accepted: 12 July 2019; Published: 16 July 2019

Abstract: We study an optimal control problem for the mathematical model that describes steady non-isothermal creeping flows of an incompressible fluid through a locally Lipschitz bounded domain. The control parameters are the pressure and the temperature on the in-flow and out-flow parts of the boundary of the flow domain. We propose the weak formulation of the problem and prove the existence of weak solutions that minimize a given cost functional. It is also shown that the marginal function of this control system is lower semi-continuous.

Keywords: non-isothermal flows; creeping flows; viscous fluid; optimal control; boundary control; pressure boundary conditions; weak solution; existence theorem; marginal function

1. Introduction and Problem Formulation

In this work, we consider the following optimal control problem for the model describing steady non-isothermal creeping flows of an incompressible fluid through a bounded domain $\Omega \subset \mathbb{R}^d$ $(d = 2, 3)$ with the locally Lipschitz boundary $\partial\Omega$:

$$\begin{cases} -\operatorname{div}[\mu(\theta)\boldsymbol{D}(\boldsymbol{u})] + \nabla p = \boldsymbol{f}(\boldsymbol{x},\theta) \ \text{ in } \Omega, \\ \operatorname{div} \boldsymbol{u} = 0 \ \text{ in } \Omega, \\ (\boldsymbol{u} \cdot \nabla)\theta - \operatorname{div}[\kappa(\theta)\nabla\theta] = \omega(\boldsymbol{x},\theta) \ \text{ in } \Omega, \\ \boldsymbol{u} = \boldsymbol{0}, \ \kappa(\theta)\dfrac{\partial\theta}{\partial\mathbf{n}} = -\alpha\theta \ \text{ on } \partial\Omega \setminus S, \\ \boldsymbol{u}_\tau = \boldsymbol{0}, \ p = \pi, \ \theta = \zeta \ \text{ on } S, \\ (\pi,\zeta) \in \mathbb{U}_1 \times \mathbb{U}_2, \\ J(\boldsymbol{u},\theta,\pi,\zeta) \to \min, \end{cases} \tag{1}$$

where $\boldsymbol{u} = \boldsymbol{u}(\boldsymbol{x})$ is the velocity of the fluid at a point $\boldsymbol{x} \in \Omega$, $\theta = \theta(\boldsymbol{x})$ is the deviation from the average temperature value, $\boldsymbol{f}(\boldsymbol{x},\theta)$ denotes the external forces, $\boldsymbol{D}(\boldsymbol{u})$ is the deformation rate tensor, $\boldsymbol{D}(\boldsymbol{u}) \stackrel{\text{def}}{=} (\nabla\boldsymbol{u} + (\nabla\boldsymbol{u})^{\mathrm{T}})/2$, the function $p = p(\boldsymbol{x})$ represents the pressure field, $\mu(\theta) > 0$ is the viscosity, $\kappa(\theta) > 0$ is the thermal conductivity, $\alpha > 0$ is a coefficient characterizing the heat transfer on solid walls of the flow domain, $\omega(\boldsymbol{x},\theta)$ stands for the heat source intensity, $\mathbf{n} = \mathbf{n}(\boldsymbol{x})$ is the unit outward normal to the surface $\partial\Omega$, S is a flat (straight for $d = 2$) portion of $\partial\Omega$ or the union of several such portions. Functions $\zeta\colon S \to \mathbb{R}$ and $\pi\colon S \to \mathbb{R}$ play the role of controls, $\mathbb{U}_1 \times \mathbb{U}_2$ is the set of admissible controls, while $J = J(\boldsymbol{u},\theta,\pi,\zeta)$ is a given cost functional. By the symbol τ we denote the tangential component of a vector, i.e., $\boldsymbol{u}_\tau \stackrel{\text{def}}{=} \boldsymbol{u} - (\boldsymbol{u} \cdot \mathbf{n})\mathbf{n}$.

The main scope of the present paper is to prove the solvability of problem (1) in the weak formulation. The proof is based on the Galerkin procedure, methods of the topological degree theory, compactness arguments, and the well-known theorem of Krasnoselskii on the continuity of a superposition operator acting in Lebesgue spaces. In addition, for control system (1), we introduce the concept of the marginal function, which shows how the optimal value of the cost functional changes with a change in the set of admissible controls. In this paper, it is proved that the marginal function of (1) is lower semi-continuous.

It should be mentioned that many mathematical works are devoted to optimization and control problems for non-isothermal flows of a viscous fluid. The results from the existing literature mainly deal with the case where the flow occurs inside a bounded domain and external body forces and/or heat sources are used as control parameters (see, for instance, [1–8]). Moreover, the Dirichlet-type boundary control by the velocity field is quite well studied (for details, see References [9–13]). However, for the analysis of real applications, it is very important to consider models of pressure driven flows, which typically occur in control problems for the heat and mass transfer in pipeline networks. For this purpose, following the approach from [14], we employ mixed boundary conditions, including the inhomogeneous Dirichlet condition for the pressure on a portion of the boundary of the flow domain; but at the same time, we formulate the optimal control problem without using the **curl** operator and boundary conditions associated with this operator. Another distinguishing feature of the present paper is that it takes into account the dependence of both the viscosity and the thermal conductivity coefficient on the temperature. This expands the range of possible applications of our results.

2. Preliminaries

In this section, we present some notations and function spaces utilized in the paper.

Let E_1, E_2 be Banach spaces. By $\mathcal{L}(E_1, E_2)$ we denote the space of all bounded linear mappings from E_1 to E_2. The space $\mathcal{L}(E_1, E_2)$ is equipped with the norm

$$\|A\|_{\mathcal{L}(E_1,E_2)} \stackrel{\text{def}}{=} \sup_{\|v\|_{E_1}\neq 0} \frac{\|A(v)\|_{E_2}}{\|v\|_{E_1}}.$$

The strong (weak) convergence in a Banach space is denoted by $\to$ ($\rightharpoonup$).

Let $\mathcal{U}$ and $\mathcal{W}$ be subsets of a Banach space E. By definition, put

$$d_E(\mathcal{U},\mathcal{W}) \stackrel{\text{def}}{=} \sup_{u\in\mathcal{U}} \inf_{w\in\mathcal{W}} \|u-w\|_E.$$

This quantity is termed as the directed Hausdorff distance (or one-sided Hausdorff distance) from the set $\mathcal{U}$ to the set $\mathcal{W}$.

Throughout this paper, boldface symbols denote vector- and matrix-valued quantities. For vectors $\boldsymbol{x}, \boldsymbol{y} \in \mathbb{R}^d$ and matrices $\mathbf{X}, \mathbf{Y} \in \mathbb{R}^{d\times d}$, by $\boldsymbol{x}\cdot\boldsymbol{y}$ and $\mathbf{X} : \mathbf{Y}$, we denote the scalar products, respectively:

$$\boldsymbol{x}\cdot\boldsymbol{y} \stackrel{\text{def}}{=} \sum_{i=1}^{d} x_i y_i, \qquad \mathbf{X}:\mathbf{Y} \stackrel{\text{def}}{=} \sum_{i,j=1}^{d} \mathrm{X}_{ij}\mathrm{Y}_{ij}.$$

As usual, $|\cdot|$ denotes the Euclidean norm:

$$|\boldsymbol{x}| \stackrel{\text{def}}{=} (\boldsymbol{x}\cdot\boldsymbol{x})^{1/2}, \qquad |\mathbf{X}| \stackrel{\text{def}}{=} (\mathbf{X}:\mathbf{X})^{1/2}.$$

The symbol ∇ stands for the gradient with respect to the variables $x_1, \ldots, x_d$. The divergence operator is defined as follows:

$$\operatorname{div} \boldsymbol{w} \stackrel{\text{def}}{=} \sum_{i=1}^{d} \frac{\partial w_i}{\partial x_i}, \qquad \operatorname{div} \boldsymbol{Q} \stackrel{\text{def}}{=} \left(\sum_{i=1}^{d} \frac{\partial Q_{i1}}{\partial x_i}, \dots, \sum_{i=1}^{d} \frac{\partial Q_{id}}{\partial x_i} \right)^{\mathrm{T}},$$

for $\boldsymbol{w}\colon \mathbb{R}^d \to \mathbb{R}^d$ and $\boldsymbol{Q}\colon \mathbb{R}^d \to \mathbb{R}^{d\times d}$.

We employ the standard notation for Lebesgue spaces, such as $L^q(\Omega)$, $L^q(\partial\Omega)$, where $q \geq 1$, and the Sobolev space $H^1(\Omega) \stackrel{\text{def}}{=} W^{1,2}(\Omega)$. When it comes to classes of $\mathbb{R}^d$-valued functions, we use bold face letters, for instance, $\boldsymbol{L}^q(\Omega) \stackrel{\text{def}}{=} L^q(\Omega)^d$, $\boldsymbol{H}^1(\Omega) \stackrel{\text{def}}{=} H^1(\Omega)^d$, etc.

Recall that the restriction of a function $w \in H^1(\Omega)$ to the surface $\partial\Omega$ is defined by the rule $w|_{\partial\Omega} \stackrel{\text{def}}{=} \gamma_{\partial\Omega} w$, where $\gamma_{\partial\Omega}\colon H^1(\Omega) \to L^q(\partial\Omega)$ is the trace operator (see, e.g., [15] Section 2.4.2), $q = 4$ if $d = 3$, and q is an arbitrary number from $[1, +\infty)$ if $d = 2$.

Let us introduce two sets:

$$\mathbb{V}_S(\Omega) \stackrel{\text{def}}{=} \{\boldsymbol{v} \in \boldsymbol{C}^\infty(\overline{\Omega}) : \ \operatorname{div} \boldsymbol{v} = 0, \ \boldsymbol{v}|_{\partial\Omega\setminus S} = \boldsymbol{0}, \ \boldsymbol{v}_\tau|_S = \boldsymbol{0}\},$$
$$\mathbb{Y}_S(\Omega) \stackrel{\text{def}}{=} \{\eta \in C^\infty(\overline{\Omega}) : \ \eta|_S = 0\},$$

and two basic spaces for handling problem (1):

$$\boldsymbol{V}_S(\Omega) \stackrel{\text{def}}{=} \text{ the closure of the set } \mathbb{V}_S(\Omega) \text{ in the space } \boldsymbol{H}^1(\Omega),$$
$$Y_S(\Omega) \stackrel{\text{def}}{=} \text{ the closure of the set } \mathbb{Y}_S(\Omega) \text{ in the space } H^1(\Omega).$$

In our study, it is convenient to use the following scalar product and norm in $\boldsymbol{V}_S(\Omega)$:

$$(\boldsymbol{v}, \boldsymbol{w})_{\boldsymbol{V}_S(\Omega)} \stackrel{\text{def}}{=} \int_\Omega \boldsymbol{D}(\boldsymbol{v}) : \boldsymbol{D}(\boldsymbol{w})\, \boldsymbol{dx}, \qquad \|\boldsymbol{v}\|_{\boldsymbol{V}_S(\Omega)} \stackrel{\text{def}}{=} (\boldsymbol{v}, \boldsymbol{v})^{1/2}_{\boldsymbol{V}_S(\Omega)}.$$

From ([16] Chap. I, Theorems 2.2 and 2.3) it follows that the scalar product $(\cdot,\cdot)_{\boldsymbol{V}_S(\Omega)}$ is well defined and the norm $\|\cdot\|_{\boldsymbol{V}_S(\Omega)}$ is equivalent to the standard $\boldsymbol{H}^1$-norm if the set $\partial\Omega \setminus S$ has the positive $(d-1)$-dimensional measure.

In the space $Y_S(\Omega)$, we introduce the scalar product and the norm as follows:

$$(\eta, \xi)_{Y_S(\Omega)} \stackrel{\text{def}}{=} \int_\Omega \nabla\eta \cdot \nabla\xi\, \boldsymbol{dx}, \qquad \|\eta\|_{Y_S(\Omega)} \stackrel{\text{def}}{=} (\eta, \eta)^{1/2}_{Y_S(\Omega)}.$$

By the Friedrichs inequality (see, e.g., [15] Section 1.1.8, Theorem 1.9), it is easy to show that the scalar product $(\cdot,\cdot)_{Y_S(\Omega)}$ is well defined and the norm $\|\cdot\|_{Y_S(\Omega)}$ is equivalent to the standard H^1-norm.

3. Main Assumptions and Some Examples of Cost Functionals

Let us assume that the following conditions are fulfilled:

- **(C1)** the inequalities $\mathrm{meas}_{d-1}(S) > 0$ and $\mathrm{meas}_{d-1}(\partial\Omega \setminus S) > 0$ hold, where $\mathrm{meas}_{d-1}(\cdot)$ denotes the Lebesgue $(d-1)$-dimensional measure of a set;
- **(C2)** the functions $\mu\colon \mathbb{R} \to \mathbb{R}$ and $\kappa\colon \mathbb{R} \to \mathbb{R}$ are continuous and there exist constants μ_i, κ_i, $i = 0, 1$, such that $0 < \mu_0 \leq \mu(y) \leq \mu_1$, $0 < \kappa_0 \leq \kappa(y) \leq \kappa_1$ for every $y \in \mathbb{R}$;
- **(C3)** the functions $f_i(\cdot, y)\colon \Omega \to \mathbb{R}$, $i = 1, \dots, d$, and $\omega(\cdot, y)\colon \Omega \to \mathbb{R}$ are measurable for every $y \in \mathbb{R}$;
- **(C4)** the functions $f_i(x, \cdot)\colon \mathbb{R} \to \mathbb{R}$, $i = 1, \dots, d$, and $\omega(x, \cdot)\colon \mathbb{R} \to \mathbb{R}$ are continuous for almost every $x \in \Omega$;
- **(C5)** there exist functions g_0 and ω_0 from the space $L^2(\Omega)$ and a positive constant M such that

  $$|\boldsymbol{f}(x, y)| \leq g_0(x) + M|y|^2, \qquad |\omega(x, y)| \leq \omega_0(x),$$

  for every $y \in \mathbb{R}$ and for almost every $x \in \Omega$;

(C6) the set $\mathbb{U}_1$ is sequentially weakly closed in the space $L^2(S)$;
(C7) the set $\mathbb{U}_2$ is closed in the space $L^2(S)$ and $0 \in \mathbb{U}_2$;
(C8) the functional $J\colon \boldsymbol{V}_S(\Omega) \times H^1(\Omega) \times L^2(S) \times L^2(S) \to \mathbb{R}$ is lower weakly semi-continuous; that is, for any sequence $\{(\boldsymbol{u}^n, \theta^n, \pi^n, \zeta^n)\}_{n=1}^{\infty}$ such that $\boldsymbol{u}^n \rightharpoonup \boldsymbol{u}^0$ weakly in $\boldsymbol{V}_S(\Omega)$, $\theta^n \rightharpoonup \theta^0$ weakly in $H^1(\Omega)$, $\pi^n \rightharpoonup \pi^0$ weakly in $L^2(S)$, and $\zeta^n \rightharpoonup \zeta^0$ weakly in $L^2(S)$ as $n \to \infty$, we have

$$J(\boldsymbol{u}^0, \theta^0, \pi^0, \zeta^0) \le \liminf_{n\to\infty} J(\boldsymbol{u}^n, \theta^n, \pi^n, \zeta^n);$$

(C9) the coercivity condition [17] holds for J; that is, for every $R > 0$ the set

$$\mathcal{G}_R \stackrel{\text{def}}{=} \{(\boldsymbol{u}, \theta, \pi, \zeta) \in \boldsymbol{V}_S(\Omega) \times H^1(\Omega) \times L^2(S) \times L^2(S) : \ J(\boldsymbol{u}, \theta, \pi, \zeta) \le R\}$$

is bounded in $\boldsymbol{V}_S(\Omega) \times H^1(\Omega) \times L^2(S) \times L^2(S)$.

Here, we also give two examples of cost functionals satisfying conditions **(C8)** and **(C9)**:

$$\begin{aligned} J = J_1(\boldsymbol{u}, \theta, \pi, \zeta) &\stackrel{\text{def}}{=} \lambda_1 \int_\Omega |\boldsymbol{u} - \boldsymbol{u}_0|^2 \, \boldsymbol{dx} + \lambda_2 \sum_{i=1}^d \int_\Omega \left| \frac{\partial(\boldsymbol{u} - \boldsymbol{u}_0)}{\partial x_i} \right|^2 \boldsymbol{dx} \\ &+ \lambda_3 \int_\Omega |\theta - \theta_0|^2 \, \boldsymbol{dx} + \lambda_4 \sum_{i=1}^d \int_\Omega \left| \frac{\partial(\theta - \theta_0)}{\partial x_i} \right|^2 \boldsymbol{dx} \\ &+ \lambda_5 \int_S |\pi|^2 \, d\sigma + \lambda_6 \int_S |\zeta|^2 \, d\sigma, \end{aligned}$$

$$\begin{aligned} J = J_2(\boldsymbol{u}, \theta, \pi, \zeta) &\stackrel{\text{def}}{=} \lambda_1 \int_\Omega |\boldsymbol{D}(\boldsymbol{u})|^2 \, \boldsymbol{dx} + \lambda_2 \int_\Omega |\nabla\theta|^2 \, \boldsymbol{dx} + \lambda_3 \int_{\partial\Omega\setminus S} |\theta|^2 \, d\sigma \\ &+ \lambda_4 \int_S |\pi - \pi_0|^2 \, d\sigma + \lambda_5 \int_S |\zeta - \zeta_0|^2 \, d\sigma, \end{aligned}$$

where $\boldsymbol{u}_0$, θ_0, π_0, ζ_0 are given functions, and $\lambda_1, \lambda_2, \ldots, \lambda_6$ are positive parameters (weight coefficients).

4. Weak Formulation of Problem (1) and Marginal Function

Definition 1. We shall say that $(\boldsymbol{u}, \theta, \pi, \zeta) \in \boldsymbol{V}_S(\Omega) \times H^1(\Omega) \times L^2(S) \times L^2(S)$ is an *admissible quadruplet* to problem (1) if $(\pi, \zeta) \in \mathbb{U}_1 \times \mathbb{U}_2$, $\theta|_S = \zeta$, and

$$\int_\Omega \mu(\theta) \boldsymbol{D}(\boldsymbol{u}) : \boldsymbol{D}(\boldsymbol{v}) \, \boldsymbol{dx} + \int_S \pi(\boldsymbol{v} \cdot \mathbf{n}) \, d\sigma = \int_\Omega \boldsymbol{f}(\boldsymbol{x}, \theta) \cdot \boldsymbol{v} \, \boldsymbol{dx}, \tag{2}$$

$$\sum_{i=1}^d \int_\Omega u_i \frac{\partial\theta}{\partial x_i} \eta \, \boldsymbol{dx} + \int_\Omega \kappa(\theta) \nabla\theta \cdot \nabla\eta \, \boldsymbol{dx} + \alpha \int_{\partial\Omega\setminus S} \theta\eta \, d\sigma = \int_\Omega \omega(\boldsymbol{x}, \theta) \eta \, \boldsymbol{dx} \tag{3}$$

for any vector-valued function $\boldsymbol{v} \in \boldsymbol{V}_S(\Omega)$ and function $\eta \in Y_S(\Omega)$.

The following lemma shows that relations (2) and (3) are natural in the weak formulation of problem (1).

Lemma 1. *Suppose $(\boldsymbol{u}, \theta, p)$ is a classic solution of boundary-value problem $(1)_1$–$(1)_5$, then equalities (2) and (3) hold.*

Proof. Let us fix an arbitrary vector-valued function $\boldsymbol{v} \in \boldsymbol{V}_S(\Omega)$. On taking the scalar product of both the left-hand and right-hand sides of equality $(1)_1$ with $\boldsymbol{v}$ and integrating over the domain Ω, we obtain

$$\underbrace{-\int_\Omega \operatorname{div}\,[\mu(\theta)\boldsymbol{D}(\boldsymbol{u})] \cdot \boldsymbol{v} \, \boldsymbol{dx}}_{I_1} + \underbrace{\int_\Omega \nabla p \cdot \boldsymbol{v} \, \boldsymbol{dx}}_{I_2} = \int_\Omega \boldsymbol{f}(\boldsymbol{x}, \theta) \cdot \boldsymbol{v} \, \boldsymbol{dx}. \tag{4}$$

By using integration by parts, one can show that

$$I_1 = \int_{\partial\Omega} [\mu(\theta)\boldsymbol{D}(\boldsymbol{u})\mathbf{n}] \cdot \boldsymbol{v}\, d\sigma - \int_{\Omega} \mu(\theta)\boldsymbol{D}(\boldsymbol{u}) : \boldsymbol{D}(\boldsymbol{v})\, \boldsymbol{dx},$$

from which, bearing in mind that $\boldsymbol{v} = (\boldsymbol{v} \cdot \mathbf{n})\mathbf{n}$ on S and $\boldsymbol{v} = \mathbf{0}$ on $\partial\Omega \setminus S$, we derive

$$I_1 = \int_S \mu(\theta)\big([\boldsymbol{D}(\boldsymbol{u})\mathbf{n}] \cdot \mathbf{n}\big)(\boldsymbol{v} \cdot \mathbf{n})\, d\sigma - \int_{\Omega} \mu(\theta)\boldsymbol{D}(\boldsymbol{u}) : \boldsymbol{D}(\boldsymbol{v})\, \boldsymbol{dx}. \tag{5}$$

Note that

$$[\boldsymbol{D}(\boldsymbol{u})\mathbf{n}] \cdot \mathbf{n} = \frac{\partial(\boldsymbol{u} \cdot \mathbf{n})}{\partial \mathbf{n}} \quad \text{on } S.$$

Since

$$\operatorname{div} \boldsymbol{u} = 0 \ \text{ in } \Omega, \qquad \boldsymbol{u}_\tau = 0 \ \text{ on } S,$$

we see that

$$\frac{\partial(\boldsymbol{u} \cdot \mathbf{n})}{\partial \mathbf{n}} = 0 \ \text{ on } S,$$

and hence,

$$[\boldsymbol{D}(\boldsymbol{u})\mathbf{n}] \cdot \mathbf{n} = 0 \ \text{ on } S.$$

Substituting the last relation into (5), we obtain, obviously,

$$I_1 = -\int_{\Omega} \mu(\theta)\boldsymbol{D}(\boldsymbol{u}) : \boldsymbol{D}(\boldsymbol{v})\, \boldsymbol{dx}. \tag{6}$$

To handle the term I_2, we again use integration by parts and get

$$I_2 = \int_{\partial\Omega} p(\boldsymbol{v} \cdot \mathbf{n})\, d\sigma - \int_{\Omega} p \operatorname{div} \boldsymbol{v}\, \boldsymbol{dx}. \tag{7}$$

Since

$$\operatorname{div} \boldsymbol{v} = 0 \ \text{ in } \Omega, \qquad \boldsymbol{v} = \mathbf{0} \ \text{ on } \Omega \setminus S, \qquad p = \pi \ \text{ on } S,$$

from (7) it follows that

$$I_2 = \int_S \pi(\boldsymbol{v} \cdot \mathbf{n})\, d\sigma. \tag{8}$$

Finally, by substituting (6) and (8) into equality (4), we arrive at (2).

Next, we take the scalar product in $L^2(\Omega)$ of each term from $(1)_3$ with a function $\eta \in Y_S(\Omega)$ and get

$$\sum_{i=1}^{d} \int_{\Omega} u_i \frac{\partial \theta}{\partial x_i} \eta\, \boldsymbol{dx} - \underbrace{\int_{\Omega} \operatorname{div}[\kappa(\theta)\nabla\theta]\eta\, \boldsymbol{dx}}_{I_3} = \int_{\Omega} \omega(\boldsymbol{x}, \theta)\eta\, \boldsymbol{dx}.$$

Integrating by parts the term I_3, we can rewrite the last equality as follows:

$$\sum_{i=1}^{d} \int_{\Omega} u_i \frac{\partial \theta}{\partial x_i} \eta\, \boldsymbol{dx} - \int_{\partial\Omega} \kappa(\theta)\frac{\partial \theta}{\partial \mathbf{n}} \eta\, d\sigma + \int_{\Omega} \kappa(\theta)\nabla\theta \cdot \nabla\eta\, \boldsymbol{dx} = \int_{\Omega} \omega(\boldsymbol{x}, \theta)\eta\, \boldsymbol{dx}.$$

Taking into account the boundary conditions

$$\kappa(\theta)\frac{\partial \theta}{\partial \mathbf{n}} = -\alpha\theta \ \text{ on } \partial\Omega \setminus S, \qquad \eta = 0 \ \text{ on } S,$$

we arrive at equality (3). This finishes the proof of Lemma 1. □

We are now ready to accurately define the concept of solutions to the considered optimal control problem. Let $\mathfrak{M}(\mathbb{U}_1, \mathbb{U}_2)$ be the set of admissible quadruplets to problem (1).

Definition 2. By a *solution* of problem (1) we mean a quadruplet $(\boldsymbol{u}_*, \theta_*, \pi_*, \zeta_*) \in \mathfrak{M}(\mathbb{U}_1, \mathbb{U}_2)$ at which the functional J attains the minimum:

$$J(\boldsymbol{u}_*, \theta_*, \pi_*, \zeta_*) = \inf_{(\boldsymbol{u},\theta,\pi,\zeta)\in\mathfrak{M}(\mathbb{U}_1,\mathbb{U}_2)} J(\boldsymbol{u}, \theta, \pi, \zeta).$$

By $\mathfrak{M}_{\mathrm{opt}}(\mathbb{U}_1, \mathbb{U}_2)$ denote the set of solutions to problem (1). Assuming that $\mathfrak{M}_{\mathrm{opt}}(\mathbb{U}_1, \mathbb{U}_2) \neq \emptyset$, we define

$$\Phi(\mathbb{U}_1, \mathbb{U}_2) \stackrel{\mathrm{def}}{=} J(\boldsymbol{u}_*, \theta_*, \pi_*, \zeta_*), \tag{9}$$

where $(\boldsymbol{u}_*, \theta_*, \pi_*, \zeta_*)$ is a quadruplet that belongs to the set $\mathfrak{M}_{\mathrm{opt}}(\mathbb{U}_1, \mathbb{U}_2)$. It is obvious that the value of $\Phi(\mathbb{U}_1, \mathbb{U}_2)$ does not depend on the choice of an element from the set $\mathfrak{M}_{\mathrm{opt}}(\mathbb{U}_1, \mathbb{U}_2)$ in the right-hand sides of (9).

Definition 3. The function Φ that is defined by equality (9) is called the *marginal function* of control system (1).

5. Main Results

The main results of the present work are summarized as follows:

Theorem 1. *Suppose conditions* **(C1)–(C9)** *hold. Then:*

(a) *optimal control problem (1) has at least one solution;*
(b) *the marginal function Φ is lower semi-continuous in the following sense: if $\mathfrak{M}_{\mathrm{opt}}(\mathbb{U}_1^n, \mathbb{U}_2^n) \neq \emptyset$, for any $n \in \mathbb{N}$, and*

$$\lim_{n\to\infty} d_{L^2(S)}(\mathbb{U}_1^n, \mathbb{U}_1) = 0, \qquad \lim_{n\to\infty} d_{L^2(S)}(\mathbb{U}_2^n, \mathbb{U}_2) = 0, \tag{10}$$

then

$$\Phi(\mathbb{U}_1, \mathbb{U}_2) \leq \liminf_{n\to\infty} \Phi(\mathbb{U}_1^n, \mathbb{U}_2^n).$$

The proof of this theorem is given in Section 7.

6. Auxiliary Propositions

For the reader's convenience, we state here some preparatory results on which the proof of Theorem 1 is based.

Proposition 1. *Suppose $\mathfrak{B}_r \stackrel{\mathrm{def}}{=} \{\boldsymbol{a} \in \mathbb{R}^N : |\boldsymbol{a}| < r\}$ and $\boldsymbol{F}\colon \overline{\mathfrak{B}}_r \times [0,1] \to \mathbb{R}^N$ is a continuous mapping such that*

(i) *$\boldsymbol{F}(\boldsymbol{a}, \lambda) \neq \boldsymbol{0}$ for any pair $(\boldsymbol{a}, \lambda) \in \partial\mathfrak{B}_r \times [0,1]$;*
(ii) *$\boldsymbol{F}(\cdot, 0)\colon \overline{\mathfrak{B}}_r \to \mathbb{R}^N$ is an odd mapping; that is, $\boldsymbol{F}(-\boldsymbol{a}, 0) = -\boldsymbol{F}(\boldsymbol{a}, 0)$ for any vector $\boldsymbol{a} \in \overline{\mathfrak{B}}_r$.*

Then, for any $\lambda \in [0,1]$, there exists a vector $\boldsymbol{a}_\lambda \in \mathfrak{B}_r$ such that $\boldsymbol{F}(\boldsymbol{a}_\lambda, \lambda) = \boldsymbol{0}$; in other words, the equation $\boldsymbol{F}(\boldsymbol{a}, \lambda) = \boldsymbol{0}$ is solvable with respect to $\boldsymbol{a}$ in the ball $\mathfrak{B}_r$.

Proof. Employing the homotopy invariance property of Brouwer's degree and condition (i), one can show that $\deg(\boldsymbol{F}(\cdot, \lambda), \mathfrak{B}_r, \boldsymbol{0})$ does not depend on $\lambda \in [0,1]$ and, consequently,

$$\deg(\boldsymbol{F}(\cdot, \lambda), \mathfrak{B}_r, \boldsymbol{0}) = \deg(\boldsymbol{F}(\cdot, 0), \mathfrak{B}_r, \boldsymbol{0}),$$

for any $\lambda \in [0,1]$. Besides, since $\boldsymbol{F}(\cdot,0)\colon \overline{\mathfrak{B}}_r \to \mathbb{R}^N$ is an odd mapping, we see that $\deg(\boldsymbol{F}(\cdot,0),\mathfrak{B}_r,\mathbf{0})$ is an odd number. This follows from Borsuk's theorem (see, e.g., [18] Chap. 1, Theorem 1.2.11). Thereby, for any $\lambda \in [0,1]$,

$$\deg(\boldsymbol{F}(\cdot,\lambda),\mathfrak{B}_r,\mathbf{0}) \neq 0.$$

Then, by [18] (Chap. 1, Theorem 1.2.6), we deduce that the equation $\boldsymbol{F}(\boldsymbol{a},\lambda) = \mathbf{0}$ has at least one solution $\boldsymbol{a}_\lambda \in \mathfrak{B}_r$. □

Proposition 2. *Suppose $G \subset \mathbb{R}^N$ is a Lebesgue measurable set and a given function $\varphi\colon G \times \mathbb{R} \to \mathbb{R}$ satisfies the following conditions:*

- *the function $\varphi(\cdot,y)\colon G \to \mathbb{R}$ is measurable for every $y \in \mathbb{R}$;*
- *the function $\varphi(\boldsymbol{x},\cdot)\colon \mathbb{R} \to \mathbb{R}$ is continuous for almost every $\boldsymbol{x} \in G$;*
- *there exist constants $q_1 \geq 1$, $q_2 \geq 1$, $K > 0$ and a function $\phi_0 \in L^{q_2}(G)$ such that*

$$|\varphi(\boldsymbol{x},y)| \leq \phi_0(\boldsymbol{x}) + K|y|^{q_1/q_2},$$

for every $y \in \mathbb{R}$ and for almost every $\boldsymbol{x} \in G$.

Then, the superposition operator T_φ defined by

$$T_\varphi\colon L^{q_1}(G) \to L^{q_2}(G), \quad T_\varphi[w](\boldsymbol{x}) \stackrel{\text{def}}{=} \varphi(\boldsymbol{x}, w(\boldsymbol{x}))$$

is a bounded and continuous mapping.

This proposition was proved by Krasnoselskii (see [19] Chap. 1).

7. Proof of Theorem 1

Let us prove the existence result (a). Our first step is to show that $\mathfrak{M}(\mathbb{U}_1,\mathbb{U}_2) \neq \emptyset$.

Fix a pair $(\pi^0,0) \in \mathbb{U}_1 \times \mathbb{U}_2$. Let $\{\boldsymbol{v}^j\}_{j=1}^\infty \subset \mathbb{V}_S(\Omega)$ be an orthonormal basis of the space $V_S(\Omega)$, $\{\eta^j\}_{j=1}^\infty \subset \mathbb{Y}_S(\Omega)$ an orthonormal basis of the space $Y_S(\Omega)$, and m an arbitrary fixed integer.

Consider the following one-parameter problem in space $\mathbb{R}^{2m}$:

Find a vector $(a_{m1},\ldots,a_{mm},b_{m1},\ldots,b_{mm})^{\mathrm{T}} \in \mathbb{R}^{2m}$ such that

$$\int_\Omega \mu(\lambda\theta^m)\boldsymbol{D}(\boldsymbol{u}^m) : \boldsymbol{D}(\boldsymbol{v}^j)\,d\boldsymbol{x} + \lambda\int_S \pi^0(\boldsymbol{v}^j\cdot\mathbf{n})\,d\sigma = \lambda\int_\Omega \boldsymbol{f}(\boldsymbol{x},\theta^m)\cdot\boldsymbol{v}^j\,d\boldsymbol{x}, \quad j = 1,\ldots,m, \tag{11}$$

$$\lambda\sum_{i=1}^d\int_\Omega u_i^m\frac{\partial\theta^m}{\partial x_i}\eta^j\,d\boldsymbol{x} + \int_\Omega \kappa(\lambda\theta^m)\nabla\theta^m\cdot\nabla\eta^j\,d\boldsymbol{x} + \lambda\alpha\int_{\partial\Omega\setminus S}\theta^m\eta^j\,d\sigma$$
$$= \lambda\int_\Omega \omega(\boldsymbol{x},\theta^m)\eta^j\,d\boldsymbol{x}, \quad j = 1,\ldots,m, \tag{12}$$

where $\boldsymbol{u}^m$ and θ^m are defined as follows:

$$\boldsymbol{u}^m \stackrel{\text{def}}{=} \sum_{j=1}^m a_{mj}\boldsymbol{v}^j, \qquad \theta^m \stackrel{\text{def}}{=} \sum_{j=1}^m b_{mj}\eta^j, \tag{13}$$

and λ is a parameter, $\lambda \in [0,1]$.

We shall establish *a priori* estimates of solutions to problem (11)–(13). Suppose a vector $(a_{m1},\ldots,a_{mm},b_{m1},\ldots,b_{mm})^{\mathrm{T}}$ satisfies (11)–(13). Let us multiply (11) by a_{mj} and add the obtained equalities for $j = 1,\ldots,m$. This yields

$$\int_\Omega \mu(\lambda\theta^m)|\boldsymbol{D}(\boldsymbol{u}^m)|^2\,d\boldsymbol{x} + \lambda\int_S \pi^0(\boldsymbol{u}^m\cdot\mathbf{n})\,d\sigma = \lambda\int_\Omega \boldsymbol{f}(\boldsymbol{x},\theta^m)\cdot\boldsymbol{u}^m\,d\boldsymbol{x}.$$

Then, using conditions **(C2)** and **(C5)**, the Cauchy–Bunyakovsky–Schwarz inequality, and $0 \le \lambda \le 1$, we derive

$$\begin{aligned}
\mu_0 \|\boldsymbol{u}^m\|^2_{\boldsymbol{V}_S(\Omega)} &= \mu_0 \int_\Omega |\boldsymbol{D}(\boldsymbol{u}^m)|^2 \, d\boldsymbol{x} \\
&\le \int_\Omega \mu(\lambda\theta^m) |\boldsymbol{D}(\boldsymbol{u}^m)|^2 \, d\boldsymbol{x} \\
&= -\lambda \int_S \pi^0 (\boldsymbol{u}^m \cdot \mathbf{n}) \, d\sigma + \lambda \int_\Omega \boldsymbol{f}(\boldsymbol{x}, \theta^m) \cdot \boldsymbol{u}^m \, d\boldsymbol{x} \\
&\le \Big(\int_S |\pi^0|^2 \, d\sigma \Big)^{1/2} \Big(\int_S |\boldsymbol{u}^m|^2 \, d\sigma \Big)^{1/2} + \Big(\int_\Omega |g_0|^2 \, d\boldsymbol{x} \Big)^{1/2} \Big(\int_\Omega |\boldsymbol{u}^m|^2 \, d\boldsymbol{x} \Big)^{1/2} \\
&\quad + M \Big(\int_\Omega |\theta^m|^4 \, d\boldsymbol{x} \Big)^{1/2} \Big(\int_\Omega |\boldsymbol{u}^m|^2 \, d\boldsymbol{x} \Big)^{1/2} \\
&= \|\pi^0\|_{L^2(S)} \|\boldsymbol{u}^m\|_{\boldsymbol{L}^2(S)} + \|g_0\|_{L^2(\Omega)} \|\boldsymbol{u}^m\|_{\boldsymbol{L}^2(\Omega)} + M \|\theta^m\|^2_{L^4(\Omega)} \|\boldsymbol{u}^m\|_{\boldsymbol{L}^2(\Omega)}.
\end{aligned} \tag{14}$$

Note that

$$\|\boldsymbol{u}^m\|_{\boldsymbol{L}^2(S)} \le \|\gamma_S\|_{\mathcal{L}(\boldsymbol{V}_S(\Omega), \boldsymbol{L}^2(S))} \|\boldsymbol{u}^m\|_{\boldsymbol{V}_S(\Omega)}, \tag{15}$$

$$\|\boldsymbol{u}^m\|_{\boldsymbol{L}^2(\Omega)} \le \|\mathcal{I}\|_{\mathcal{L}(\boldsymbol{V}_S(\Omega), \boldsymbol{L}^2(\Omega))} \|\boldsymbol{u}^m\|_{\boldsymbol{V}_S(\Omega)}, \tag{16}$$

$$\|\theta^m\|_{L^4(\Omega)} \le \|\mathcal{I}\|_{\mathcal{L}(Y_S(\Omega), L^4(\Omega))} \|\theta^m\|_{Y_S(\Omega)}, \tag{17}$$

where γ_S is the trace operator, the symbol $\mathcal{I}$ stands for the identity mapping. Combining (14) with (15)–(17), we get

$$\begin{aligned}
\mu_0 \|\boldsymbol{u}^m\|^2_{\boldsymbol{V}_S(\Omega)} \le & \|\pi^0\|_{L^2(S)} \|\gamma_S\|_{\mathcal{L}(\boldsymbol{V}_S(\Omega), \boldsymbol{L}^2(S))} \|\boldsymbol{u}^m\|_{\boldsymbol{V}_S(\Omega)} \\
& + \|g_0\|_{L^2(\Omega)} \|\mathcal{I}\|_{\mathcal{L}(\boldsymbol{V}_S(\Omega), \boldsymbol{L}^2(\Omega))} \|\boldsymbol{u}^m\|_{\boldsymbol{V}_S(\Omega)} \\
& + M \|\mathcal{I}\|^2_{\mathcal{L}(Y_S(\Omega), L^4(\Omega))} \|\mathcal{I}\|_{\mathcal{L}(\boldsymbol{V}_S(\Omega), \boldsymbol{L}^2(\Omega))} \|\theta^m\|^2_{Y_S(\Omega)} \|\boldsymbol{u}^m\|_{\boldsymbol{V}_S(\Omega)}.
\end{aligned}$$

It immediately follows that

$$\begin{aligned}
\|\boldsymbol{u}^m\|_{\boldsymbol{V}_S(\Omega)} \le & \mu_0^{-1} \|\pi^0\|_{L^2(S)} \|\gamma_S\|_{\mathcal{L}(\boldsymbol{V}_S(\Omega), \boldsymbol{L}^2(S))} + \mu_0^{-1} \|g_0\|_{L^2(\Omega)} \|\mathcal{I}\|_{\mathcal{L}(\boldsymbol{V}_S(\Omega), \boldsymbol{L}^2(\Omega))} \\
& + M \mu_0^{-1} \|\mathcal{I}\|^2_{\mathcal{L}(Y_S(\Omega), L^4(\Omega))} \|\mathcal{I}\|_{\mathcal{L}(\boldsymbol{V}_S(\Omega), \boldsymbol{L}^2(\Omega))} \|\theta^m\|^2_{Y_S(\Omega)}.
\end{aligned} \tag{18}$$

Now multiply (12) by b_{mj} and add the results for $j = 1, \ldots, m$; this gives

$$\underbrace{\lambda \sum_{i=1}^d \int_\Omega u_i^m \frac{\partial \theta^m}{\partial x_i} \theta^m \, d\boldsymbol{x}}_{I_4} + \int_\Omega \kappa(\lambda\theta^m) |\nabla \theta^m|^2 \, d\boldsymbol{x} + \lambda\alpha \int_{\partial\Omega \setminus S} |\theta^m|^2 \, d\sigma = \lambda \int_\Omega \omega(\boldsymbol{x}, \theta^m) \theta^m \, d\boldsymbol{x}. \tag{19}$$

Using the integration by parts formula, it is easy to show that the term I_4 vanishes. Indeed,

$$\begin{aligned}
I_4 &= \frac{1}{2} \sum_{i=1}^d \int_\Omega u_i^m \frac{\partial |\theta^m|^2}{\partial x_i} \, d\boldsymbol{x} \\
&= \frac{1}{2} \sum_{i=1}^d \int_{\partial\Omega} u_i^m n_i |\theta^m|^2 \, d\sigma - \frac{1}{2} \sum_{i=1}^d \int_\Omega \frac{\partial u_i^m}{\partial x_i} |\theta^m|^2 \, d\boldsymbol{x} \\
&= \frac{1}{2} \int_S (\boldsymbol{u}^m \cdot \mathbf{n}) \underbrace{|\theta^m|^2}_{=0} \, d\sigma + \frac{1}{2} \int_{\partial\Omega \setminus S} \underbrace{(\boldsymbol{u}^m \cdot \mathbf{n})}_{=0} |\theta^m|^2 \, d\sigma - \frac{1}{2} \int_\Omega \underbrace{\operatorname{div} \boldsymbol{u}^m}_{=0} |\theta^m|^2 \, d\boldsymbol{x} \\
&= 0.
\end{aligned}$$

Therefore, equality (19) can be rewritten as follows:

$$\int_\Omega \kappa(\lambda\theta^m)|\nabla\theta^m|^2\,\boldsymbol{dx} = -\lambda\alpha\int_{\partial\Omega\setminus S}|\theta^m|^2\,d\sigma + \lambda\int_\Omega \omega(\boldsymbol{x},\theta^m)\theta^m\,\boldsymbol{dx},$$

from which, by using conditions **(C2)** and **(C5)**, the Cauchy–Bunyakovsky–Schwarz inequality, and $0 \le \lambda \le 1$, we derive

$$\begin{aligned}\kappa_0\|\theta^m\|^2_{Y_S(\Omega)} =& \kappa_0\int_\Omega|\nabla\theta^m|^2\,\boldsymbol{dx}\\ \le& \int_\Omega\kappa(\lambda\theta^m)|\nabla\theta^m|^2\,\boldsymbol{dx}\\ =& -\lambda\alpha\int_{\partial\Omega\setminus S}|\theta^m|^2\,d\sigma + \lambda\int_\Omega\omega(\boldsymbol{x},\theta^m)\theta^m\,\boldsymbol{dx}\\ \le& \left(\int_\Omega|\omega_0|^2\,\boldsymbol{dx}\right)^{1/2}\left(\int_\Omega|\theta^m|^2\,\boldsymbol{dx}\right)^{1/2}\\ =& \|\omega_0\|_{L^2(\Omega)}\|\theta^m\|_{L^2(\Omega)}\\ \le& \|\omega_0\|_{L^2(\Omega)}\|\mathcal{I}\|_{\mathcal{L}(Y_S(\Omega),L^2(\Omega))}\|\theta^m\|_{Y_S(\Omega)},\end{aligned}$$

whence

$$\|\theta^m\|_{Y_S(\Omega)} \le \kappa_0^{-1}\|\omega_0\|_{L^2(\Omega)}\|\mathcal{I}\|_{\mathcal{L}(Y_S(\Omega),L^2(\Omega))}. \tag{20}$$

With this majoration, we deduce from (18) that

$$\begin{aligned}\|\boldsymbol{u}^m\|_{\boldsymbol{V}_S(\Omega)} \le& \mu_0^{-1}\|\pi^0\|_{L^2(S)}\|\gamma_S\|_{\mathcal{L}(\boldsymbol{V}_S(\Omega),\boldsymbol{L}^2(S))} + \mu_0^{-1}\|g_0\|_{L^2(\Omega)}\|\mathcal{I}\|_{\mathcal{L}(\boldsymbol{V}_S(\Omega),\boldsymbol{L}^2(\Omega))}\\ &+ M\mu_0^{-1}\kappa_0^{-2}\|\omega_0\|^2_{L^2(\Omega)}\|\mathcal{I}\|^2_{\mathcal{L}(Y_S(\Omega),L^4(\Omega))}\|\mathcal{I}\|_{\mathcal{L}(\boldsymbol{V}_S(\Omega),\boldsymbol{L}^2(\Omega))}\|\mathcal{I}\|^2_{\mathcal{L}(Y_S(\Omega),L^2(\Omega))}.\end{aligned} \tag{21}$$

Furthermore, since $\{\boldsymbol{v}^j\}_{j=1}^\infty$ is an orthonormal basis of the space $\boldsymbol{V}_S(\Omega)$ and $\{\eta^j\}_{j=1}^\infty$ is an orthonormal basis of the space $Y_S(\Omega)$, from (13) it follows that

$$\|\boldsymbol{u}^m\|^2_{\boldsymbol{V}_S(\Omega)} = \sum_{j=1}^m a_{mj}^2, \qquad \|\theta^m\|^2_{Y_S(\Omega)} = \sum_{j=1}^m b_{mj}^2.$$

Taking into account (20) and (21), we obtain

$$\begin{aligned}\sum_{j=1}^m a_{mj}^2 \le& \Big\{\mu_0^{-1}\|\pi^0\|_{L^2(S)}\|\gamma_S\|_{\mathcal{L}(\boldsymbol{V}_S(\Omega),\boldsymbol{L}^2(S))} + \mu_0^{-1}\|g_0\|_{L^2(\Omega)}\|\mathcal{I}\|_{\mathcal{L}(\boldsymbol{V}_S(\Omega),\boldsymbol{L}^2(\Omega))}\\ &+ M\mu_0^{-1}\kappa_0^{-2}\|\omega_0\|^2_{L^2(\Omega)}\|\mathcal{I}\|^2_{\mathcal{L}(Y_S(\Omega),L^4(\Omega))}\|\mathcal{I}\|_{\mathcal{L}(\boldsymbol{V}_S(\Omega),\boldsymbol{L}^2(\Omega))}\|\mathcal{I}\|^2_{\mathcal{L}(Y_S(\Omega),L^2(\Omega))}\Big\}^2,\\ \sum_{j=1}^m b_{mj}^2 \le& \kappa_0^{-2}\|\omega_0\|^2_{L^2(\Omega)}\|\mathcal{I}\|^2_{\mathcal{L}(Y_S(\Omega),L^2(\Omega))}.\end{aligned}$$

Note that these estimates do not depend on $m \in \mathbb{N}$ as well as $\lambda \in [0,1]$. Therefore, we can apply Proposition 1 to justify the solvability of problem (11)–(13). Here, of course, we used Proposition 2 and conditions **(C2)–(C5)** to establish the continuity of the corresponding mappings.

Let $(\widetilde{a}_{m1},\dots,\widetilde{a}_{mm},\widetilde{b}_{m1},\dots,\widetilde{b}_{mm})$ be a solution to (11)–(13) with $\lambda = 1$. Setting

$$\widetilde{\boldsymbol{u}}^m \overset{\text{def}}{=} \sum_{j=1}^m \widetilde{a}_{mj}\boldsymbol{v}^j, \qquad \widetilde{\theta}^m \overset{\text{def}}{=} \sum_{j=1}^m \widetilde{b}_{mj}\eta^j,$$

we have

$$\int_\Omega \mu(\widetilde{\theta}^m)\boldsymbol{D}(\widetilde{\boldsymbol{u}}^m) : \boldsymbol{D}(\boldsymbol{v}^j)\,\boldsymbol{dx} + \int_S \pi^0(\boldsymbol{v}^j \cdot \mathbf{n})\,d\sigma = \int_\Omega \boldsymbol{f}(\boldsymbol{x},\widetilde{\theta}^m)\cdot \boldsymbol{v}^j\,\boldsymbol{dx}, \quad j = 1,\ldots,m, \tag{22}$$

$$\sum_{i=1}^d \int_\Omega \widetilde{u}_i^m \frac{\partial \widetilde{\theta}^m}{\partial x_i}\eta^j\,\boldsymbol{dx} + \int_\Omega \kappa(\widetilde{\theta}^m)\nabla\widetilde{\theta}^m \cdot \nabla\eta^j\,\boldsymbol{dx} + \alpha\int_{\partial\Omega\setminus S}\widetilde{\theta}^m\eta^j\,d\sigma = \int_\Omega \omega(\boldsymbol{x},\widetilde{\theta}^m)\eta^j\,\boldsymbol{dx},$$
$$j = 1,\ldots,m. \tag{23}$$

Consider the sequence $\{(\widetilde{\boldsymbol{u}}^m,\widetilde{\theta}^m)\}_{m=1}^\infty$. It is clear that estimate (21) remains valid if we replace $\boldsymbol{u}^m$ with $\widetilde{\boldsymbol{u}}^m$. Consequently, the sequence $\{\widetilde{\boldsymbol{u}}^m\}_{m=1}^\infty$ is bounded in the space $\boldsymbol{V}_S(\Omega)$. Likewise, taking into account estimate (20), we see that the sequence $\{\widetilde{\theta}^m\}_{m=1}^\infty$ is bounded in the space $Y_S(\Omega)$. Therefore, there exists a pair $(\widetilde{\boldsymbol{u}}^0,\widetilde{\theta}^0) \in \boldsymbol{V}_S(\Omega)\times Y_S(\Omega)$ such that $\widetilde{\boldsymbol{u}}^{m_k}$ converges to $\widetilde{\boldsymbol{u}}^0$ weakly in $\boldsymbol{V}_S(\Omega)$ and $\widetilde{\theta}^{m_k}$ converges to $\widetilde{\theta}^0$ weakly in $Y_S(\Omega)$, for some subsequence $m_k \to \infty$ as $k \to \infty$. Without loss of generality, it can be assumed that

$$\widetilde{\boldsymbol{u}}^m \rightharpoonup \widetilde{\boldsymbol{u}}^0 \text{ weakly in } \boldsymbol{V}_S(\Omega) \text{ as } m \to \infty, \tag{24}$$

$$\widetilde{\theta}^m \rightharpoonup \widetilde{\theta}^0 \text{ weakly in } Y_S(\Omega) \text{ as } m \to \infty. \tag{25}$$

Since the trace operator $\gamma_{\partial\Omega}\colon H^1(\Omega) \to L^2(\partial\Omega)$ is compact (see [15] Section 2.6.2, Theorem 6.2), we have

$$\widetilde{\theta}^m|_{\partial\Omega\setminus S} \to \widetilde{\theta}^0|_{\partial\Omega\setminus S} \text{ strongly in } L^2(\partial\Omega\setminus S) \text{ as } m \to \infty. \tag{26}$$

Moreover, from (24) and (25) and the compactness theorem for the identity mapping $\mathcal{I}\colon H^1(\Omega) \to L^4(\Omega)$ (see [15] Section 2.6.1, Theorem 6.1), it follows that

$$\widetilde{\boldsymbol{u}}^m \to \widetilde{\boldsymbol{u}}^0 \text{ strongly in } \boldsymbol{L}^4(\Omega) \text{ as } m \to \infty, \tag{27}$$

$$\widetilde{\theta}^m \to \widetilde{\theta}^0 \text{ strongly in } L^4(\Omega) \text{ as } m \to \infty. \tag{28}$$

Using (24)–(28) and Proposition 2, we can pass to the limit $m \to \infty$ in (22) and (23); this gives

$$\int_\Omega \mu(\widetilde{\theta}^0)\boldsymbol{D}(\widetilde{\boldsymbol{u}}^0) : \boldsymbol{D}(\boldsymbol{v}^j)\,\boldsymbol{dx} + \int_S \pi^0(\boldsymbol{v}^j \cdot \mathbf{n})\,d\sigma = \int_\Omega \boldsymbol{f}(\boldsymbol{x},\widetilde{\theta}^0)\cdot \boldsymbol{v}^j\,\boldsymbol{dx}, \tag{29}$$

$$\sum_{i=1}^d \int_\Omega \widetilde{u}_i^0 \frac{\partial \widetilde{\theta}^0}{\partial x_i}\eta^j\,\boldsymbol{dx} + \int_\Omega \kappa(\widetilde{\theta}^0)\nabla\widetilde{\theta}^0 \cdot \nabla\eta^j\,\boldsymbol{dx} + \alpha\int_{\partial\Omega\setminus S}\widetilde{\theta}^0\eta^j\,d\sigma = \int_\Omega \omega(\boldsymbol{x},\widetilde{\theta}^0)\eta^j\,\boldsymbol{dx}, \tag{30}$$

for each $j \in \mathbb{N}$. Because $\{\boldsymbol{v}^j\}_{j=1}^\infty$ is a basis of $\boldsymbol{V}_S(\Omega)$ and $\{\eta^j\}_{j=1}^\infty$ is a basis of $Y_S(\Omega)$, equalities (29) and (30) remain valid if we replace $\boldsymbol{v}^j$ and η^j with arbitrary vector function $\boldsymbol{v} \in \boldsymbol{V}_S(\Omega)$ and function $\eta \in Y_S(\Omega)$, respectively:

$$\int_\Omega \mu(\widetilde{\theta}^0)\boldsymbol{D}(\widetilde{\boldsymbol{u}}^0) : \boldsymbol{D}(\boldsymbol{v})\,\boldsymbol{dx} + \int_S \pi^0(\boldsymbol{v} \cdot \mathbf{n})\,d\sigma = \int_\Omega \boldsymbol{f}(\boldsymbol{x},\widetilde{\theta}^0)\cdot \boldsymbol{v}\,\boldsymbol{dx},$$

$$\sum_{i=1}^d \int_\Omega \widetilde{u}_i^0 \frac{\partial \widetilde{\theta}^0}{\partial x_i}\eta\,\boldsymbol{dx} + \int_\Omega \kappa(\widetilde{\theta}^0)\nabla\widetilde{\theta}^m \cdot \nabla\eta\,\boldsymbol{dx} + \alpha\int_{\partial\Omega\setminus S}\widetilde{\theta}^0\eta\,d\sigma = \int_\Omega \omega(\boldsymbol{x},\widetilde{\theta}^0)\eta\,\boldsymbol{dx}.$$

Thus, we established that $(\widetilde{\boldsymbol{u}}^0,\widetilde{\theta}^0,\pi^0,0)$ is an admissible quadruplet to problem (1) and hence $\mathfrak{M}(\mathbb{U}_1,\mathbb{U}_2) \neq \varnothing$.

Now, consider a sequence $\{(\widehat{\boldsymbol{u}}^s,\widehat{\theta}^s,\widehat{\pi}^s,\widehat{\zeta}^s)\}_{s=1}^\infty \subset \mathfrak{M}(\mathbb{U}_1,\mathbb{U}_2)$ such that

$$\lim_{s\to\infty} J(\widehat{\boldsymbol{u}}^s,\widehat{\theta}^s,\widehat{\pi}^s,\widehat{\zeta}^s) = \inf_{(\boldsymbol{u},\theta,\pi,\zeta)\in\mathfrak{M}(\mathbb{U}_1,\mathbb{U}_2)} J(\boldsymbol{u},\theta,\pi,\zeta). \tag{31}$$

Owing to coercivity condition **(C9)**, we deduce from (31) that the set $\{(\widehat{\boldsymbol{u}}^s, \widehat{\theta}^s, \widehat{\pi}^s, \widehat{\zeta}^s)\}_{s=1}^\infty$ is bounded in the space $\boldsymbol{V}_S(\Omega) \times H^1(\Omega) \times L^2(S) \times L^2(S)$. Hence, there exists a subsequence $\{s_k\}_{k=1}^\infty$ such that

$$\widehat{\boldsymbol{u}}^{s_k} \rightharpoonup \boldsymbol{u}_* \text{ weakly in } \boldsymbol{V}_S(\Omega) \text{ as } k \to \infty, \tag{32}$$

$$\widehat{\theta}^{s_k} \rightharpoonup \theta_* \text{ weakly in } Y_S(\Omega) \text{ as } k \to \infty, \tag{33}$$

$$\widehat{\pi}^{s_k} \rightharpoonup \pi_* \text{ weakly in } L^2(S) \text{ as } k \to \infty, \tag{34}$$

$$\widehat{\zeta}^{s_k} \rightharpoonup \zeta_* \text{ weakly in } L^2(S) \text{ as } k \to \infty, \tag{35}$$

for some quadruplet $(\boldsymbol{u}_*, \theta_*, \pi_*, \zeta_*) \in \boldsymbol{V}_S(\Omega) \times H^1(\Omega) \times L^2(S) \times L^2(S)$.

The inclusion $\{(\widehat{\boldsymbol{u}}^{s_k}, \widehat{\theta}^{s_k}, \widehat{\pi}^{s_k}, \widehat{\zeta}^{s_k})\}_{k=1}^\infty \subset \mathfrak{M}(\mathbb{U}_1, \mathbb{U}_2)$ implies that

$$\int_\Omega \mu(\widehat{\theta}^{s_k}) \boldsymbol{D}(\widehat{\boldsymbol{u}}^{s_k}) : \boldsymbol{D}(\boldsymbol{v})\, \boldsymbol{dx} + \int_S \widehat{\pi}^{s_k} (\boldsymbol{v} \cdot \mathbf{n})\, d\sigma = \int_\Omega \boldsymbol{f}(\boldsymbol{x}, \widehat{\theta}^{s_k}) \cdot \boldsymbol{v}\, \boldsymbol{dx}, \quad \forall \boldsymbol{v} \in \boldsymbol{V}_S(\Omega),$$

$$\sum_{i=1}^d \int_\Omega \widehat{u}_i^{s_k} \frac{\partial \widehat{\theta}^{s_k}}{\partial x_i} \eta\, \boldsymbol{dx} + \int_\Omega \kappa(\widehat{\theta}^{s_k}) \nabla \widehat{\theta}^{s_k} \cdot \nabla \eta\, \boldsymbol{dx} + \alpha \int_{\partial\Omega \setminus S} \widehat{\theta}^{s_k} \eta\, d\sigma = \int_\Omega \omega(\boldsymbol{x}, \widehat{\theta}^{s_k}) \eta\, \boldsymbol{dx}, \quad \forall \eta \in Y_S(\Omega),$$

from which, by (32)–(35), one can derive the following equalities:

$$\int_\Omega \mu(\theta_*) \boldsymbol{D}(\boldsymbol{u}_*) : \boldsymbol{D}(\boldsymbol{v})\, \boldsymbol{dx} + \int_S \pi_* (\boldsymbol{v} \cdot \mathbf{n})\, d\sigma = \int_\Omega \boldsymbol{f}(\boldsymbol{x}, \theta_*) \cdot \boldsymbol{v}\, \boldsymbol{dx}, \quad \forall \boldsymbol{v} \in \boldsymbol{V}_S(\Omega),$$

$$\sum_{i=1}^d \int_\Omega u_{*i} \frac{\partial \theta_*}{\partial x_i} \eta\, \boldsymbol{dx} + \int_\Omega \kappa(\theta_*) \nabla \theta_* \cdot \nabla \eta\, \boldsymbol{dx} + \alpha \int_{\partial\Omega \setminus S} \theta_* \eta\, d\sigma = \int_\Omega \omega(\boldsymbol{x}, \theta_*) \eta\, \boldsymbol{dx}, \quad \forall \eta \in Y_S(\Omega).$$

Note that

$$\widehat{\zeta}^{s_k} = \widehat{\theta}^{s_k}|_S \to \theta_*|_S \text{ strongly in } L^2(S) \text{ as } k \to \infty.$$

Comparing this convergence with (35), we conclude that $\theta_*|_S = \zeta_*$. Moreover, since $\{\widehat{\zeta}^{s_k}\}_{k=1}^\infty \subset \mathbb{U}_2$ and the set $\mathbb{U}_2$ is closed in the space $L^2(S)$, we see that the function ζ_* belongs to $\mathbb{U}_2$. Next, taking into account condition **(C6)**, the inclusion $\{\widehat{\pi}^{s_k}\}_{k=1}^\infty \subset \mathbb{U}_1$, and convergence (34), we get $\pi_* \in \mathbb{U}_1$. Thus, $(\boldsymbol{u}_*, \theta_*, \pi_*, \zeta_*) \in \mathfrak{M}(\mathbb{U}_1, \mathbb{U}_2)$.

In addition, by condition **(C8)** and (31)–(35), we can obtain

$$\begin{aligned} J(\boldsymbol{u}_*, \theta_*, \pi_*, \zeta_*) &\le \liminf_{k\to\infty} J(\widehat{\boldsymbol{u}}^{s_k}, \widehat{\theta}^{s_k}, \widehat{\pi}^{s_k}, \widehat{\zeta}^{s_k}) \\ &= \inf_{(\boldsymbol{u},\theta,\pi,\zeta) \in \mathfrak{M}(\mathbb{U}_1, \mathbb{U}_2)} J(\boldsymbol{u}, \theta, \pi, \zeta). \end{aligned}$$

This means that the quadruplet $(\boldsymbol{u}_*, \theta_*, \pi_*, \zeta_*)$ belongs to the set $\mathfrak{M}_{\rm opt}(\mathbb{U}_1, \mathbb{U}_2)$, and hence $\mathfrak{M}_{\rm opt}(\mathbb{U}_1, \mathbb{U}_2) \neq \varnothing$. Therefore, we have shown that optimal control problem (1) is solvable in the sense of Definition 2.

We now turn to proving assertion (b). Assume the converse. Then, there exists a subsequence $\{n_k\}_{k=1}^\infty$ such that

$$\lim_{k\to\infty} \Phi(\mathbb{U}_1^{n_k}, \mathbb{U}_2^{n_k}) < \Phi(\mathbb{U}_1, \mathbb{U}_2). \tag{36}$$

Consider a sequence $\{(\boldsymbol{u}_*^{n_k}, \theta_*^{n_k}, \pi_*^{n_k}, \zeta_*^{n_k})\}_{k=1}^\infty$ such that

$$(\boldsymbol{u}_*^{n_k}, \theta_*^{n_k}, \pi_*^{n_k}, \zeta_*^{n_k}) \in \mathfrak{M}_{\rm opt}(\mathbb{U}_1^{n_k}, \mathbb{U}_2^{n_k}), \quad \forall k \in \mathbb{N}.$$

By Definition 1, we get

$$\int_\Omega \mu(\theta_*^{n_k})\boldsymbol{D}(\boldsymbol{u}_*^{n_k}) : \boldsymbol{D}(\boldsymbol{v})\, \boldsymbol{dx} + \int_S \pi_*^{n_k}(\boldsymbol{v}\cdot\mathbf{n})\, d\sigma = \int_\Omega \boldsymbol{f}(\boldsymbol{x},\theta_*^{n_k})\cdot\boldsymbol{v}\, \boldsymbol{dx}, \quad \forall \boldsymbol{v} \in \boldsymbol{V}_S(\Omega), \tag{37}$$

$$\sum_{i=1}^{d}\int_\Omega u_{*i}^{n_k}\frac{\partial\theta_*^{n_k}}{\partial x_i}\eta\, \boldsymbol{dx} + \int_\Omega \kappa(\theta_*^{n_k})\nabla\theta_*^{n_k}\cdot\nabla\eta\, \boldsymbol{dx} + \alpha\int_{\partial\Omega\setminus S}\theta_*^{n_k}\eta\, d\sigma = \int_\Omega \omega(\boldsymbol{x},\theta_*^{n_k})\eta\, \boldsymbol{dx}, \quad \forall\eta\in Y_S(\Omega). \tag{38}$$

Moreover, by Definition 3, we have

$$\Phi(\mathbb{U}_1^{n_k},\mathbb{U}_2^{n_k}) = J(\boldsymbol{u}_*^{n_k},\theta_*^{n_k},\pi_*^{n_k},\zeta_*^{n_k}), \quad \forall k\in\mathbb{N}. \tag{39}$$

It follows from (36) and (39) that there exists a number k_0 such that

$$J(\boldsymbol{u}_*^{n_k},\theta_*^{n_k},\pi_*^{n_k},\zeta_*^{n_k}) < \Phi(\mathbb{U}_1,\mathbb{U}_2)$$

for each $k \geq k_0$. Hence,

$$\{(\boldsymbol{u}_*^{n_k},\theta_*^{n_k},\pi_*^{n_k},\zeta_*^{n_k})\}_{k=k_0}^{\infty} \subset \mathcal{G}_{|\Phi(\mathbb{U}_1,\mathbb{U}_2)|}.$$

In view of condition **(C9)**, the set $\mathcal{G}_{|\Phi(\mathbb{U}_1,\mathbb{U}_2)|}$ is bounded in $\boldsymbol{V}_S(\Omega)\times H^1(\Omega)\times L^2(S)\times L^2(S)$. Therefore, without loss of generality, it can be assumed that

$$\boldsymbol{u}_*^{n_k} \rightharpoonup \boldsymbol{u}_*^0 \text{ weakly in } \boldsymbol{V}_S(\Omega) \text{ as } k\to\infty, \tag{40}$$

$$\theta_*^{n_k} \rightharpoonup \theta_*^0 \text{ weakly in } Y_S(\Omega) \text{ as } k\to\infty, \tag{41}$$

$$\pi_*^{n_k} \rightharpoonup \pi_*^0 \text{ weakly in } L^2(S) \text{ as } k\to\infty, \tag{42}$$

$$\zeta_*^{n_k} \rightharpoonup \zeta_*^0 \text{ weakly in } L^2(S) \text{ as } k\to\infty, \tag{43}$$

for some quadruplet $(\boldsymbol{u}_*^0,\theta_*^0,\pi_*^0,\zeta_*^0)\in\boldsymbol{V}_S(\Omega)\times H^1(\Omega)\times L^2(S)\times L^2(S)$.

Besides, since both the identity mapping $\mathcal{I}\colon H^1(\Omega)\to L^4(\Omega)$ and the trace operator $\gamma_{\partial\Omega}\colon H^1(\Omega)\to L^2(\partial\Omega)$ are compact, the following convergences hold:

$$\boldsymbol{u}_*^{n_k} \to \boldsymbol{u}_*^0 \text{ strongly in } \boldsymbol{L}^4(\Omega) \text{ as } k\to\infty, \tag{44}$$

$$\theta_*^{n_k} \to \theta_*^0 \text{ strongly in } L^4(\Omega) \text{ as } k\to\infty, \tag{45}$$

$$\zeta_*^{n_k} = \theta_*^{n_k}|_S \to \theta_*^0|_S \text{ strongly in } L^2(S) \text{ as } k\to\infty. \tag{46}$$

Comparing (43) with (46), one can infer that $\theta_*^0|_S = \zeta_*^0$ and

$$\zeta_*^{n_k} \to \zeta_*^0 \text{ strongly in } L^2(S) \text{ as } k\to\infty. \tag{47}$$

In view of (10), there exist sequences $\{\overline{\pi}^{n_k}\}_{k=1}^{\infty}\subset\mathbb{U}_1$ and $\{\overline{\zeta}^{n_k}\}_{k=1}^{\infty}\subset\mathbb{U}_2$ such that

$$\lim_{k\to\infty}\|\pi_*^{n_k}-\overline{\pi}^{n_k}\|_{L^2(S)} = 0, \qquad \lim_{k\to\infty}\|\zeta_*^{n_k}-\overline{\zeta}^{n_k}\|_{L^2(S)} = 0. \tag{48}$$

From (42), (47), and (48), it follows that

$$\overline{\pi}^{n_k} \rightharpoonup \pi_*^0 \text{ weakly in } L^2(S) \text{ as } k\to\infty,$$

$$\overline{\zeta}^{n_k} \to \zeta_*^0 \text{ strongly in } L^2(S) \text{ as } k\to\infty,$$

and, by conditions **(C6)** and **(C7)**, we get $(\pi_*^0,\zeta_*^0)\in\mathbb{U}_1\times\mathbb{U}_2$.

Next, taking into account the convergence results (40)–(42), (44), and (45), we pass to the limit $k \to \infty$ in equalities (37) and (38) and obtain

$$\int_\Omega \mu(\theta^0_*)\boldsymbol{D}(\boldsymbol{u}^0_*) : \boldsymbol{D}(\boldsymbol{v})\, \boldsymbol{dx} + \int_S \pi^0_*(\boldsymbol{v}\cdot\mathbf{n})\, d\sigma = \int_\Omega \boldsymbol{f}(\boldsymbol{x},\theta^0_*)\cdot\boldsymbol{v}\, \boldsymbol{dx}, \qquad \forall \boldsymbol{v} \in \boldsymbol{V}_S(\Omega),$$

$$\sum_{i=1}^{d}\int_\Omega u^0_{*i}\frac{\partial\theta^0_*}{\partial x_i}\eta\, \boldsymbol{dx} + \int_\Omega \kappa(\theta^0_*)\nabla\theta^0_*\cdot\nabla\eta\, \boldsymbol{dx} + \alpha\int_{\partial\Omega\setminus S}\theta^0_*\eta\, d\sigma = \int_\Omega \omega(\boldsymbol{x},\theta^0_*)\eta\, \boldsymbol{dx}, \qquad \forall \eta \in Y_S(\Omega).$$

Therefore, we have established that the quadruplet $(\boldsymbol{u}^0_*, \theta^0_*, \pi^0_*, \zeta^0_*)$ belongs to the set $\mathfrak{M}(\mathbb{U}_1, \mathbb{U}_2)$. This is a key point of the proof of assertion (b). Indeed, using condition **(C8)** and (39)–(43), we can deduce the following relations:

$$\begin{aligned}
\Phi(\mathbb{U}_1, \mathbb{U}_2) &= \inf_{(\boldsymbol{u},\theta,\pi,\zeta)\in\mathfrak{M}(\mathbb{U}_1,\mathbb{U}_2)} J(\boldsymbol{u},\theta,\pi,\zeta) \\
&\le J(\boldsymbol{u}^0_*, \theta^0_*, \pi^0_*, \zeta^0_*) \\
&\le \liminf_{k\to\infty} J(\boldsymbol{u}^{n_k}_*, \theta^{n_k}_*, \pi^{n_k}_*, \zeta^{n_k}_*) \\
&= \lim_{k\to\infty} \Phi(\mathbb{U}^{n_k}_1, \mathbb{U}^{n_k}_2),
\end{aligned}$$

thus contradicting inequality (36). This contradiction concludes the proof.

8. Concluding Remarks

In this study, we considered an optimal control problem for the system of nonlinear equations describing steady non-isothermal creeping flows of an incompressible fluid through a locally Lipschitz bounded domain. Using the discussion presented above, we established the existence of weak solutions that minimize a given cost functional. For the considered control system, we also introduced the concept of the marginal function and proved that this function is lower semi-continuous. This means that it is impossible to achieve a significant improvement in the optimal value of the cost functional by small changes in the set of admissible controls.

Author Contributions: Conceptualization, E.S.B.; methodology, E.S.B.; writing—original draft, E.S.B. and A.A.D.; writing—review and editing, M.A.A.

Funding: This research received no external funding.

Conflicts of Interest: The authors declare no conflict of interest.

References

1. Lee, H.C.; Imanuvilov, O.Y. Analysis of optimal control problems for the 2-D stationary Boussinesq equations. *J. Math. Anal. Appl.* **2000**, *242*, 191–211. [CrossRef]
2. Lee, H.-C.; Shin, B.C. Piecewise optimal distributed controls for 2D Boussinesq equations. *Math. Meth. Appl. Sci.* **2000**, *23*, 227–254. [CrossRef]
3. Li, S.; Wang, G. The time optimal control of the Boussinesq equations. *Numer. Funct. Anal. Optim.* **2003**, *24*, 163–180. [CrossRef]
4. Trenchea, C. Periodic optimal control of the Boussinesq equation. *Nonlinear Anal.* **2003**, *53*, 81–96. [CrossRef]
5. Lee, H.-C.; Kim, S. Finite element approximation and computations of optimal Dirichlet boundary control problems for the Boussinesq equations. *J. Korean Math. Soc.* **2004**, *41*, 681–715. [CrossRef]
6. Guerrero, S. Local exact controllability to the trajectories of the Boussinesq system. *Ann. Inst. Henri Poincaré Anal. Non Linéaire* **2006**, *23*, 29–61. [CrossRef]
7. Hinze, M.; Matthes, U. Optimal and model predictive control of the Boussinesq approximation. In *Control of Coupled Partial Differential Equations*; Kunisch, K., Leugering, G., Sprekels, J., Tröltzsch, F., Eds.; International Series of Numerical Mathematics; Birkhäuser Verlag AG: Basel, Switzerland, 2007; Volume 155, pp. 149–174, ISBN 978-3-7643-7720-5.

8. Korotkii, A.I.; Kovtunov, D.A. Optimal boundary control of a system describing thermal convection. *Proc. Steklov Inst. Math.* **2011**, *272* (Suppl. 1), S74–S100. [CrossRef]
9. Alekseev, G.V. Solvability of stationary boundary control problems for heat convection equations. *Sib. Math. J.* **1998**, *39*, 844–858. [CrossRef]
10. Fursikov, A.V.; Imanuvilov, O.Y. Exact controllability of the Navier–Stokes and Boussinesq equations. *Russ. Math. Surv.* **1999**, *54*, 565–618. [CrossRef]
11. Alekseev, G.V. Solvability of inverse extremal problems for stationary heat and mass transfer equations. *Sib. Math. J.* **2001**, *42*, 811–827. [CrossRef]
12. Alekseev, G.V.; Tereshko, D.A. Boundary control problems for stationary equations of heat convection. In *New Directions in Mathematical Fluid Mechanics*; Fursikov, A.V., Galdi, G., Pukhnachev, V.V., Eds.; The Alexander V. Kazhikhov Memorial Volume; Birkhäuser Verlag: Basel, Switzerland, 2010; pp. 1–21, ISBN 978-3-0346-0151-1.
13. Alekseev, G.V.; Tereshko, D.A. Stability of optimal controls for the stationary Boussinesq equations. *Int. J. Differ. Equ.* **2011**, *2011*. [CrossRef]
14. Conca, C.; Murat, F.; Pironneau, O. The Stokes and Navier–Stokes equations with boundary conditions involving the pressure. *Jpn. J. Math.* **1994**, *20*, 279–318. [CrossRef]
15. Nečas, J. *Direct Methods in the Theory of Elliptic Equations*; Springer: Heidelberg, Germany, 2012.
16. Litvinov, V.G. *Motion of a Nonlinear-Viscous Fluid*; Nauka: Moscow, Russia, 1982.
17. Fursikov, A.V. *Optimal Control of Distributed Systems: Theory and Applications*; AMS: Providence, RI, USA, 2000.
18. O'Regan, D.; Cho, Y.J.; Chen, Y.Q. *Topological Degree Theory and Applications*; Taylor & Francis Group: Boca Raton, FL, USA, 2006.
19. Krasnoselskii, M.A. *Topological Methods in the Theory of Nonlinear Integral Equations*; Pergamon Press: New York, NY, USA, 1964.

fluids

MDPI

Article

A New Model for Thermodynamic Characterization of Hemoglobin

Francesco Farsaci [1], Ester Tellone [2,*], Antonio Galtieri [2] and Silvana Ficarra [2]

1 Institute for Chemical and Physical Processes (IPCF-C.N.R.), Via Ferdinando Stagno d'Alcontres 37, Faro Superiore, 98158 Messina, Italy

2 Department of Chemical, Biological, Pharmaceutical and Environmental Sciences, University of Messina, Viale Ferdinando Stagno d'Alcontres 31, 98166 Messina, Italy

* Correspondence: etellone@unime.it

Received: 14 June 2019; Accepted: 5 July 2019; Published: 17 July 2019

Abstract: In this paper, we formulate a thermodynamic model of hemoglobin that describes, by a physical point of view, phenomena favoring the binding of oxygen to the protein. Our study is based on theoretical methods extrapolated by experimental data. After some remarks on the non-equilibrium thermodynamic theory with internal variables, some thermodynamic functions are determined by the value of the complex dielectric constant. In previous papers, we determined the explicit expression of a dielectric constant as a function of a complex dielectric modulus and frequency. The knowledge of these functions allows a new characterization of the material and leads to the study of new phenomena that has yet to be studied. In detail, we introduce the concept of "hemoglobe", a model that considers the hemoglobin molecule as a plane capacitor, the dielectric of which is almost entirely constituted by the quaternary structure of the protein. This model is suggested by considering a phenomenological coefficient of the non-equilibrium thermodynamic theory related to the displacement polarization current. The comparison of the capacity determined by the mean of this coefficient, and determined by geometrical considerations, gives similar results; although more thermodynamic information is derived by the capacity determined considering the aforementioned coefficient. This was applied to the normal human hemoglobin, homozygous sickle hemoglobin, and sickle cell hemoglobin C disease. Moreover, the energy of the capacitor of the three hemoglobin was determined. Through the identification of displacement currents, the introduction of this model presents new perspectives and helps to explain hemoglobin functionality through a physical point of view.

Keywords: hemoglobin; biological capacitor; non-equilibrium thermodynamics; hemoglobe capacitor; thermodynamic capacitor

1. Introduction

From an evolutionary point of view, the respiratory pigment known as hemoglobin (Hb), appeared very early. Its presence in the symbiotic nitrogen-fixing microorganisms of the *Leguminosae* plant family, where Hb, quickly capturing the oxygen (O_2), plays a "protective" role towards the nitrogenase enzyme complex poisoned by O_2 [1]; in invertebrates, in some bivalve mollusk, such as *Scapharca inaequivalvis*, the role of O_2 transport is played by a more primitive hemocyanin pigment [2]. Once the Hb appears, first in cartilaginous fishes and then in teleosts, it preserves an identical molecular architecture at all levels of the vertebrate evolutionary scale: Amphibians, reptiles, birds, terrestrial, and marine mammals. This unique, extraordinary, and complex respiratory pigment, responsible for the correct and adequate distribution of O_2 at all tissue levels, is a conjugated protein with a tetrameric structure $\alpha_2\beta_2$ that has a fine, and complex functional regulation that defines Hb as the allosteric protein par excellence [3]. Several modulators (effectors) influence oxygen binding to Hb such as the classic

2,3-bisphosphoglycerate (BPG), the chloride anion, carbon dioxide (CO_2), hydrogen ions responsible for the Bohr effect, temperature (i.e., the enthalpy of oxygenation ΔH), and inositol hexaphosphate and adenosine triphosphate depending on the evolutionary levels. The O_2, Hb preferential ligand, is also responsible for the complex homotropic regulation defined as cooperative, because it develops "mechanically" through a conformational change. This conformational change is due to the O_2 binding to the heme iron of one Hb subunit that in turn causes the sliding of the other subunits on each other. It is truly intriguing and surprising to note that for millions of years Hb has managed to maintain the tertiary and quaternary characteristics that characterize it, although the primary structure of its subunits has considerably varied. This peculiar architecture is basic and indispensable for the proper performance of Hb's delicate function, related to the binding, transport and O_2 release, which are finely and smartly regulated even in very different environments. Even more surprising is the fact that, through contact free of electronic transfers, a small molecule like O_2 can cause such an intramolecular disarrangement in a protein with about 3000 atoms (almost 1500 times larger than O_2). In fact, O_2 binding to the porphyrinic iron, causes a sort of intramolecular transduction that leads to a 15° rotation of the $\alpha_1\beta_1$ protomer with respect to the $\alpha_2\beta_2$. This results in a T-R transition makes the central cavity of Hb smaller by 50% (site of BPG binding, in the T state). In conclusion, the importance of the quaternary structure of Hb, preserved throughout evolution, and the disconcerting structural variation due the O_2 binding to heme suggest the existence of a driving force that is still not well identified and characterized. In detail, the "strength" of breathing work, has arisen from a Hb "intrinsic characteristic" certainly linked to its three-dimensional geometry (strictly maintained). This extraordinary machine, a slightly flattened globe with a central cavity in which the polar charges of the surface amino acid residues are perfectly oriented towards the (polar) medium and helps to stabilize the molecule's architecture, is a unique intrinsically well-designed "tool" to perform its "job" in an almost perfect way (perfect protein?). Moreover, it is a surprising three-dimensional morphology; the distribution and the variation of its charges on the internal and external surfaces that derive surprising thermodynamic responses [4], have led us to consider hemoglobin as an efficient electric machine. An ancestral biological capacitor with variable geometry whose model, in our opinion, needs to be characterized with complex thermodynamic models in order to obtain potential answers which may aid in clarifying that which is still ambiguous.

2. Thermodynamic Considerations

In this section, we remark on some aspects of the Hb capacitor model that justifies our point of view. In particular, we want to show this model as it is related to the non-equilibrium thermodynamics fundamental laws with internal variables and how it is a "consequence" of entropy production and therefore of the phenomenological equations. We observed that entropy is an important function in the thermodynamic description of the phenomenon. Entropy assumes a determinant role in the irreversible processes in which its production is different from zero. In fact, null entropy production is synonymous with reversibility. The choice of the variables on which the entropy depends leads to different developments of non-equilibrium thermodynamics. In all our researches, we chose to use the non-equilibrium thermodynamics with hidden internal variables formulated by Klutenberg in its theoretical aspects [5–9] and further developed by us [10–18]. This choice was made because we think this theory is more suitable in the study of biological systems [12] and it is a more physical–mathematical approach in regard to the classical meaning of this term. Here we consider only electrodynamics phenomena because the experiments on which we based the formulation of our Hb capacitor model was carried out only considering the dielectric measurements at constant temperature T. In this context, we assume the following functional dependence for the specific entropy s:

$$s = s\left(u, \underline{p}, \underline{\Omega}\right) \quad (1)$$

where u is the specific internal energy, p the specific polarization vector, and Ω a hidden vectorial internal variable. It can be shown that the introduction of Ω allows splitting of the polarization p into two parts $p^{(0)}$ and $p^{(1)}$ [7–9], such that:

$$\underline{p}^{(0)} = \underline{p} - \underline{p}^{(1)} \tag{2}$$

After this, the functional dependence in Equation (1) can be rewritten [7–9]:

$$s = s\left(u, p, \underline{p}^{(1)}\right) \tag{3}$$

and the entropy production $\sigma^{(s)}$ is:

$$\sigma^{(s)} = \frac{1}{T}\left[\rho \underline{E}^{(irr)} \frac{d\underline{p}}{dt} + \rho \underline{E}^{(1)} \frac{d\underline{p}^{(1)}}{dt}\right] \tag{4}$$

where by defining the equilibrium electric vector $\underline{E}^{(eq)}$ as:

$$\underline{E}^{(eq)} = -T \frac{\partial s\left(u, \varepsilon_{ik}, p, \underline{p}^{(1)}\right)}{\partial \underline{p}} \tag{5}$$

one has:

$$\underline{E}^{(ir)} = \underline{E} - \underline{E}^{(eq)} \tag{6}$$

$$\underline{E}^{(1)} = -T\rho \frac{\partial s\left(u, \varepsilon_{ik}, p, \underline{p}^{(1)}\right)}{\partial \underline{p}^{(1)}} \tag{7}$$

ρ is the mass density and $\underline{E}$ is the electric field that appears in the Maxwell's equations. Note that Equation (4) is a sum of the inner products of vectors, and therefore, according with the non equilibrium thermodynamics set of linear relations among these quantities can be obtained. For our purpose, we consider only one of these equations (a complete treatment of these questions can be studied in depth in References [7–9]); it is:

$$\rho \frac{dp_i^{(1)}}{dt} = \rho L_{ik}^{(1,0)} \frac{dp_k}{dt} + L_{ik}^{(1,1)} E_k^{(1)} \tag{8}$$

We here adopt the Einstein convection for the index. $L_{ik}^{(1,0)}$ that is a phenomenological tensor that takes into account possible cross effects between the relaxation phenomena described by Equation (8) and other ones described by an equation that we do not consider because it is not necessary for our purpose. $L_{ik}^{(1,1)}$ is the conductivity phenomenological tensor associated to polarization current $dP^{(1)}/dt$. Tensors were assumed as constant in time, but they can depend on a parameter characterizing the perturbation and this is the frequency of perturbation. It is important to emphasize that these tensors are a phenomenological characterization of the medium object of the study. For our purpose, it is enough to limit our consideration to the case of isotropic media neglecting the mentioned cross effects represented by the $L_{ik}^{(1,0)}$ tensor. In this case, one has:

$$L_{ik}^{(1,1)} = L^{(1,1)} \delta_{ik} \tag{9}$$

where δ_{ik} is the Kronecker's tensor. With these specification the Equation (8) can be rewritten:

$$\rho \frac{dp_i^{(1)}}{dt} = L^{(1,1)} \delta_{ik} E_k^{(1)} \tag{10}$$

By putting:

$$P_i^{(i)} = \rho p_i^{(i)} \tag{11}$$

and assuming ρ = constant (in time), Equation (10) becomes:

$$\frac{dP_i^{(1)}}{dt} = L^{(1,1)} \delta_{ik} E_k^{(1)} \tag{12}$$

The position ρ = constant is justified by the short time necessary for the measurement. We are recalling only the concept that we used to introduce our model, especially taking into account the hypothesis under which the dielectric experiment has been carried out. The Hb capacitor model was conceived, starting by Equation (12) and it has been deduced directly by entropy production (4) and in linear approximation, as we will explain in the next section. The experimental apparatus for dielectric measurements on which we based our model allows us to consider only one component of Equation (12) so as to write:

$$\frac{dP^{(1)}}{dt} = L^{(1,1)} E^{(1)} \tag{13}$$

This is a phenomenological equation connecting the affinity electric field $E^{(1)}$ (see Equation (7)) and the time variation of the polarization $p^{(1)}$ that is ascribed to the rotation of polar molecules according the Debye's model. $L^{(1,1)}$ is our keystone to introduce our model, as we will show in the next section.

3. Hemoglobe Model

In our previous paper, we formulated a non equilibrium thermodynamic characterization of the Hb molecule [4]. In particular, we studied the thermodynamic behavior of human hemoglobin (AA) comparing with homozygous sickle hemoglobin (SS) and its heterozygous (SC) originating sickle cell hemoglobin C disease, deducing biochemical properties associated to each one and relative differences. Indeed, we referred to dielectric experimental data reported by Laogun et al. (1997) [19]. Starting from these assumptions, we calculated the thermodynamic functions, which can be considered, so to speak, as the identity card of each Hb in solution. As mentioned in the introduction, the analysis of the so obtained functions, in particular of the $L^{(1,1)}$ coefficient of the $dP^{(1)}/dt$ function and the geometrical aspect of the Hb moleculeallow to evaluate the possibility of consider the Hb as a capacitor with quaternary structure as dielectric. Thus, we consider the Hb as a capacitor (see Figure 1) two walls of which are schematized as capacitors plates and whose dielectric is constituted by almost all the quaternary structure α_1, β_1, α_2, and β_2. Moreover, the three dimensionality of the Hb and a good capacitor functionality suggest the distance between the two plates as the lesser among the three dimension provided by literature. For our purpose, Hb will be only considered has a dipole moment and it is not necessary to evaluate the dipole moment or how data were obtained [20–22]. We indicate the total dipole moment as the vectorial sum of the four dipole moments of each subunits and calculate these last. The model that describes the hemoglobe can answer in an efficient way to three essential requirements of the T-R transition; they are: (1) How is the binding of the first oxygen facilitated? (2) How is the cooperativity generated? (3) How is the energy transmitted during the T-R transition?

The hemoglobe like a capacitor has a set of charges on each plate (the density of which is described by the induction vector $\underline{D}$). A set of opposite polarization charges correspond to these set of charges on adjacent dielectric represented by a side of the four subunities. Somehow there is a strong external mechanical obstruction that is an obstacle for the entry of external agents. Therefore, in order to carry out the T-R transition, it is necessary to use certain energy to overcome the aforementioned obstruction. During the journey in the bloodstream, this energy may be supplied to Hb by the different environments between tissues and lungs. These differences are especially characterized by a shift in: pH, temperature, BPG concentrations bound to Hb, and levels of oxygen and CO_2. For simplicity the explanation of our model will consider only the difference in pH and pO_2 (further differences of BPG bound to Hb and temperature confirm the model). In detail, pH variation (between tissues and lungs) causes a charging change on the external plate of the hemoglobe. This leads to a change on the internal plate charges and consequently a rearrangement of the structure of the four subunits (dielectric). This change is induced by displacement current processes deactivating ionic pairs, which were a cause of stabilization of the T state. This weakening and the influence of oxygen pressure, for example in the lungs, is approximately in the order of 100 mmHg, push the oxygen inside the Hb. The charges variation on the external plates due to pH changes, is similar to a voltage variation on the capacitor in an electric circuit. In more detail, on the capacitor plates, the change of charges causes the polarization of the dielectric in which displacement current appears. Additionally, this allows for a correlation of the energy propagation inside the Hb to a polarization displacement current. Since the change in pH is of small entity (7.4 in the lungs and 7.2 in tissues), it causes a small variation of charges on the hemoglobe plates. However, in the lungs (as in tissues where the conditions are simply opposite, but of equal entity), this small variation, accompanied by the high pO_2, allow the binds of the first oxygen to Hb. In turn, the oxygen binding causes the chemical rearrangement of the molecule that, by displacement current effects, influences the internal polarization charges allowing for facilitation of the binding of the other oxygen molecules. Further and so on until the T-R transition is complete. It is easy to notice that our model fits perfectly with the sequential model that describes the Hb cooperative behavior and somehow hemoglobe theory provides further explanations on the dynamics of oxygen binding to Hb. In fact, the hemoglobe model gives a physical description of some aspects of the T-R transition based on thermodynamic considerations so we may define it as a "Physical Mathematical Model of Hb". All this is supported by mathematical calculations in Sections 2 and 4.

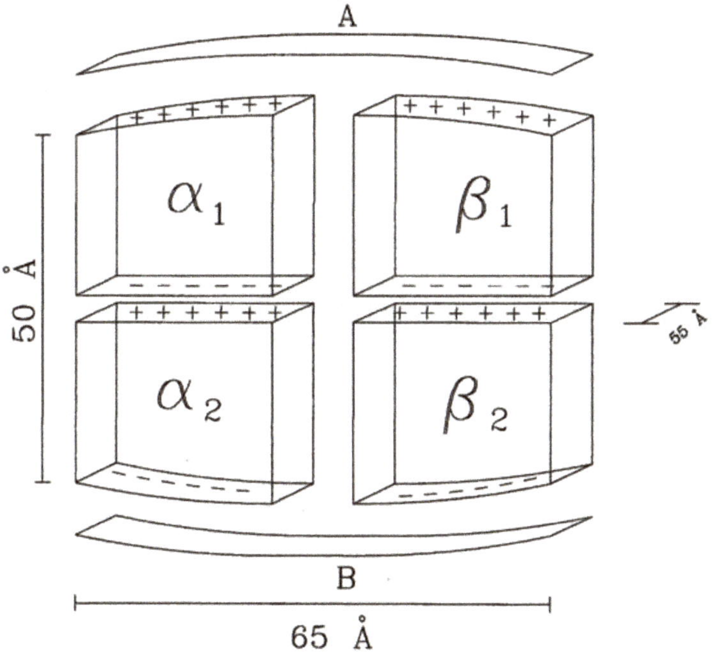

Figure 1. Hemoglobe: schematic model of the four subunits of the Hb assembled to form a capacitor with two faces A and B and surface S (armatures) with relative charges and characteristic dielectric (great part of the internal mass is the quaternary structure).

4. Physical-Mathematics Deductions

Based on the geometrical aspect of the Hb molecule and on the idea of Hb as a capacitor in which the quaternary structure is the dielectric, an important role is assigned to $L^{(1,1)}$ coefficient that has the dimension of a conductivity. From now on, $L^{(1,1)}$ will be called Displacement Equivalent Conductivity Coefficient (DECC) since it is related to $dP^{(1)}/dt$. $dP(1)/dt$ has the dimension of a current, due to orientation polarization $P^{(1)}$, so we call it: Polarization Displacement Equivalent Current (PDEC). The term "equivalent" is justified by observing that a change of the polarization state implies a movement of charges and so it is equivalent to a current. The connection between PDEC and DECC is represented by Equation (13). The PDEC appearing in Equation (13) is only defined inside the matter (unlike the term $d\underline{D}/dt$, where $\underline{D}$ is the displacement vector) and in the Debye's model inside the matter with polar molecules. Recall that, in an electric circuit RLC [23] (Resistance, Inductance, and Capacitance) subjected to harmonic perturbation of frequency ω we introduce a quantity, called impedance, which is analogous to the Ohmic resistance, in the case of constant perturbation; this results from the sum of three terms, one Ohmic, one inductive, and one capacitive. The capacitance component of the impedance is:

$$Z_c = \frac{1}{i\omega C}$$

and therefore, it has the dimensions of a resistance linked to capacitive factors. Since the reciprocal of a resistance is a conductivity, it follows that ωC has the conductivity dimensions per unit of length. It was written from:

$$i\omega C$$

From this, it follows that ωC_e (conductivity for unitary length) in the MKSQ system has the following dimension:

$$[\omega c] = \frac{1}{s}\frac{Q^2 s^2}{Kg\ m^2} = \frac{Q^2 s}{Kg\ m^2} \tag{14}$$

while

$$\left[L^{(1,1)}\right] = \frac{Q^2 s}{Kg\ m^3} \tag{15}$$

This dimensional analysis and by observing that $L^{(1,1)}$ refers to Hb, now considered as a capacitor of defined size, suggest to us the following equality (in agreement with Somerfeld point of view on dimensional analysis) [23]:

$$L^{(1,1)}(meter) = \omega C_e \tag{16}$$

From which:

$$C_e = \frac{L^{(1,1)}(meter)}{\omega} \tag{17}$$

Since, from Equation (17), C_e is connected to DECC, we call it Displacement Equivalent Capacity (DEC). Obviously it results in:

$$C_e = C_e(w) \tag{18}$$

The position (16) is legitimate if we consider capacitor with polar dielectric (Debye). Figure 1 shows Hb like a parallelepiped with curved faces, since the protein is a permanent dipole, it can be considered like a plane capacitor of A and B plates, and a great part of the quaternary structure as dielectric. We are conscious that A and B sides are not plane, but we can deduce that they do not have an important curvature. Indeed, the Hb molecule is not a well defined figure, but it certainly is a permanent dipole with a dipole moment [20,21]. The non-angular shape allows us to coin a new term, well representing a geometrical image of the Hb model introduced by us. Thus, we call Hemoglobe the plane capacitor with A and B plates of area S placed to a distance d.

It will be remembered a capacitor is made up of two plates placed at a very small distance, so as to have distribution of charges, whose density is σ. The field near a plate is (see Somerfeld) [23]:

$$E = \frac{\sigma}{\varepsilon} \tag{19}$$

and capacity C (i.e., the possibility of the system to assume charges) is:

$$C = \frac{S}{d}\varepsilon \tag{20}$$

being S the surface area of the plate and d the distance between the two plates; ε is the dielectric constant of the medium interposed between the frames. If the charges on the surface remain constant over time, then C will be constant because ε is constant too; this occurs when the capacitor is subjected to a constant perturbation. If the perturbation is not constant and of the harmonic type like the perturbation used in the paper [23], then dispersive phenomena can be observed and a complex dielectric function is introduced that will vary with the perturbation frequency. It will be written [23]:

$$\varepsilon = \varepsilon_1 - i\varepsilon_2 \tag{21}$$

where ε_1 is linked to storage phenomena and ε_2 to loss phenomena. Replacing Equation (21) in (20) one has:

$$C = \frac{S}{d}\varepsilon_1 - i\frac{S}{d}\varepsilon_2 \tag{22}$$

therefore, the capacity can be considered as a complex function depending on the perturbation frequency ω and the real part of C is introduced:

$$C_1 = \frac{S}{d}\varepsilon_1 \tag{23}$$

and the imaginary one is:

$$C_2 = \frac{S}{d}\varepsilon_2 \tag{24}$$

so Equation (4) is:

$$C(\omega) = C_1(\omega) - iC_2(\omega) \tag{25}$$

The imaginary part of C is related to dissipation phenomena since this occur for ε_2. Taking into account that a Hb molecule has the following dimension: 64Å*55Å*50Å, we have S = 64*55Å and d = 50Å [22]. If $\varepsilon(\omega)$ is the dielectric loss of the Hb one has:

$$C_2 = \varepsilon_2(w) \times (64 \times 55)/50\text{Å} = \varepsilon_2(w)70\text{Å} \tag{26}$$

C_2 is the capacity of the hemoglobe. Both Equations (17) and (26) are two capacities expressed as functions of the perturbation frequency ω. While Equation (17) is obtained by thermodynamic considerations (we use the coefficient $L^{(1,1)}$), Equation (26) is the geometrical definitions of capacity for a plane capacitor. We note that Equations (17) and (26) should give similar values. This is true as can be seen in the Figures 2–4, that will be commented in the next section.

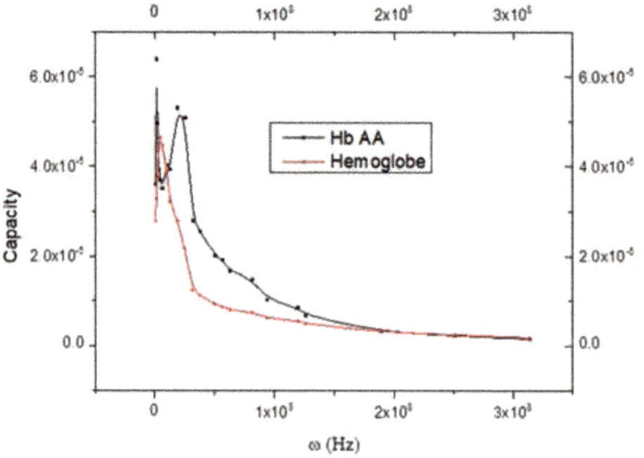

Figure 2. Comparison between HbAA (black line) and Hemoglobe (red line) capacity expressed in p-Farad (pF) calculated by Equations (30) and (31), respectively, as a function of frequency.

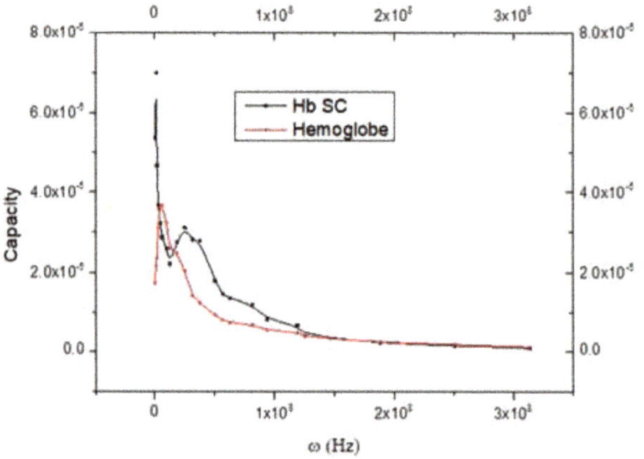

Figure 3. Comparison between HbSC (black line) and Hemoglobe (red line) capacity expressed in p-Farad (pF) calculated by Equations (30) and (31), respectively, as a function of frequency.

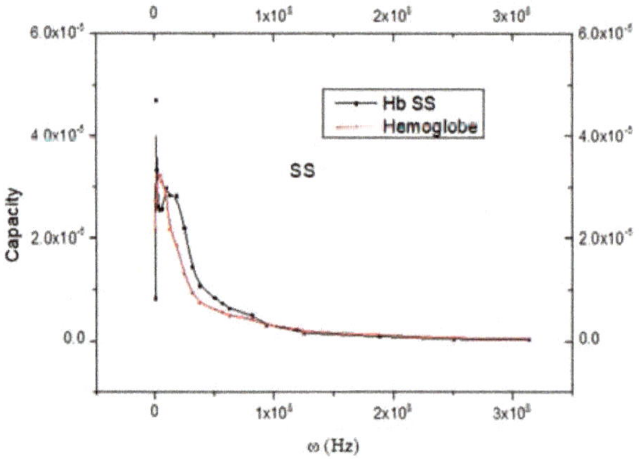

Figure 4. Comparison between HbSS (black line) and Hemoglobe (red line) capacity expressed in p-Farad (pF) calculated by Equations (30) and (31), respectively, as a function of frequency.

5. Hemoglobin as a Capacitor

From the non-equilibrium thermodynamics point of view, why is Hb regarded as a capacitor? The answer lies in the comparison between the following positions:

$$L^{(1,1)}(meter) = \omega C_e \tag{27}$$

from which it derives:

$$C_e = \frac{L^{(1,1)}(meter)}{\omega} \tag{28}$$

and comparing with capacity:

$$C_2 = \frac{S}{d}\varepsilon_2 \tag{29}$$

it is possible to deduct that both positions (28 and 29) have a similar behavior with different values for frequency of about 2.5×10^7 Hz. The comparison between these two curves (3 and 4) is not a fit. It must be considered as only a guideline to the concept that a flat capacitor with Hb structure as dielectric has a capacity with orders of magnitude similar to the capacity calculated using the equality (27). Furthermore, as an experimental support in a "Hypotheses non fingo" perspective we neglect the edge effects for the shape of the Hemoglobe. In a frequency range between 2.5×10^7 Hz and 10^8 Hz, the difference between the two curves (28) and (29) is due to a type of phenomena emerging when the thermodynamic coefficient $L^{(1,1)}$ is used. In line with our studies [4,5,10–18], the introduced coefficients (not only $L^{(1,1)}$) show phenomena not detectable with ε, because ε is the sum of many phenomena. The above on $L^{(1,1)}$ and ε is reported in the previous paper [4]. In this range of frequency, $L^{(1,1)}$ shows a fluctuating trend while ε shows a "regular" trend. The question arises, which of the calculated capacities is the correct one. Hb is a polar molecule, so the orientation polarization $P^{(1)}$ is predominant and only the displacement conductivity $L^{(1,1)}$ associated with $P^{(1)}$ will be considered. The correct Hb conductivity, or rather the most reliable, is that calculated with (28), where only displacement phenomena are considered as it should be done in a dielectric. In the ε_2 value, which appears only in Equation (29), a series of phenomena are present, because ε_2 is not selective as the $L^{(1,1)}$. Hb is not a perfect dielectric, because it presents dispersive phenomena. Then, from a capacitive point of view, it is very important to select displacement phenomena. In these circumstances, Equation (29) is not very suitable to calculate the capacity; position (29) is only considered because it can give an idea of the order of magnitude of the capacity. It should be remembered that these statements arise from the desire to study dielectric relaxation phenomena resulting from a harmonic perturbation. In the physiological reality, this does not happen and therefore these affirmations have the role to support the idea that Hb may be considered as a capacitor when perturbed by a harmonic action. Now, some considerations, further justifying what has been said so far, will be made. Let us start by pointing out that our deductions are founded on experimental data reported by Laogun et al. (1997) [19]. The analysis of the experiments described in this paper allow us to state that each point of the solutions of Hb AA, SC, and SS is subjected to the same perturbation and therefore each Hb molecule is subjected to the same field generated by the surface change on the plates [19]. Let us not forget that we are in linear approximation. Moreover, Equation (17) implies that conductivity phenomena have a capacitive character, therefore not inductive and not resistive. This statement is reasonable if the experiment is analyzed and especially the range of frequencies investigated. Even if $L^{(1,1)}$ refers to the solution, Equation (13) induces us to write position (16) so that DECC has a capacitive character, peculiar of the displacement current in a capacitor. In the range of frequency under investigation, the only elements of the solution irrefutably influencing $L^{(1,1)}$ are the Hb molecules, whence the approach is to a capacitor. Obviously, the "Hemoglobe" is the first approach to a capacitor model: a curve is substituted with a plane side, even if the difference is small; moreover, we only study the dielectric relaxation of the very complex phenomena occurring in the Hb molecule. However, we think that, by the development of this approach, it can be possible to conceive the model more and more near the reality. After these clarifications, for a Hb molecule, Equation (17) is:

$$C_e = (L^{(1,1)}/\omega) \times 56.3\text{Å}$$

from which, in the MKSQ system, this becomes:

$$C_e = (L^{(1,1)}/\omega) \times 56.3 \times 10^2\text{pF} \tag{30}$$

where 56.3 is the medium value of the three Hb dimensions. Equation (26) become:

$$C_2 = \varepsilon_2(\omega) \times 70\text{Å} = 8.856 \times 10^{-12} \times 70 \times 10^{-10}\ \varepsilon_{2r}{}'(w)10^{12}\text{pF} \tag{31}$$

where $\varepsilon_{2r}(\omega)$ is the relative dielectric constant [19]. By substituting, respectively, $L^{(1,1)}$ calculated in our previous paper [4] in Equation (30) and $\varepsilon_{2r}(\omega)$ measured by Laogun et al. (1997) [19] in Equation (31), we obtain two curves as a function of the frequencies for the Hb AA, SC, and SS plotted in Figures 2–4. The capacity (26) shows characteristics only obtained by taking into account $\varepsilon_{2r}(\omega)$, while the capacity (17) shows more details that are deductible by the theory developed by us, as is shown in the Figures 2–4. This is to confirm that the developed approach is able to evidence and detect some phenomena not considered by all other approaches. Extending this idea to the real physiological case, in the next paragraph, some aspects of the hemoglobin T-R transition will be described considering the protein as a "plane capacitor" not subjected to harmonic actions, but to a variation of surface charges induced by physiological phenomena.

5.1. Observations

The in vitro measurements do not give any information about the Hb functions when the protein is inside its biological place (inside the blood). Thus, all the processes that occur in human body cannot be studied with a Hb solution in vitro. Nevertheless, our theoretical deductions, based on a Hb solution perturbed with an electric field (dielectric measurements), allows to hazard the hypothesis that somehow processes occur because it was stimulated by the field. It is as to partially stimulate the Hb molecule to work as an electrical machine. Physical considerations induced us to consider this machine as a capacitor and to think that this behavior can be preserved even in the human body. Better yet, can this model be used to explain some energetic phenomena? In other words: If the Hb shows behaviors similar to a capacitor in vitro, why can it not work as a capacitor even in vivo, inside the human body? This will explain, qualitatively (but not quantitatively because we have not experimental data), some energetic balance processes occurring in vivo; this is in agreement with a characteristic of the capacitor: To accumulate energy and release it at the appropriate time and with a certain frequency.

5.2. Capacitor Model

The variation of the surface charge density of the Hb during its travel from the tissues to the lungs occurs in a chaotic way, we would say randomly, because it is due to turbulence of the components inside the Hb and to the influence of the blood flow that carries it. This disorder can be considered as a background sensor to the phenomenon that somehow directs this randomness towards a well defined causal order: The T-R transition. The phenomenon to which we refer is linked to the potential binding of an oxygen molecule (and then the other two) to iron by penetrating the surface barrier of Hb. We believe that the oxygen binding to Hb may be favored by the generation of displacement currents through the breaking of chemical bonds inside the Hb. Now let us see how the introduced model takes into account the generation of these displacement currents. Comparing the Hb to a capacitor, we assume that a randomly variable charge is formed on its walls, during the journey in the blood stream. For the reasons mentioned above, we consider the contribution of this charge to the generation of displacement currents irrelevant, which will be random and with random directions too. We cannot quantify the magnitude of this current, but we believe that it is slight because it was caused by factors internal to the Hb. When Hb reaches the pulmonary environment, the influence of charges generated on the walls of the hemoglobin capacitor is different. Here, the different environmental conditions cause the generation of polarization charges in a very short time. Given the rapidity with which this occurs, we think that these charges can be represented with an exponential function, which we indicate with σ.

$$\sigma = \sigma_0 e^{\pm Rt} \tag{32}$$

where σ_0 is a constant, whose dimensions are of a superficial charge density and R is a constant, whose dimensions are t^{-1} and it is characteristic of Hb. The sign $\pm$ indicates an increase or a decrease in surface charges, what is important is the charge variation because this is the cause of the generation of displacement currents occurring in this time frame. In other words, the variation of s over time (very short) generates the displacement currents. Only the variations in occurring when the Hb reaches the lungs, will be taken into account like favoring factor to the binding of the first oxygen molecule to Hb. Therefore, the function (32) is defined in a time interval T indicated with $\Delta t = t_1 - t_0$ and that will also be written as:

$$t_0 < t < t_1$$

being t_0 and t_1 the instants in which the Hb arrives in the lungs and the oxygen binds to the iron, respectively. A change of σ on the "Hb wall" generates an electric field E on the "inner wall", given by:

$$E = \frac{\sigma}{\varepsilon}$$

i.e.,:

$$E = \frac{\sigma_0 e^{\pm Rt}}{\varepsilon} \tag{33}$$

where ε is the dielectric constant of the quaternary structure of the Hb. ε will be a function of time as the displacement currents act by breaking the chemical bonds. Therefore, there will be orientation polarization currents identified by the function $P^{(1)}$; the deformation polarization will be neglected because Hb is a polar molecule so the effects due to permanent dipoles prevail. Neglecting the deformation polarization, the field $E^{(1)}$ related to the orientation polarization $P^{(1)}$ is [4]:

$$E^{(1)} = \left(a^{(0,0)} - a^{(1,1)}\right)P^{(1)} \tag{34}$$

this field is equal to the one expressed by (33) for what was said before:

$$E^{(1)} = \frac{\sigma_0 e^{\pm Rt}}{\varepsilon} \tag{35}$$

The displacement currents $dP^{(1)}/dt$ due to orientation polarization phenomena are expressed by [4]:

$$\frac{dP^{(1)}}{dt} = L^{(1,1)}E^{(1)}$$

in which, substituting Equation (35):

$$\frac{dP^{(1)}}{dt} = L^{(1,1)}\frac{\sigma_0}{\varepsilon}e^{\pm Rt} \tag{36}$$

The integration of this differential equation allows for calculation of the orientation polarization in the function of the time. Unfortunately we do not know the expression of $\varepsilon(t)$ because there are no experimental data, so we cannot integrate it. From Equation (36):

$$P^{(1)}(t) = \int L^{(1,1)} \frac{\sigma_0}{\varepsilon} e^{\pm Rt} dt + c \tag{37}$$

C is an integration constant. From the Equation (32), σ_0 is the density of charge for $t = 0$, so assuming $t_0 = 0$ as the time in which the Hb reaches the lungs, σ_0 will be the value of the surface charge density on the Hb wall at the moment the protein reaches the lungs.

Recalling Equations (32) and (37), we have:

$$\sigma = \sigma_0 e^{\pm Rt} \tag{38}$$

$$P^{(1)}(t) = \int L^{(1,1)} \frac{\sigma_0}{\varepsilon} e^{\pm Rt} dt + c \tag{39}$$

since σ_0 and $L^{(1,1)}$ are temporal constants, Equation (39) can be written as:

$$P^{(1)}(t) = L^{(1,1)} \sigma_0 \int \frac{e^{\pm Rt}}{\varepsilon} dt + c \tag{40}$$

The problem to calculate this integral arises from the lack of analytical knowledge of $\varepsilon(t)$. ε derives from the dielectric phenomena occurring inside the Hb and these are affected by the displacement currents generated at $t = 0$ (due to the pulmonary environment) so ε may also vary exponentially, but with a time lag with respect to density expressed by Equation (38). Therefore, the dielectric function can be expressed in the following form:

$$\varepsilon = \varepsilon_0 e^{st+\phi} \tag{41}$$

s is a constant whose dimensions are t^{-1} and ϕ is the time delay with respect to the variation of σ. Substituting Equation (41) in (40) is:

$$P^{(1)}(t) = \frac{L^{(1,1)} \sigma_0}{\varepsilon_0} \int e^{\pm Rt} e^{-st-\phi} dt + c$$

$$P^{(1)}(t) = \frac{L^{(1,1)} \sigma_0 e^{-\phi}}{\varepsilon_0} \int e^{(\pm R - s)t} dt + c \tag{42}$$

$$P^{(1)}(t) = \frac{L^{(1,1)} \sigma_0 e^{-\phi}}{\varepsilon_0 (\pm R - s)} e^{(\pm R - s)t} + c \tag{43}$$

This is the expression of orientation polarization. Only the variations of the phenomena for $t > 0$ (i.e., from when the Hb reaches the lungs) were considered, so $P^{(1)}$ expressed by Equation (43) will be null for $t = 0$ and therefore:

$$P^{(1)}(t) = \frac{L^{(1,1)} \sigma_0 e^{-\phi}}{\varepsilon_0 (\pm R - s)} + c = 0$$

i.e.,:

$$c = -\frac{L^{(1,1)} \sigma_0 e^{-\phi}}{\varepsilon_0 (\pm R - s)} \tag{44}$$

and Equation (43) is:

$$P^{(1)}(t) = \frac{L^{(1,1)} \sigma_0 e^{-\phi}}{\varepsilon_0 (\pm R - s)} \left[e^{(\pm R - s)t} - 1 \right] \tag{45}$$

this is the final expression of $P^{(1)}(t)$.

5.3. *Hemoglobe Energy In Vitro*

It is well known that all capacitors have an intrinsic energy given by:

$$U = \frac{1}{2} \times Q^2 / C_0 \tag{46}$$

where Q is the charge in a side of the Hb and C_0 is its capacity. Now Q is calculated. We remember that:

$$E = E^{(eq)} + E^{(ir)} \tag{47}$$

and

$$Q = D \times S \tag{48}$$

where D is the dielectric displacement from equation:

$$D = \varepsilon E \tag{49}$$

and from Equation (48), it follows:

$$U = \frac{1}{2} \times E^2 S^2 \times \left(E^{(eq)} + E^{(ir)}\right)^2 / C_0 \tag{50}$$

By substituting Equations (17), (26), and (27) become, respectively:

$$U_G = \frac{1}{2} \times ESd \times \left(E^{(eq)} + E^{(ir)}\right)^2 \tag{51}$$

$$U_T = \frac{1}{2} \times E^2 S^2 \left(E^{(eq)} + E^{(ir)}\right)^2 / L^{(1,1)} \tag{52}$$

By considering the explicit expression of $E^{(eq)}$ and $E^{(ir)}$ [13], Equations (51) and (52) become:

$$U_G = \frac{1}{2} \varepsilon s d P_0^2 \left(\Gamma_1 \sin \omega t + L^{(0,0)} \omega \cos \omega t\right)^2 \tag{53}$$

$$U_T = \frac{1}{2} \frac{\varepsilon^2 s^2 P_0^2 \left(\Gamma_1 \sin \omega t + L^{(0,0)} \omega \cos \omega t\right)^2 \omega}{L^{(1,1)} (meter)} \tag{54}$$

where we remember that $P = P^{(0)}$ sen ωt, and $P = P^{(0)} + P^{(1)}$.

In both Equations (53) and (54) P_0 is an unknown value since we do not know the field values applied in the experiment. From Equation (53), one obtains:

$$P_0 = \sqrt{\frac{2U_g}{\varepsilon s d}} \left(\Gamma_1 \sin \omega t + L^{(0,0)} \omega \cos \omega t\right) \tag{55}$$

and from Equation (54):

$$P_0 = \sqrt{\frac{L^{(1,1)} (meter)}{\omega} \frac{2U_T}{\varepsilon s \left(\Gamma_1 \sin \omega t + L^{(0,0)} \omega \cos \omega t\right)}} \tag{56}$$

These two expressions are able to assign the values of P_0 if we know the energy U_G and U_T, that in this case are the same; they are calculated with other techniques. Other terms are known.

5.4. Hemoglobe Energy In Vivo

It is well known that all capacitors have an intrinsic energy given by

$$U = \frac{1}{2} \times Q^2/C \tag{57}$$

where Q, taking into account Equation (32), is:

$$Q = s \times S = S\sigma_0 e^{\pm Rt} \tag{58}$$

Taking into account Equations (20) and (57) become:

$$U = \frac{1}{2}\frac{Sd\sigma_0^2 e^{\pm 2Rt}}{\varepsilon} \tag{59}$$

This is the in vivo energy of Hb considered as a capacitor.

6. Curve Description

Figures 2–4 show the capacity of hemoglobe evaluated by the mean of Equation (31); it is possible to note a sufficiently regular course of the capacity and an asymptotically decreasing trend for the three Hbs. By observing Equation (30) and the $L^{(1,1)}$ trend, in this range, we say that they are in agreement, as can be noted by the $L^{(1,1)}$ curve shown in Farsaci et al. (2019) [4]. Apart of this, the curves evaluated by Equations (30) and (31) show a sufficient similarity. It is important to emphasize that, by Equation (30), results were obtained that are not shown by Equation (31). This means that the "thermodynamic" evaluation of the capacity (by Equation (30)) gives information that cannot be obtained in another way. Figures 5 and 6 show a comparison of the three Hbs, on the trend of thermodynamic capacity and energy, evaluated by Equations (30) and (54), respectively. More careful analysis of all curves reveals significant differences, certainly attributed to primary structural substitutions on β6 residues in pathological HbSS (homozygous) and HbSC (heterozygous) with respect to HbA. In detail, on the shape of the curves the substitutions with loss of charge (as in HbSS, where two negative charges are replaced by two residues without charge) have more significant effects than those with an opposite charge (as in HbSC, where two negative charges are replaced with two positive). All this strongly supports the idea of the capacitor, which electrical and thermodynamic functionality is due to electric charges and is expressed through them.

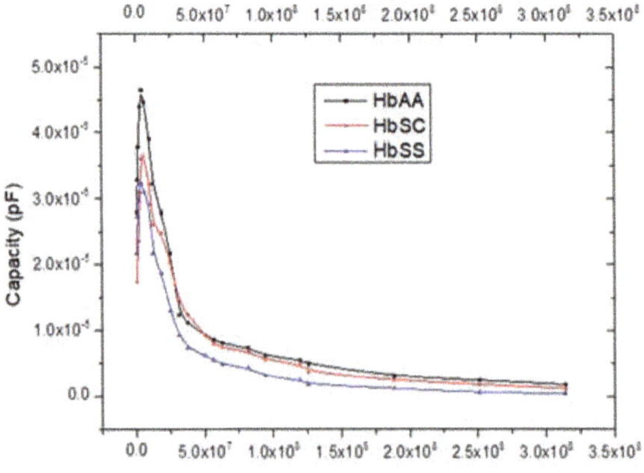

Figure 5. Comparison between HbAA (black line), HbSC (red line) and HbSS (blue line) capacity expressed in p-Farad (pF) calculated by Equation (30), as a function of frequency.

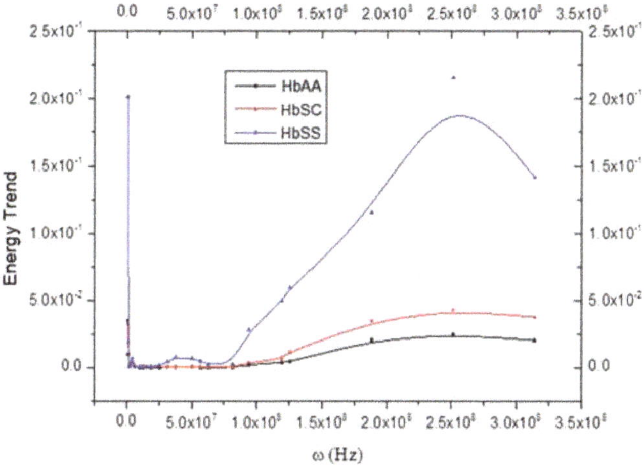

Figure 6. Comparison between HbAA (black line), HbSC (red line) and HbSS (blue line) Energy trend calculated by Equation (54), as a function of frequency.

7. Conclusions

The characterization of the Hb, trough the non equilibrium thermodynamic with internal variables allowed for reaching the goal of our study: The formulation of a physical mathematical model of Hb, the "Hemoglobe" a plane capacitor "with non-plane faces" and with dielectric characters. This simple model, which we (etymologically) call hemoglobe, reports the Hb as a geometric figure with the dimensions of 64 Å × 55 Å × 50 Å, very close to its complex biological reality. In a comparative context with the pathological hemoglobins HbSS and HbSC, the capacity that is a function of the angular frequency and the energy of the hemoglobe have been calculated. The phenomenon of conductivity, represented by the term $L^{(1,1)}$, has been interesting and determinative, because it revealed a capacitive character, therefore neither inductive nor Ohmic and peculiar of the displacement currents in a capacitor. Furthermore, at the molecular level and at the frequencies considered, the experimental logics described, allowed for the evaluation of the Hb molecules as the only "agents" of the solution that could influence $L^{(1,1)}$, from which the association almost unquestionable to a plane capacitor. The introduction of the hemoglobe discloses a new vision of the T-R transition of the Hb, which does not contrast with the existing theories but rather strengthens them. Because, our thermodynamic results provide useful information, or at least points of correlation, between mechanical movements of a certain entity, such as the rotation of Hb protomers, and structures where production and storage of the energy, required for the T-R transition, occurs. Through the physical mathematical model of hemoglobe, we can explain some aspects of the T-R transition and cooperativity of the Hb in terms of capacitor functionality, looking at the oxygenation-deoxygenation cycle of the protein in the blood circle as a voltage variation on the capacitor in an electric circuit. In other words, the theory of "hemoglobe" reinforces the importance of the peculiar quaternary structure of the protein whose molecular architecture is preserved at all levels of the vertebrate evolutionary scale. The protein appears constructed in such a way to full maximize its functionality as an oxygen carrier, so that to do this, the molecule does not only exploit the chemistry of its bonds, but also the mechanics of its structure and the thermodynamics of its charges. We can state that, during the vertebrate evolution, Hb is perhaps the only molecule, that although significantly changes its primary structure manages to maintain its functionality, almost ignoring the dogma that correlates each structural variation to the functional one. An architecture of quaternary structure ables of generating and accumulating energy and of being able to release appropriately it at the moment of the respiratory transition of the protein. Concluding the capacitor (hemoglobe) from simple and elementary electric machine becomes a complex and unusual, but effective electric machine. In fact, as thermodynamically highlighted, the protein is able not only to properly store and release energy but also to strategically produce it for a "mechanical consequence" induced by the O_2 binding. In fact, the breaking of the salt bridges that stabilize the structure T, generates new electric charges on the capacitor and these provide new capacity and new energy charge that are stored. An oxy-deoxy cycle of the Hb is carried out through the T-R and R-T transitions that produce and conserve energy, so Hb is able to generate energy (under the oxygen mechanical thrust) and at the same time, being a capacitor, it can conserve and release energy.

Author Contributions: F.F. was responsible for physical-mathematical processing, S.F. and E.T. were responsible for biological interpretation of theoretical results, A.G. conceived the idea of research.

Funding: This research received no external funding.

Conflicts of Interest: The authors declare no conflict of interest.

Abbreviations

Hemoglobin (Hb); Oxygen (O_2); 2,3-bisphosphoglycerate (BPG); carbon dioxide (CO_2); Human Hemoglobin (HbAA); Homozygous Sickle Hemoglobin (HbSS); Sickle Cell Hemoglobin C Disease (HbSC); Displacement Equivalent Conductivity Coefficient (DECC); Polarization Displacement Equivalent Current (PDEC); Displacement Equivalent Capacity (DEC).

References

1. Appleby, C.A. The origin and function of haemoglobin in plants. *Sci. Prog.* **1994**, *76*, 365–398.
2. Van-Holde, K.K.E.; Miller, K.I.; Decker, H. Hemocyanins and invertebrate evolution. *J. Biol. Chem.* **2001**, *276*, 15563–15566. [CrossRef] [PubMed]
3. Perutz, M.F. Haemoglobin as a model of an allosteric protein. *Symp. Soc. Exp. Biol.* **1973**, *27*, 1–3. [PubMed]
4. Farsaci, F.; Tellone, E.; Galtieri, A.; Ficarra, S. A new model with internal variables for theoretical thermodynamic characterization of hemoglobin: Entropy determination and comparative study. *J. Mol. Liq.* **2019**, *279*, 632–639. [CrossRef]
5. Farsaci, F.; Tellone, E.; Galtieri, A.; Ficarra, S. Is a dangerous blood clot formation a reversible process? Introduction of new characteristic parameter for thermodynamic clot blood characterization: Possible molecular mechanisms and pathophysiologic applications. *J. Mol. Liq.* **2018**, *262*, 345–353. [CrossRef]
6. De Groot, S.R.; Mazur, P. *Non Equilibrium Thermodynamics*; Dover Publication: New York, NY, USA, 1984.
7. Kluitenberg, G.A. On dielectric and magnetic relaxation phenomena and non-equilibrium thermodynamics. *Physica* **1973**, *68*, 75–82. [CrossRef]
8. Kluitenberg, G.A. On dielectric and magnetic relaxation phenomena and vectorial internal degrees. *Physica* **1977**, *87*, 302–330. [CrossRef]
9. Kluitenberg, G.A. On vectorial internal variables and dielectric and magnetic relaxation phenomena. *Physica* **1981**, *109*, 91–122. [CrossRef]
10. Farsaci, F.; Tellone, E.; Galtieri, A.; Ficarra, S. Molecular characterization of a peculiar blood clot fluidification by theoretical thermodynamic models and entropy production study. *J. Mol. Liq.* **2018**, *26*, 457–462. [CrossRef]
11. Farsaci, F.; Tellone, E.; Cavallaro, M.; Russo, A.; Ficarra, S. Low frequency dielectric characteristics of human blood: A non-equilibrium thermodynamic approach. *J. Mol. Liq.* **2013**, *188*, 113–119. [CrossRef]
12. Farsaci, F.; Ficarra, S.; Russo, A.; Galtieri, A.; Tellone, E. Dielectric properties of human diabetic blood: Thermodynamic characterization and new prospective for alternative diagnostic techniques. *J. Adv. Dielectr.* **2015**, *5*, 3. [CrossRef]
13. Farsaci, F.; Russo, A.; Ficarra, S.; Tellone, E. Dielectric properties of human normal and malignant liver tissue: A non-equilibrium thermodynamics approach. *Open Access Libr. J.* **2015**, *2*, e1395. [CrossRef]
14. Farsaci, F.; Ficarra, S.; Russo, A.; Galtieri, A.; Tellone, E. On evaluation of electric conductivity by mean of non equilibriumthermodynamic approach with internal variables. An application to human erythrocyte suspension for metabolic characterizations. *J. Mol. Liq.* **2016**, *224*, 1181–1188. [CrossRef]

15. Farsaci, F.; Tellone, E.; Galtieri, A.; Ficarra, S. Expanding the repertoire of dielectric fractional models: A comprehensive development and functional applications to predict metabolic alterations in experimentally-inaccessible cells or tissues. *Fluids* **2018**, *3*, 9. [CrossRef]
16. Farsaci, F.; Tellone, E.; Galtieri, A.; Russo, A.; Ficarra, S. Evaluation of the human blood entropy production: A new thermodynamic approach. *J. Ultrasound* **2016**, *19*, 265–273. [CrossRef]
17. Farsaci, F.; Tellone, E.; Russo, A.; Galtieri, A.; Ficarra, S. Rheological properties of human blood in the network of non-equilibrium thermodynamic with internal variables by means of ultrasound wave perturbation. *J. Mol. Liq.* **2017**, *231*, 206–212. [CrossRef]
18. Farsaci, F.; Ficarra, S.; Galtieri, A.; Tellone, E. New non-equilibrium thermodynamic fractional visco-inelastic model to predict experimentally inaccessible processes and investigate pathophysiological cellular structures. *Fluids* **2017**, *2*, 59. [CrossRef]
19. Laogun, A.A.; Agba, E.H.; Ajayi, N.O. A comparison of the dielectric behaviour of human haemoglobin SC with SS and AA in solution. *Phys. Med. Biol.* **1997**, *42*, 707–715. [CrossRef]
20. Takashima, S. Use of protein database for the computation of the dipole moments of normal and abnormal hemoglobins. *J. Biophys. Soc.* **1993**, *64*, 1550–1558. [CrossRef]
21. Antosiewicz, J.; Porschke, D. Electrostatics of Hemoglobins from Measurements of the Electric Dichroism and Computer Simulations. *Biophys. J.* **1995**, *68*, 655–664. [CrossRef]
22. Huang, Y.X.; Wu, Z.J.; Huang, B.T.; Luo, M. Pathway and Mechanism of pH Dependent Human Hemoglobin Tetramer-Dimer-Monomer Dissociations. *PLoS ONE* **2013**, *8*, e81708. [CrossRef] [PubMed]
23. Sommerfeld, A. *Electrodynamics: Lectures on Theoretical Physics*; Academic Press: Cambridge, MA, USA, 1956; Volume III.

fluids

MDPI

Article

Investigation of Hydrodynamically Dominated Membrane Rupture, Using Smoothed Particle Hydrodynamics–Finite Element Method

Hossein Asadi [1,*,†], Mohammad Taeibi-Rahni [1], Amir Mahdi Akbarzadeh [2], Khodayar Javadi [1] and Goodarz Ahmadi [3]

1 Department of Aerospace Engineering, Sharif University of Technology, Azadi Ave, P.O. Box: 11365-11155, Tehran, Iran
2 J. Mike Walker '66 Department of Mechanical Engineering, Texas A&M University, College Station, TX 77843, USA
3 Department of Mechanical and Aeronautical Engineering, Clarkson University, Potsdam, NY 13699, USA
* Correspondence: hasadi@tamu.edu
† Current address: J. Mike Walker '66 Department of Mechanical Engineering, Texas A&M University, College Station, TX 77843, USA.

Received: 28 June 2019; Accepted: 30 July 2019; Published: 3 August 2019

Abstract: The rupturing process of a membrane, located between two fluids at the center of a three-dimensional channel, is numerically investigated. The smoothed particle hydrodynamics (SPH) and the finite element method (FEM) are used, respectively, for modeling the fluid and solid phases. A range of pressure differences and membrane thicknesses are studied and two different rupturing processes are identified. These processes differ in the time scale of the rupture, the location of the rupture initiation, the level of destruction and the driving mechanism.

Keywords: smoothed particle hydrodynamics (SPH); meshless; fluid-solid interaction (FSI); membrane; rupture; SPH-FEM

1. Introduction

Predicting the exact values of impact loads exerted from fluids on structures is of great importance due to its many applications, for example, in cardiovascular systems [1–7] and industrial flows [8,9]. In particular, simulation of fluid-solid interaction (FSI) between a membrane and two fluids have many applications in petrochemical [10] and aerospace industries. Better understanding of membrane-fluid interactions helps mitigate failures of heat exchangers [9], fuel cells, ship tankers and aircraft fuel containers, which hinge on fluid pressure difference on the sides of membranes. Due to complex nature of FSI, progress in developing effective numerical methods have been slow. However, with the advances in meso-scale methods, for example the Lattice Boltzmann method (LBM) and smoothed particle hydrodynamics (SPH), satisfactory computational schemes have been emerged for handling the complex physics associated with FSI.

The SPH method was first introduced by Monaghan and Gingold [11] for solving astrophysical problems. In this method, the fluid is discretized to many particles, which are tracked in both time and space. The most important advantage of the SPH is the absence of typical errors of Eulerian methods, such as numerical diffusion occurring in high gradient regions of the flow.

The SPH method was first employed in 1990 for simulation of solid materials [12]. It was also used to study the failure of structures. Subsequently, several important developments in SPH were achieved. Swegle et al. [13] analyzed various numerical features of the SPH and explored its stability margins, zero energy modes, interface modeling and artificial viscosity to overcome the numerical challenges of the SPH method. Further developments were obtained by optimizing kernel functions [14] and symmetrical boundaries with extra particles, that is, ghost particles [15].

In early twenty first century, the SPH method was widely used in computational fluid dynamics (CFD), including simulation of compressible flows [16], incompressible flows [17], two-phase flows [18], flows in porous media [19], explosion modeling [20], and non-Newtonian flows [21]. The SPH method has been used as a suitable context to simulate viscoelastic fluid flow by implementing Maxwell [22] and cross models [23]. In addition, Hashemi et al. [24] studied the effect of fluid elasticity on the migration mechanism of solid bodies in Oldroyd-B shear flows. Xu and Ouyang [25] used SPH to simulate the challenging jet buckling and rod-climbing effects of viscoelastic fluids. SPH has shown good results for molding flows of Non-Newtonian fluids for both isothermal [26] and non-isothermal [27] conditions. Recently, Venkatesan and Ganesan [28] employed SPH to investigate viscoelasticity on droplet spreading dynamics on hydrophilic surfaces.

With advancements in simulation of fluid flow and growing interest in FSI, the smoothed particle hydrodynamics was used to simultaneously simulate the fluids and structures. In 2008, Maurel and Combescure [29] provided a formulation for SPH, which enabled it to simulate behaviors of shell structures. This formulation was capable of simulating elastic and plastic deformations of solids until failure. Concurrently, Potapov et al. [30] made an effort to simulate FSI in a problem in which random collision of fast-moving fluid to a structure caused major deformation of the solid.

In 2013, Caleyron et al. [31] modeled the rupture of a shell structure using the SPH method. He considered a cylinder and a piston system, in which a shell was placed at one end of the cylinder. The shell underwent large deformations after a weight was collided with the piston, generating extreme pressure waves. Recently, Faucher et al. [32] implemented a high resolution adaptive framework for fast transient FSI with interfaces and structural failure for application to failing tanks under impact. However, this work was focused on accurate evaluation of crack propagation under fluid pressure but did not analyze the structure failure modes. The present work aims to systematically identify the hydrodynamically dominated membrane rupture processes.

In this study, the transient FSI problem of membrane ruputre is investigated, in which the flow and the solid deformations are computed using, respectively, the SPH and finite element method (FEM). Initially, a membrane is located at the middle of a vertical three-dimensional rectangular channel between two weakly compressible liquids with different properties. By increasing the pressure difference, the membrane experiences large deformation and finally ruptures when the pressure difference becomes sufficiently large. The main contribution of the present study is in identifying different membrane failure modes, which are governed by different mechanisms, namely, the pressure waves and steady pressure force field. The rupture initiation time, the affected area, pressure difference sensitivity, thickness sensitivity and the level of destruction are the factors that influence different modes of membrane rupture.

2. Governing Equations

In this section, the governing equations of fluid flow and the solid phase are described. In addition, the damage model of the solid phase is briefly discussed. Note that the isothermal condition is assumed throughout this study and therefore, there was no need to sue the energy equation in fluid and solid phases. In the SPH method, the fluid dynamics are derived in a Lagrangian framework, similar to the lattice Boltzmann and MD [33], which obviates the need for a mesh to solve the governing equations, in contrast to the traditional CFD methods [34–37].

2.1. Fluid Phase

In this work, the fluid phase is modeled using the SPH method that is capable of handling free-surface flows involving ruptures and multiphase flows with small density differences [38]. The corresponding governing equations are the Lagrangian form of the Navier-Stokes equations, which in SPH context are given as,

$$\frac{D\rho_i}{Dt} = \sum_{j=1}^{N} m_j v_{ij}^{\alpha} \frac{\partial W_{ij}}{\partial x_i^{\alpha}},$$
$$\frac{Dv_i^{\alpha}}{Dt} = -\sum_{j=1}^{N} m_j \left(\frac{\sigma_i^{\alpha\beta}}{\rho_i^2} + \frac{\sigma_j^{\alpha\beta}}{\rho_j^2}\right) \frac{\partial W_{ij}}{\partial x_i^{\beta}} + F_i^{\alpha}, \tag{1}$$

where, the superscripts α and β are coordinate direction indicators and the subscripts i and j denote the central particle and its neighboring particles, respectively. In addition, ρ is density, m is mass, W is kernel function, v is velocity, σ is total stress tensor, F is external force and $v_{ij} = v_i - v_j$ [39,40]. The continuity equation for a weakly compressible fluid is given as,

$$\frac{\partial \rho}{\partial t} + \rho(\nabla . u) = 0. \tag{2}$$

Note, in Equation (2), $u.\nabla\rho$ that appears in the compressible continuity equation has been ignored. This assumption is valid when density variation is orthogonal to the velocity vector. This occurs when the maximum velocity of the fluid is negligible in comparison with the speed of sound [17]. Using Equation (2), for calculation of pressure p, one can use the following equation,

$$dp = c^2 d\rho \tag{3}$$

Since the flow is considered to be inviscid, the viscous effects are also ignored throughout this study. Thus, the momentum equation becomes:

$$\frac{Dv^{\alpha}}{Dt} = -\frac{1}{\rho}\frac{\partial p}{\partial x^{\alpha}} + F^{\alpha} \tag{4}$$

Substituting Equations (2) and (4) into Equation (1) leads to:

$$\left(\frac{D\rho}{Dt}\right)_i = \rho_i \sum_{j=1}^{N} \frac{m_j}{\rho_j} (v_{ij}^{\alpha}) \frac{\partial W_{ij}}{\partial x_i^{\alpha}},$$
$$\frac{Dv_i^{\alpha}}{Dt} = -\sum_{j=1}^{N} m_j \left(\frac{p_i}{\rho_i^2} + \frac{p_j}{\rho_j^2} + \prod_{ij} \delta^{\alpha\beta}\right) \frac{\partial W_{ij}}{\partial x_i^{\beta}} + F_i^{\alpha}, \tag{5}$$

where W denotes a cubic B-spline kernel function and $\delta^{\alpha\beta}$ is Kronecker's delta function [41,42]. In Equation (5), artificial viscosity terms ($\prod_{ij}$) are added to the momentum equation to enhance the stability of the method for the computation of discontinuities [43]. Note that the reason for including the fluid phase in this study, rather than a simple pressure loading on the membrane, is that we are interested in evaluation of flow after the rupture of the membrane which is considered for the future work.

2.2. Solid Phase

The behavior of the materials used in the solid phase are assumed to be independent of strain rate and thus von Mises' elastic-plastic laws are utilized. The Lemaitre-Chaboche damage laws are also coupled with Von Mises' elastic-plastic laws to model the rupturing mechanism of the membrane. The Lemaitre-Chaboche failure criterion (f_D) is presented in the following form [44],

$$f_D = [\frac{2}{3}(1+\nu) + 3(1-2\nu)(\frac{\sigma_h}{\sigma_{eq}})^2]P - \epsilon_p^s \tag{6}$$

where ν is Poisson's ratio, σ_h is the hydrostatic stress, σ_{eq} is Von Mises' equivalent stress and p is the isotropic pressure. Also, ϵ_p^s denotes the threshold equivalent plastic strain, beyond which microcrackes start to grow. This criterion gives the damage evolution ($\dot{D}_{nc}$) in the following form,

$$\dot{D}_{nc} = \frac{D_c}{\epsilon_p^c - \epsilon_p^s}[\frac{2}{3}(1+\nu) + 3(1-2\nu)(\frac{\sigma_h}{\sigma_{eq}})^2]\dot{P} \tag{7}$$

where ϵ_p^c is the critical equivalent plastic strain, beyond which fracture of material occurs. Also, D_c is the volume fraction of defects at the fracture point [31].

Bazant and Belytschko [45] and Needleman [46] showed that the damage localization problem for models similar to Equation (7) reduces to transformation of the original hyperbolic problem to an ill-posed elliptic one, when material softens. Therefore, the majority of proposed solutions emphasizes on preserving the hyperbolic nature of equations. Mostly suggested solutions utilize a characteristic time or length to moderate the interactions between the elements. similar to recent work in Reference [31], we compute damage rate ($\dot{D}$) as [47],

$$\dot{D} = \frac{1}{\tau_c}(1 - exp(-a|\dot{D}_{nc} - D|)), \tag{8}$$

where $\dot{D}_{nc}$ is calculated form Equation (7) and τ_c represents the inverse of maximum damage rate. Also, a denotes a coefficient related to the material properties of the membrane.

3. Numerical Modeling

In this section, the material characteristics and element topologies of the fluid phase are first described and then, the features of the employed materials and elements of solid phase are presented. Finally, the initial and boundary conditions that includes the utilized FSI interface are discussed. The entire numerical modeling is performed using the open version of EUROPLEXUS, which is a transient FSI computer code [48].

3.1. Fluid Phase

For fluid discretization, spherical fluid particles with a radius of 650 μm are employed. To decide on the size of the particles, dependency studies are performed, which are presented in the next section. Each of these particles have 6 degrees of freedom and a total of 211,146 particles are utilized for both high and low pressure fluids. In Figure 1, the blue spheres show high pressure liquid (water) and gold spheres represent low pressure liquid (oil). Also, the pink plane placed between the two liquids is the membrane under study. As it is shown in this figure, the lower bottom of the channel is closed by a rigid wall, while the upper end is open and thus, a free surface of low pressure fluid is created. The characteristics of the fluid phase are presented in Table 1.

Table 1. Fluid phase characterestics.

Fluid	Density (Kg/m^3)	Speed of Sound (m/s)
Water	1000	1450
Diesel Oil	900	1250

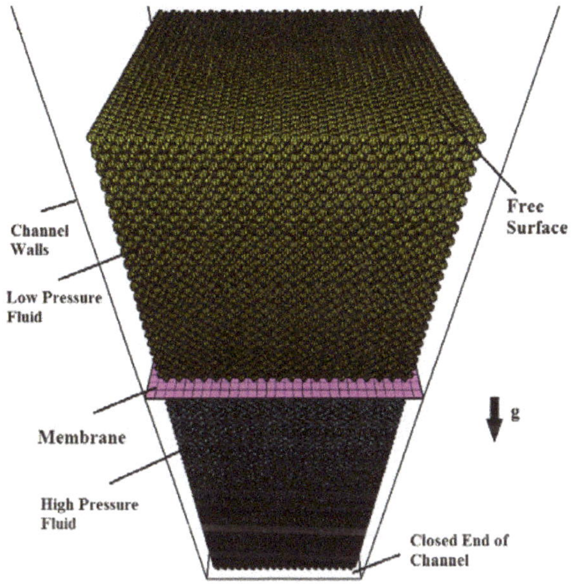

Figure 1. Particles representing fluids with gold balls for low pressure and blue balls for high pressure liquids (particles were magnified in size for better illustration).

3.2. *Solid Phase*

The solid phase is composed of two main parts, namely: channel walls and the membrane placed in the middle of the channel, both of which are modeled using FEM. The three-dimensional channel has a length of 20 cm, which is placed vertically. It is closed from the bottom but its top end is open. It has a cross section of 5 cm × 5 cm square. In simulating the behavior of the structural components, 16,200 two-dimensional plane elements are used with the optimal dimensions of 1.6 cm × 1.6 mm. Each of these elements has 20 Gaussian points for stress calculations and every 4 Gaussian points are placed in 5 different layers of its thickness. Three different thickness length are considered for the membrane, ranging from 1 to 3 mm. Given the fact that the components of the structure are expected to show different behaviors, we utilize two different materials. The channel walls are assumed to be made of iron with perfectly-elastic behavior. The features of this component are given in Table 2. For the membrane, von Mises' elastic-plastic laws are employed, coupled by the Lemaitre-Chaboche damage law with XC 38 steel material characteristics. Features of this component are illustrated in Tables 2 and 3 and also in Figure 2.

Table 2. Solid phase characteristics.

Material	Application	Density [Kg/m^3]	Young's Modulus [GPa]	Poisson's Ratio	Elastic Limit [MPa]
Iron	Channel wall	7860	200	0.33	Not applicable
XC 38 steel	Membrane	7860	200	0.33	166.5

Table 3. Required parameters for damage modeling of XC38 steel [44,49].

Parameter	ϵ_p^c	ϵ_p^s	D_c	$\tau_c[ns]$	a
Value	0.0	0.56	0.22	10	1

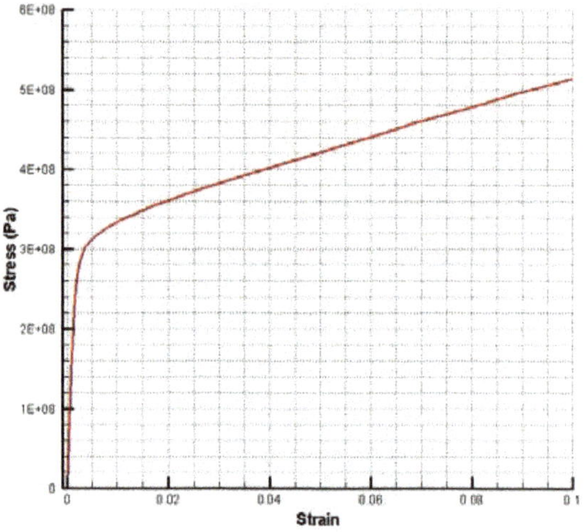

Figure 2. Strain-stress curve for XC 38 steel, employed for membrane component.

3.3. *Initial and Boundary Conditions*

As noted before, Figure 1 displays the initial conditions of the problem, wherein the pressure on both sides of the membrane is fixed at 1 KPa. For the region placed beneath the membrane, a high pressure is suddenly applied to the fluid. The magnitude of applied pressure varies depending on the test cases as will be discussed later. The free surface of the low-pressure fluid is modeled by periodic boundary condition, in order to let the high-pressure fluid to flow towards the low-pressure side after rupturing the membrane. All the membrane elements, except the ones attached to the wall, are treated using the master-slave method [50] for detection of interpenetration between the fluid and solid parts. Whenever interpenetration of two bodies is detected, proper contact forces are applied to them by the Lagrangian multiplier method [51].

4. Grid Resolution Study

The size of both SPH fluid particles and FEM solid elements are altered to study their effect on the displacement of the membrane central node (see Table 4). In all of these cases, a pressure difference of 250 MPa is exerted to a membrane and the final displacement of the central node of the membrane is measured.

As shown in Table 4, by reducing the size of the elements, relative error, which is defined as the ratio of the changes in central node displacement to its initial displacement, is decreased from 4.4% to 0.4%. In this table, only the last relative errors of the last two columns (for the study of element size) and the last two rows (for the study of particle radius) are presented for the sake of brevity. To be conservative, a 1.66 mm × 1.66 mm element was used.

Reducing the particle radius does not show similar effects due to the immobility of the particles before rupturing. For this reason, we resorted to employ equivalent plastic strain to study the effects of particle radius. According to Figure 3, it is clear that for particle sizes of 5 and 2.5 mm equivalent plastic strain patterns are highly asymmetric and thus incorrect, as far as the particles' impacting area to the membrane is seen. These figures also show that by altering the particle size from 0.83 to 0.65 mm equivalent plastic strains do not change. Consequently, a relatively conservative choice is SPH particles of radius 0.65 mm.

Table 4. Displacement of membrane central node for different particle and element sizes (all sizes are in mm).

Element Side/Particle Radius	5	2.5	1.25	0.83	0.65	Relative Error (%) (Effect of Element Size)
10	4.72	6.00	5.74	5.62	5.37	4.4
5	4.61	5.99	6.38	6.73	6.53	3
2.5	4.33	5.42	6.66	7.21	7.10	1.5
1.66	4.36	5.09	6.4	7.16	7.19	0.4
Relative error (%) (effect of particle radius)	0.7	6.1	4	0.7	1.3	-

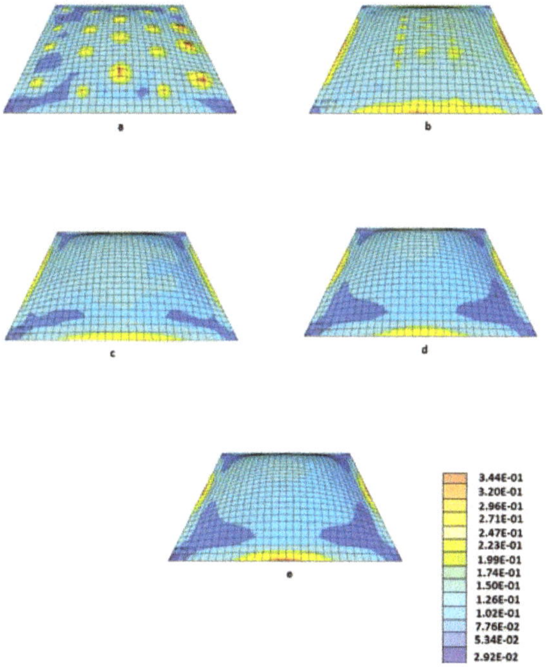

Figure 3. Effect of particle radius on equivalent plastic strain for radii (**a**) 5, (**b**) 2.5, (**c**) 1.25, (**d**) 0.83 and (**e**) 0.65 mm.

5. Code Validation

Three related test cases were previously studied to validate the EUROPLEXUS code. First, acoustic oscillations of a fluid contained in a cylinder with a piston was modeled using the SPH method and the results are compared with the FE-FV method [52]. The second case investigated the bending of a plate, which was modeled using FEM, under the impact of a column of water modeled with the SPH method [30]. In this case, the capabilities of the SPH-FEM, included in the code for FSI simulations, were studied. Third, the SPH results were compared to experimental data for simulation of the leakage rate of a metal shell at the bottom of water column, impacted by a weight at its top piston [52]. All these test cases gave satisfactory results for SPH-FEM implemented in the EUROPLEXUS code for simulating different FSI problems.

Here, it is necessary to satisfy two critical conditions. First, momentum transport from fluid particles to the membrane elements must be modeled properly. Also, the structural modeling of the membrane must be of high accuracy to simulate material behavior under external forces. Due to the lack of experimental data for von Mises' elastic-plastic material interacting with fluids, we resorted to empirical data of materials with hyperelastic behavior for validation purposes. For code validation, the experimental data of a FEM modeling of a rubber membrane under multi-axial stress is used [53]. For modeling natural rubber, we use the Mooney-Rivlin model. The fluid is modeled using the SPH method. Also, the Master-Slave method is employed for the FSI interfaces. The comparison of SPH-FEM results with the experimental data and a full FEM method is presented in Figure 4, showing satisfactory results (maximum error of 4%).

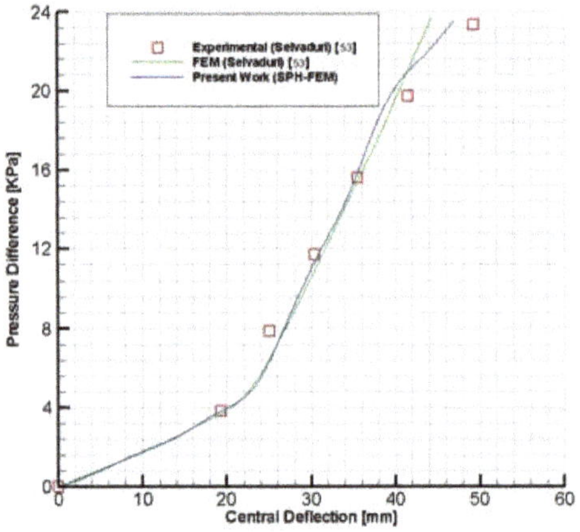

Figure 4. Comparison of Smoothed Particle Hydrodynamics–Finite Element Method (SPH-FEM) results with experimental data.

6. Results and Discussion

In this section, we investigated the occurrences of the start of membrane deformation until complete failure in two stages:

- membrane deformation (at this stage, instantaneous rise in water pressure causes membrane deformation) and,
- membrane rupture (the mechanism which controls the initiation and propagation of cracks).

Note that throughout this work, we used three different membrane thicknesses, namely: thin membrane (1 mm thickness), membrane with intermediate thickness (2 mm thickness) and thick membrane (3 mm thickness). Therefore, the deformation and rupture results are presented for these membrane thicknesses.

6.1. Membrane Deformation

The effects of of pressure difference on the membrane deflection is first examined. Then, the time variation of deformation of the membrane subjected to the imposed pressure difference is studied. Clearly, by applying higher pressure differences on the membrane, its deflection increases, until micro cracks start to emerge at a critical pressure. In order to provide quantitative results, the behavior of a particular point on the membrane is examined. Here the deflection of the central point is studied, since it is more sensitive to pressure differences. Figure 5 shows the variation of membrane central node deflection versus imposed pressure difference. It is from this figure that the central point deflection increases linearly with pressure difference.

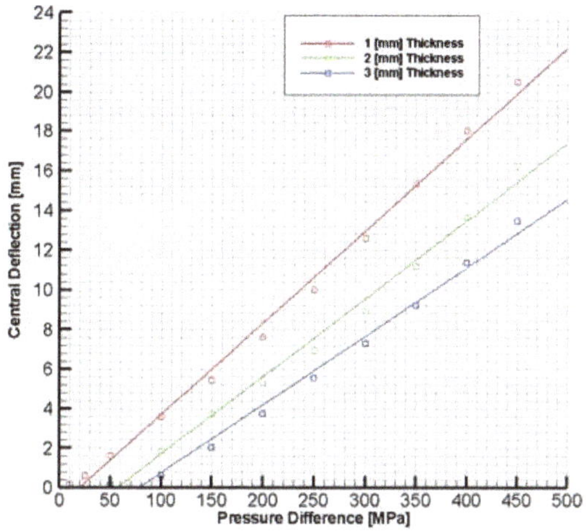

Figure 5. Variations of membrane central point deflection versus pressure difference for different thicknesses.

The equivalent plastic strain contours of a thin membrane subjected to a pressure difference was also studied and the results are shown in Figure 6. It is important to note that the maximum plastic strain occur at the center and at the edges of the membrane. Strain concentration on the edges of the membrane is because in those areas one side of the membrane is fixed to the wall. Therefore, the membrane deformation in the vicinity of the edges leads to large amount of strain. At the center of the membrane, since maximum displacement occurs, strain is large. Consequently, these areas are the potential points for onset of membrane rupture.

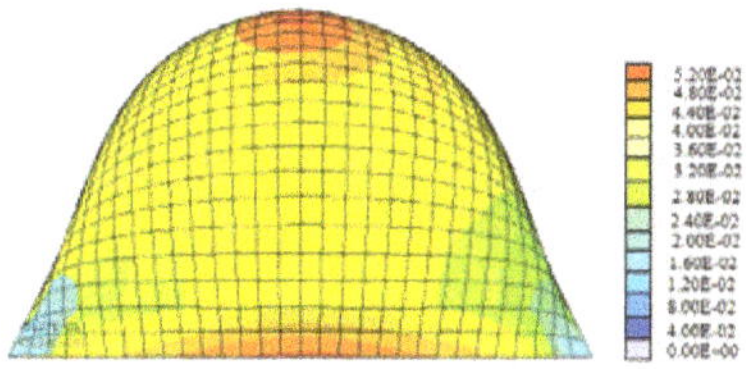

Figure 6. Equivalent plastic strain contours of the 1 mm thickness membrane under pressure difference of 400 MPa.

Another important parameter is the time variation of membrane deformation after imposing a pressure difference. The time histories of the membrane central point displacement for different thicknesses under various imposed pressure differences are examined. Figure 7a shows the corresponding time evolutions of the central point deflection for different thickness subject to a pressure difference of 450 MPa. It is seen that the deflection increases with time and reaches a saturation level. Also as the membrane thickness increases, the slope of the time evolution curve and the peak deformation decrease. It is also seen that all curves reach to their maximum displacement almost at the same time of about 0.12 ms.

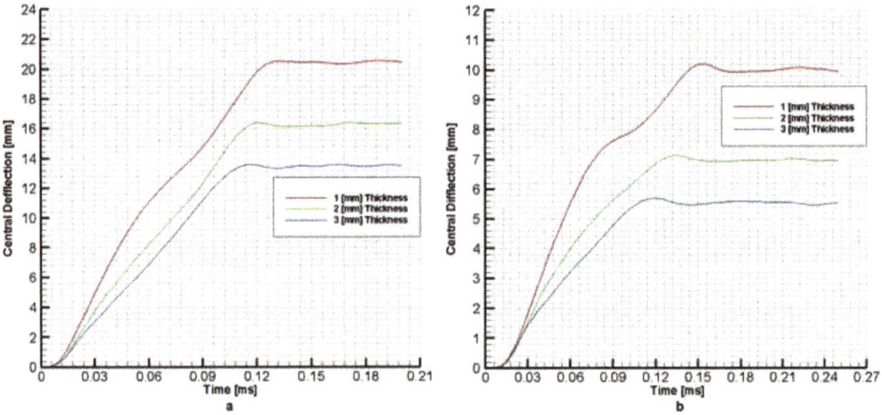

Figure 7. Time history of membrane central point displacements for different thicknesses and for pressure differences of (**a**) 450 MPa and (**b**) 250 MPa.

Similarly, Figure 7b shows similar trends for a pressure difference of 250 MPa. Note, for this pressure difference, the steady state reaches about 0.15 ms for all membrane thicknesses. It seems that the time required for reaching the membrane steady state deflection is a function of the applied pressure difference.

In order to investigate the time history of membrane deformation under a given pressure difference, the equivalent plastic strain contours for a 1 mm thickness membrane under a 450 MPa pressure difference are shown in Figure 8. These contour plots show that initially, the edges of the membrane experience the maximum strain, that is, the pressure pulse exerted on the membrane generates higher stress around fixed boundaries of the membrane. The strain in the central regions increases with time and eventually exceeds the levels at the edges. These trends are important for properly predicting the onset of membrane rupture. That is, at high pressure differences, the rupture is initiated at the membrane edges, while at low pressure difference, the rupture is initiated from the central point. These points will be further discussed in the next subsection.

6.2. Membrane Rupture

Different modes of membrane rupture are discussed in this section. As mentioned before, membrane failure is simulated using the Lemaitre-Chaboche method. First, the failure initiation time for different pressure differences is investigated. Then, two different rupturing modes are introduced. One objective of this study is to predict the minimum pressure difference that leads to the membrane rupture. In addition, the failure time as a function of pressure difference is also evaluated.

The failure time of several membranes are presented in Figure 9. This figure illustrates the first sign of rupture for several membrane thicknesses and pressure differences. The criterion utilized to interpret the first sign of membrane rupture is the failure of all 20 Gaussian points of a single membrane element. The simulation data in Figure 9 are fitted by power laws that are shown by solid lines in this figure. For a given membrane thickness, it is seen that there is a lower limit for pressure difference below which the rupture does not occur. The time required to reach the minimum pressure difference needed for rupture, appears to be roughly the same for all membranes with different thickness. We refer to the time for failure under minimum pressure difference as the "critical time", which is roughly independent of membrane thickness but perhaps is a function of the size and shape of the membrane. For the present membrane, the critical time is about 0.21 ms (varies between 0.204 to 0.218) and is illustrated in Figure 9 by a dashed line. Thus, if the membrane does not fail before the critical time, rupturing will not occur, regardless of the thickness of the membrane or the exerted pressure difference.

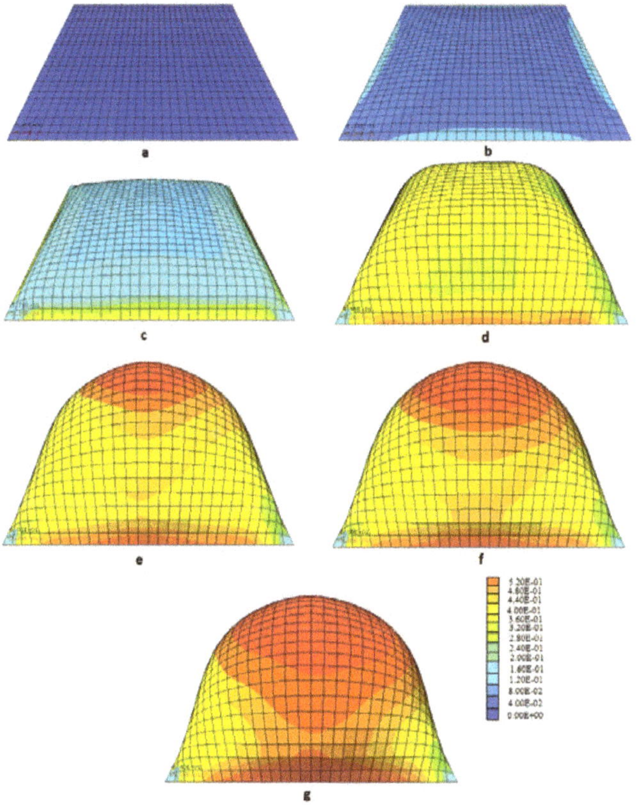

Figure 8. Resultant equivalent plastic strain contours for a 1 mm membrane under an imposed pressure difference of 450 MPa at times (**a**) 0, (**b**) 0.02, (**c**) 0.04, (**d**) 0.08, (**e**) 0.12, (**f**) 0.16 and (**g**) 0.2 ms.

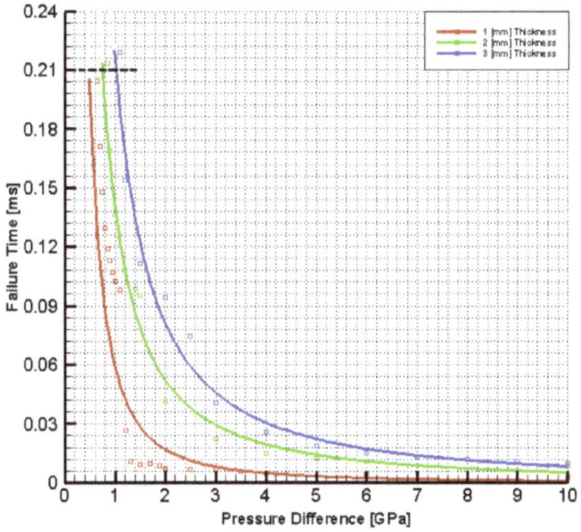

Figure 9. Failure time as a function of pressure difference for three different membrane thicknesses.

Figure 9 shows that the failure time for low values of pressure difference is highly sensitive to the value of membrane thickness. On the other hand, for high pressure differences, the failure time is not as sensitive to changes in the membrane thickness. In this context, membrane rupture initiation can be categorized into slow rupture (thickness/pressure-sensitive) and fast rupture (thickness/pressure-less sensitive) and also maybe a mixed mode between the two, namely: medium speed rupture (mildly sensitive to thickness/pressure). Membrane rupture evolution and the amount of failure over the surface of the membrane are addressed next. We first discuss slow rupturing process and then consider the fast one.

6.2.1. Slow Rupture

Figure 10 shows top-view of membrane failure by slow rupture for a pressure difference of 700 MPa (at six time instants). It is seen that for slow rupture mode, the equivalent plastic strain profiles vary smoothly. The pattern of plastic strain distribution at time 30 μs (Figure 10b) is similar to the last strain patterns of the fast rupturing mode before rupturing occurs (see Figure 11b). However, in the slow mode, strain magnitudes do not reach their critical values and thus, rupturing does not occur. Consequently, peak strain has sufficient time to develop towards central part of the membrane. In other word, the effective destructive mechanism in this mode is the pressure force magnitude (and not the pressure pulse). As shown in Figure 10c, at time 80 μs, the central parts of the membrane experience the growth of plastic strain. As the intensity of strain in the bulk of the membrane increases with time, the magnitude of strain in the central areas becomes larger than the values of the stain at edges that are relaxing time at 120 μs (Figure 10d). Then, at 190 μs, the first element at the center of the membrane fails (Figure 10e), causing its neighboring elements to fail and reach the final rupturing state at 250 μs (Figure 10f).

6.2.2. Fast Rupture

As mentioned before, the fast rupturing mode takes place for high pressure differences. In order to model fast rupturing, a pressure difference of 4 GPa is imposed suddenly on a 1 mm thick membrane. Figure 11 shows the initiation, progress and complete destruction of the membrane by the fast rupturing model at six different times. The non-smoothness of equivalent plastic strain contours particularly for the initial times is clearly seen in this figure, which is due to the highly unsteady nature of this kind of rupturing. The first signs of fast failure is observed in about 5 μs after imposing the pressure difference (Figure 11b). Figure 11c illustrates that the edges of the membrane are the first areas that the rupturing is initiated and progresses. The displacement of the central parts of the membrane in this mode is not significant. That is, due to the fast nature of this rupturing mode, the membrane does not have sufficient time to propagate the strain to its central area. At 12 μs, the membrane remains attached to the duct walls only through its corners (Figure 11d). As the rupturing process continuing, as shown in Figure 11e, at 18 μs the membrane completely detached from the duct walls with only some small pieces of the membrane which have more constraints remaining on the duct walls.

To summarize the findings of this section, different features of the membrane rupturing modes are presented in Table 5. It is clear that magnitude of the pressure difference is a major factor that controls the rupturing mode. As noted before, there is a mixed rupturing mode which occurs between the slow and fast modes. However, because of the transitional nature of the mixed mode, its detailed study is left for a future work.

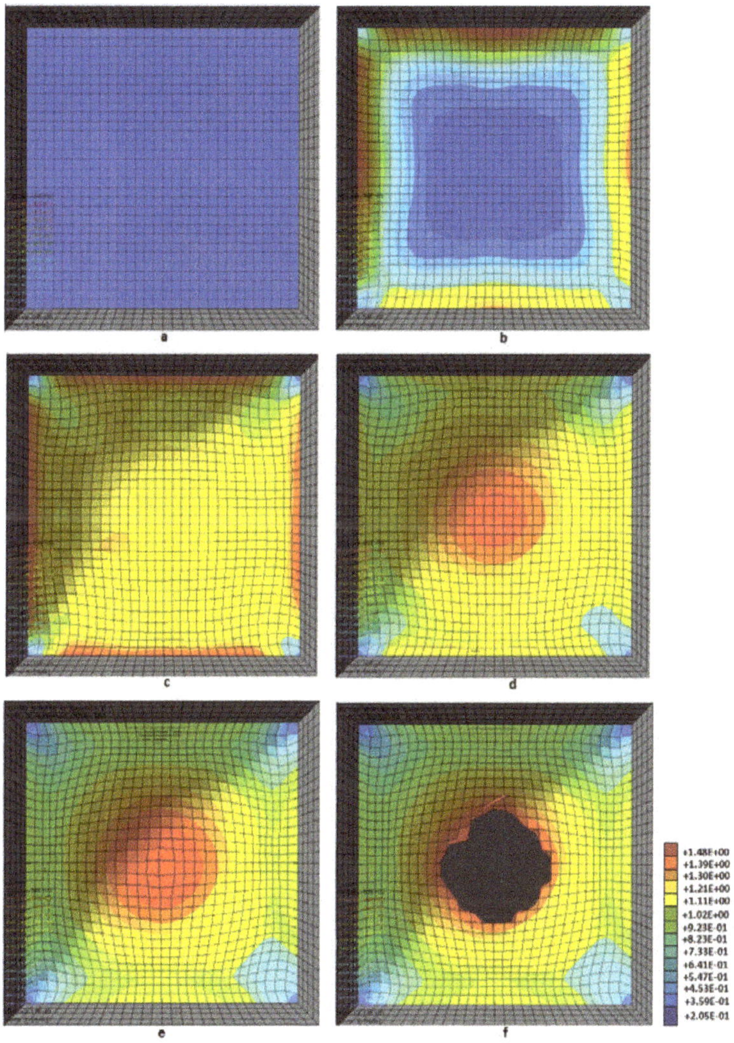

Figure 10. Equivalent plastic strain contours for the slow rupturing mode for times (**a**) 0, (**b**) 30, (**c**) 80, (**d**) 120, (**e**) 190 and (**f**) 250 μs.

Table 5. Features of fast and slow rupturing modes.

Feature	Fast Rupturing	Slow Rupturing
Initiation Time	Order of μs	Order of ms
Initiation Area	Edges	Center
Thickness/Pressure Sensitivity	low sensitivity	Sensitive
Destruction Level	Smaller Destruction per Area with Complete Separation from Channel Walls	High Destruction per Area without Separation from Channel Walls
Destructive Mechanism	Pressure pulse	Steady Pressure Field Forces

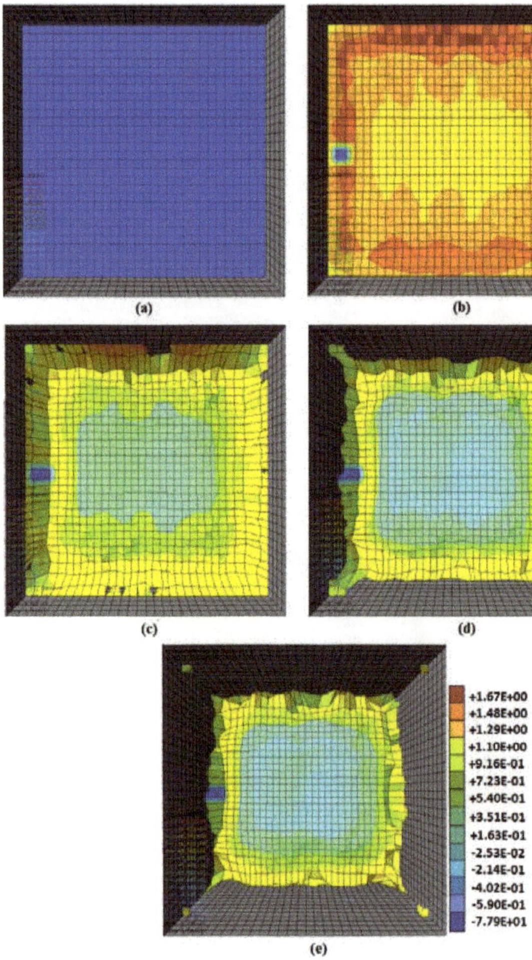

Figure 11. Equivalent plastic strain contours during the fast rupturing mode for times (**a**) 0, (**b**) 5, (**c**) 10, (**d**) 12 and (**e**) 18 μs.

7. Concluding Remarks

This study utilized the SPH and FEM for modeling of fluid and solid phases, respectively. Based on the presented results, the following conclusions are drawn:

- The maximum deflection of the membrane occurs at its center. The deflection is also a linear function of the applied pressure difference.
- The time required to reach the final deformation of the membrane is only a function of the applied pressure difference and the effect of membrane thickness is negligible.
- The maximum plastic strain on the membrane, which is first seen at the constrained edges, move to the center of the membrane with time.
- The minimum pressure difference required for initiation of membrane rupture varies with membrane thickness but the rupturing time (referred to as critical time) is roughly the same for all membrane thicknesses.
- Simulations of the rupturing time as a function of pressure difference suggest three different rupturing modes, namely, thickness/pressure-sensitive mode (slow rupturing), mild

thickness/pressure-sensitive mode (medium speed rupturing) and low thickness/pressure-sensitive mode (fast rupturing).

- Slow rupturing causes cracks in the central parts of the membrane that spreads to other areas.
- In fast rupturing, failure first occurs at the edges of the membrane and quickly detaches the membrane from the duct walls.

Given the potential of the SPH method for simulating FSI and multiphase flows, as well as its applicability to handling different types of Non-Newtonian flows, for future work, we plan to study membrane rupturing in Non-Newtonian flows and investigate the effect of shear thinning and shear thickening as well as the viscoelastic effects on the process, particularly on the onset of rupture.

Author Contributions: Conceptualization, M.T.-R.; methodology, M.T.-R. and H.A.; software, H.A.; validation, H.A.; formal analysis, H.A., M.T.-R., K.J., and G.A.; investigation, H.A., M.T.-R., and K.J.; resources, H.A.; data curation, H.A.; writing—original draft preparation, H.A., M.T.-R., K.J., and G.A.; writing—review and editing, H.A., A.M.A., M.T.-R., K.J., and G.A.; visualization, H.A.; supervision, M.T.-R. and K.J.; project administration, M.T.-R. and K.J.

Funding: This research received no external funding.

Conflicts of Interest: The authors declare no conflict of interest.

References

1. Takizawa, K.; Bazilevs, Y.; Tezduyar, T.E. Space–time and ALE-VMS techniques for patient-specific cardiovascular fluid–structure interaction modeling. *Arch. Comput. Methods Eng.* **2012**, *19*, 171–225. [CrossRef]
2. Shahidian, A.; Hassankiadeh, A.G. Stress analysis of internal carotid artery with low stenosis level: The effect of material model and plaque geometry. *J. Mech. Med. Biol.* **2017**, *17*, 1750098. [CrossRef]
3. Hedayat, M.; Asgharzadeh, H.; Borazjani, I. Platelet activation of mechanical versus bioprosthetic heart valves during systole. *J. Biomech.* **2017**, *56*, 111–116. [CrossRef] [PubMed]
4. Sharzehee, M.; Khalafvand, S.S.; Han, H.C. Fluid-structure interaction modeling of aneurysmal arteries under steady-state and pulsatile blood flow: A stability analysis. *Comput. Methods Biomech. Biomed. Eng.* **2018**, *21*, 219–231. [CrossRef] [PubMed]
5. Azar, D.; Ohadi, D.; Rachev, A.; Eberth, J.F.; Uline, M.J.; Shazly, T. Mechanical and geometrical determinants of wall stress in abdominal aortic aneurysms: A computational study. *PLoS ONE* **2018**, *13*, e0192032. [CrossRef] [PubMed]
6. Asgharzadeh, H.; Asadi, H.; Meng, H.; Borazjani, I. A non-dimensional parameter for classification of the flow in intracranial aneurysms. II. Patient-specific geometries. *Phys. Fluids* **2019**, *31*, 031905. [CrossRef] [PubMed]
7. Hedayat, M.; Borazjani I. Comparison of platelet activation through hinge vs bulk flow in bileaflet mechanical heart valves. *J. Biomech.* **2019**, *83*, 280–290. [CrossRef]
8. Kamakoti, R.; Shyy, W. Fluid–structure interaction for aeroelastic applications. *Prog. Aerosp. Sci.* **2004**, *40*, 535–558. [CrossRef]
9. Ennis, C.J.; Botros, K.K.; Patel, C. Dynamic model for a heat exchanger tube rupture discharging a high-pressure flashing liquid into a low-pressure liquid-filled shell. *J. Loss Prev. Process. Ind.* **2011**, *24*, 111–121. [CrossRef]
10. Ravaji, B.; Mashadizade, S.; Hashemi, A. Introducing optimized validated meshing system for wellbore stability analysis using 3D finite element method. *J. Nat. Gas Sci. Eng.* **2018**, *53*, 74–82. [CrossRef]
11. Gingold, R.A.; Monaghan, J.J. Smoothed particle hydrodynamics: Theory and application to non-spherical stars. *Mon. Not. R. Astron. Soc.* **1977**, *181*, 375–389. [CrossRef]
12. Libersky, L.D.; Petschek, A.G. Smooth particle hydrodynamics with strength of materials. In *Advances in the Free-Lagrange Method including Contributions on Adaptive Gridding and the Smooth Particle Hydrodynamics Method;* Springer: Berlin/Heidelberg, Germany, 1991; pp. 248–257.
13. Swegle, J.W.; Attaway, S.W.; Heinstein, M.W.; Mello, F.J.; Hicks, D.L. *An Analysis of Smoothed Particle Hydrodynamics;* Sandia National Labs: Albuquerque, NM, USA, 1994.
14. Chen, J.S.; Pan, C.; Wu, C.T.; Liu, W.K. Reproducing kernel particle methods for large deformation analysis of non-linear structures. *Comput. Methods Appl. Mech. Eng.* **1996**, *139*, 195–227. [CrossRef]

15. Randles, P.W.; Libersky, L.D. Smoothed particle hydrodynamics: Some recent improvements and applications. *Comput. Methods Appl. Mech. Eng.* **1996**, *139*, 375–408. [CrossRef]
16. Welton, W.C.; Pope, S.B. PDF model calculations of compressible turbulent flows using smoothed particle hydrodynamics. *J. Comput. Phys.* **1997**, *134*, 150–168. [CrossRef]
17. Morris, J.P.; Fox, P.J.; Zhu, Y. Modeling low Reynolds number incompressible flows using SPH. *J. Comput. Phys.* **1997**, *136*, 214–226. [CrossRef]
18. Hu, X.Y.; Adams, N.A. An incompressible multi-phase SPH method. *J. Comput. Phys.* **2007**, *227*, 264–278. [CrossRef]
19. Zhu, Y.; Fox, P.J. Simulation of Pore-Scale Dispersion in Periodic Porous Media Using Smoothed Particle Hydrodynamics. *J. Comput. Phys.* **2002**, *182*, 622–645. [CrossRef]
20. Liu, M.B.; Liu, G.R.; Zong, Z.; Lam, K.Y. Computer simulation of high explosive explosion using smoothed particle hydrodynamics methodology. *Comput. Fluids* **2003**, *32*, 305–322. [CrossRef]
21. Hosseini, S.M.; Manzari, M.T.; Hannani, S.K. A fully explicit three-step SPH algorithm for simulation of non-Newtonian fluid flow. *Int. J. Numer. Methods Heat Fluid Flow* **2007**, *17*, 715–735. [CrossRef]
22. Ellero, M.; Kroger, M.; Hess, S. Viscoelastic flows studied by smoothed particle hydrodynamics. *J. Non-Newton. Fluid Mech.* **2002**, *105*, 35–51. [CrossRef]
23. Shao, S.; Lo, E.Y.M. Incompressible SPH method for simulating Newtonian and non-Newtonian flows with a free surface. *Adv. Water Resour.* **2003**, *26*, 787–800. [CrossRef]
24. Hashemi, M.R.; Fatehi, R.; Manzari, M.T. SPH simulation of interacting solid bodies suspended in a shear flow of an Oldroyd-B fluid. *J. Non-Newton. Fluid Mech.* **2011**, *166*, 1239–1252. [CrossRef]
25. Xu, X.; Ouyang, J. A SPH-based particle method for simulating 3D transient free surface flows of branched polymer melts. *J. Non-Newton. Fluid Mech.* **2013**, *202*, 54–71. [CrossRef]
26. Fan, X.J.; Tanner, R.I.; Zheng, R. Smoothed particle hydrodynamics simulation of non-Newtonian moulding flow. *J. Non-Newton. Fluid Mech.* **2010**, *165*, 219–226. [CrossRef]
27. Ren, J.; Ouyang, J.; Jiang, T. An improved particle method for simulation of the non-isothermal viscoelastic fluid mold filling process. *Int. J. Heat Mass Transf.* **2015**, *85*, 543–560. [CrossRef]
28. Venkatesan, J.; Ganesan, S. Computational modeling of impinging viscoelastic droplets. *J. Non-Newton. Fluid Mech.* **2019**, *263*, 42–60. [CrossRef]
29. Maurel, B.; Combescure, A. An SPH shell formulation for plasticity and fracture analysis in explicit dynamics. *Int. J. Numer. Methods Eng.* **2008**, *76*, 949–971. [CrossRef]
30. Potapov, S.; Maurel, B.; Combescure, A.; Fabis, J. Modeling accidental-type fluid–structure interaction problems with the SPH method. *Comput. Struct.* **2009**, *87*, 721–734. [CrossRef]
31. Caleyron, F.; Combescure, A.; Faucher, V.; Potapov, S. SPH modeling of fluid–solid interaction for dynamic failure analysis of fluid-filled thin shells. *J. Fluids Struct.* **2013**, *39*, 126–153. [CrossRef]
32. Faucher, V.; Casadei, F.; Valsamos, G.; Larcher, M. High resolution adaptive framework for fast transient fluid-structure interaction with interfaces and structural failure–Application to failing tanks under impact. *Int. J. Impact Eng.* **2019**, *127*, 62–85. [CrossRef]
33. Kheirabadi, A.M.; Moosavi, A.; Akbarzadeh, A.M. Nanofluidic transport inside carbon nanotubes. *J. Phys. D Appl. Phys.* **2014**, *47*. [CrossRef]
34. Asadi, H.; Asgharzadeh, H.; Borazjani, I. On the scaling of propagation of periodically generated vortex rings. *J. Fluid Mech.* **2018**, *853*, 150–170. [CrossRef]
35. Akbarzadeh, A.; Borazjani, I. A numerical study on controlling flow separation via surface morphing in the form of backward traveling waves. In Proceedings of the AIAA Aviation 2019 Forum, Dallas, TX, USA, 17–21 June 2019; p. 3589.
36. Asadi, H. Two-Dimensional Numerical Investigation of Oscillatory Shear-Driven Flows in Slip Flow Regime between Two Microscale Concentric Cylinders. *Appl. Mech. Mater.* **2014**, *704*, 299–304. [CrossRef]
37. Akbarzadeh, A.M.; Moosavi, A.; Kheirabadi, A.M. Dewetting of evaporating thin films over nanometer-scale topographies. *Phys. Rev. E* **2014**, *90*, 012409. [CrossRef]
38. Monaghan, J.J.; Cas, R.A.; Kos, A.M.; Hallworth, M. Gravity currents descending a ramp in a stratified tank. *J. Fluid Mech.* **1999**, *379*, 39–69. [CrossRef]
39. Liu, M.B.; Liu, G.R. Smoothed Particle Hydrodynamics (SPH): An Overview and Recent Developments. *Arch. Comput. Methods Eng.* **2010**, *17*, 25–76. [CrossRef]

40. Liu, G.R.; Liu, M.B. *Smoothed Particle Hydrodynamics: A Meshfree Particle Method*; World Scientific: Singapore, 2003.
41. Shadloo, M.S.; Oger, G.; Touzé, D.L. Smoothed particle hydrodynamics method for fluid flows, towards industrial applications: Motivations, current state, and challenges. *Comput. Fluids* **2016**, *136*, 11–34. [CrossRef]
42. Gray, J.P.; Monaghan, J.J.; Swift, R.P. SPH elastic dynamics. *Comput. Methods Appl. Mech. Eng.* **2001**, *190*, 6641–6662. [CrossRef]
43. Monaghan, J.J.; Gingold, R.A. Shock simulation by the particle method SPH. *J. Comput. Phys.* **1983**, *52*, 374–389. [CrossRef]
44. Lemaitre, J.; Chaboche, J.L. *Mechanics of Solid Materials*; Cambridge University Press: Cambridge, UK, 1994.
45. Bažant, Z.P.; Belytschko, T.B. Wave Propagation in a Strain-Softening Bar: Exact Solution. *J. Eng. Mech.* **1985**, *111*, 381–389. [CrossRef]
46. Needleman, A. Material rate dependence and mesh sensitivity in localization problems. *Comput. Methods Appl. Mech. Eng.* **1988**, *67*, 69–85. [CrossRef]
47. Allix, O.; Deü, J.F. Delayed-damage modelling for fracture prediction of laminated composites under dynamic loading. *Eng. Trans.* **1997**, *45*, 29–46.
48. EUROPLEXUS. User's Manual. 2005. Available online: https://europlexus.jrc.ec.europa.eu (accessed on 1 August 2015).
49. Caleyron, F.; Combescure, A.; Faucher, V.; Potapov, S. Dynamic simulation of damage-fracture transition in smoothed particles hydrodynamics shells. *Int. J. Numer. Methods Eng.* **2012**, *90*, 707–738. [CrossRef]
50. Hallquist, J.O.; Goudreau, G.L.; Benson, D.J. Sliding interfaces with contact-impact in large-scale Lagrangian computations. *Comput. Methods Appl. Mech. Eng.* **1985**, *51*, 107–137. [CrossRef]
51. Casadei, F; Potapov, S. Permanent fluid–structure interaction with non-conforming interfaces in fast transient dynamics. *Comput. Methods Appl. Mech. Eng.* **2004**, *193*, 4157–4194. [CrossRef]
52. Maurel, B.; Potapov, S.; Fabis, J.; Combescure, A. Full SPH fluid-shell interaction for leakage simulation in explicit dynamics. *Int. J. Numer. Methods Eng.* **2009**, *80*, 210–234. [CrossRef]
53. Selvadurai, A.P.S.; Shi, M. Fluid pressure loading of a hyperelastic membrane. *Int. J. Non-Linear Mech.* **2012**, *47*, 228–239. [CrossRef]

fluids

MDPI

Article

The Influence of Bubbles on Foamed Cement Viscosity Using an Extended Stokesian Dynamics Approach

Eilis Rosenbaum [1,*,†], Mehrdad Massoudi [1,2,†] and Kaushik Dayal [3,4,†]

1 National Energy Technology Laboratory, 626 Cochrans Mill Road, P.O. Box 10940, Pittsburgh, PA 15236, USA; mehrdad.massoudi@netl.doe.gov
2 Department of Biomedical Engineering, Carnegie Mellon University, 5000 Forbes Avenue, Pittsburgh, PA 15213, USA
3 Department of Civil and Environmental Engineering, Carnegie Mellon University, Pittsburgh, PA 15213, USA; kaushik.dayal@cmu.edu
4 Center for Nonlinear Analysis, Carnegie Mellon University, Pittsburgh, PA 15213, USA
* Correspondence: eilis.rosenbaum@netl.doe.gov
† These authors contributed equally to this work.

Received: 7 August 2019; Accepted: 3 September 2019; Published: 6 September 2019

Abstract: We want to study the influence of bubbles on the viscosity of suspensions with a computational approach that also accounts for the arrangement of the bubbles due to shearing flow. This requires a large number of bubbles to properly simulate and requires a large amount of computational resources. Here we develop a set of equations to define the viscosity ratio from the simulation results to show the influence of the bubbles on the viscosity as a function of the volume fraction. One application of this work has been used to study a specific type of cement that has bubbles injected into the slurry while it is still fluid. The bubbles are added to reduce the density but they also improve the properties of the cement with the increase in viscosity. We show that the computed results match the few experimental results that have been reported.

Keywords: stokesian dynamics; dense suspension; rheology; bubble suspension; suspension viscosity

1. Introduction

The primary motivation for this work is to study a specialized wellbore cement that has bubbles dispersed and suspended in it. The cement is referred to as foamed cement and the cement is foamed to reduce its density while still maintaining compressive strength. The cement density is designed to keep the cement gradient between the fracture and pore pressures [1,2] and current technologies allow the foamed cement density or quality (gas volume fraction) to be controlled and optimized during operations [3]. Foamed cements offer an economically viable low-density option that still maintains compressive strength. Foaming the cement also offers other improved properties compared to conventional cement or alternative light-weight cement [1,4–7].

In drilling and completing a well, cement is placed in the annulus between the well casing and the formation to prevent fluid migration between the formation layers and to prevent fluids from migrating to the surface. The cement also provides support to the steel casing pipe in the center of the well. A successful cement job will provide complete "zonal isolation" [8–10], where fluids in the wellbore remain isolated. Any means for fluids to migrate through the cement via fluid pathways is undesirable and could lead to severe issues.

Foamed cement is typically used in wells with weak formations or anywhere a low density cement is required but it has been used successfully in other adverse environments where the foaming provides better performance, over conventional or unfoamed cements [11]. It has been used for high-temperature/high-pressure wells where stress cracking is of concern due to the cycling pressures and temperatures [4,12]. Its higher ductility when compared to conventional cements makes it more resilient to stress cracking [1,4,5,7]. With its ability to withstand cyclical conditions, foamed cement is well suited for geothermal wells [5]. Foamed cement reduces lost circulation [5,8], where some of the cement flows into the formation, since foamed cement can be maintained below the formation fracture gradient during placement. Foamed cement offers improved mud displacement and controls gas migration better than conventional cements [1,5,7] due to the presence of the bubbles in the cement slurry. Even with the specialized foaming equipment, overall costs can be reduced by using foamed cement and, due in part to using less material, foamed cement has a lower environmental impact [5].

Foamed cement is 'designed' for the downhole conditions where it is to be placed in the well [11]. The foam is created when the base cement slurry flows into a tee met at a right angle by pressurized nitrogen, which is dispersed into the cement slurry via some form of atomizer. Stabilizers and additives are generally pumped into the cement slurry prior to foaming [4]. The design of the cement is based on the midpoint of cement placement depth so the changing hydrostatic pressure gradient and temperature changes occurring as the foamed cement is placed in the well must also be considered in the final cement design [3].

The American Petroleum Institute (API) recommends a foam quality (the volume fraction of added gas) below 35% for a stable foam that will maintain mechanical integrity and proper zonal isolation [3]. Above 35% volume of gas in the cement, the permeability may increase to unacceptable levels, and the compressive strength and ability to support the casing could be compromised. The foam should also have well dispersed bubbles that remain suspended during placement in the well without significant clustering, coalescence, or other undesirable configuration changes until the cement fully hardens [11]. It is thought that the bubble sizes and extent to which the bubbles remain well dispersed affects the overall strength and performance of the cement. Coalescence or clustering of the bubbles could provide migration pathways for fluids once the cement is fully hydrated and could reduce the compressive strength of the cement sheath.

2. Methods

2.1. Model Assumptions

We used a computation method to calculate the viscosity ratio (also called the relative viscosity to describe the change in viscosity due to the suspended bubbles). The computation method that we developed [13] extends Stokesian Dynamics and Fast Lubrication Dynamics. To employ the computation methods, several assumptions must be made. The bubbles are assumed to be spherical and discrete and remain spherical throughout the simulation. Foamed cement samples that have been produced in the laboratory and collected as part of a collaborative field effort show spherical, stable, discrete bubbles [14,15]. Though the interaction considers spherical bubbles, the surface is allowed to deflect in the region of bubble pair interaction [16]. Another assumption is that the bubbles are neutrally buoyant. This has also been observed experimentally [14,15,17] and were observed to remain in place during curing under pressure [15]. We also must assume that the suspending fluid is Newtonian. Cement is known to have a yield stress and nonlinear behavior but this assumption must be made to maintain linear equations.

2.2. Stokesian Dynamics and Fast Lubrication Dynamics

Stokesian Dynamics [18] and Fast Lubrication Dynamics (FLD) [19,20] include both close-range and long-range interactions to simulate spherical objects suspended in a base fluid. They both employ methods to incorporate these ranges of interactions in more efficient dynamics so that large numbers of spherical objects can be simulated at any volume fraction of objects. Even in dilute suspensions, long-range effects need to be considered, because in a suspension, the motion of each suspended object (to also include bubbles) is transmitted through the base fluid. In the quasi-static (creeping flow) limit, this is felt immediately throughout the entire system by all the other objects. For a dilute suspension of objects, the objects are far apart so the detailed shape and structure of the suspended objects does not matter to leading-order. The velocity disturbance of a suspended object decays like a point force as $1/r^2$, where r is the radial distance from the object [18]. When the bubbles are close together, the interaction of each pair is dominated by a pairwise force which comes from lubrication theory, with specific surface properties representative of a bubble [21]. The Stokesian Dynamics method accounts for the far-field interaction through multipole expansions and for the near-field interactions through pairwise interactions [18]. The FLD method further increases the efficiency by using fast approximate methods for the far-field interaction [19,20]. We employ an extended FLD method with specific pairwise near-field interactions for bubbles [16].

2.3. Governing Equations

For the Stokes flow regime, the viscous forces dominate and the inertial forces have negligible effects. The movement of a particle is governed by the equation [22,23]:

$$\boldsymbol{m} \cdot \frac{\mathrm{d}\boldsymbol{U}}{\mathrm{d}t} = \boldsymbol{F}^H \tag{1}$$

Here, $\boldsymbol{U}$ is the generalized velocity vector of the particles with entries corresponding to both linear and angular velocities. The velocity vector therefore has 6 components, corresponding to linear and angular velocities along the Cartesian directions, for each particle. Correspondingly, $\boldsymbol{m}$ is the generalized moment of inertia matrix containing entries for mass as well as the rotational inertia and, $\boldsymbol{F}$ is the generalized force vector with entries for both force as well as moment. Contained in the hydrodynamic force vector, $\boldsymbol{F}^H$, are then the 6 Cartesian components of the force and torque for each particle.

The hydrodynamic forces can be determined from the generalized equation relating the velocities and particle positions, $\boldsymbol{X}$, to the forces.

$$\left(\boldsymbol{F}^H\right) = \mathcal{R}(\boldsymbol{X}) \cdot \left(\boldsymbol{U}\right) \tag{2}$$

$\mathcal{R}$ is the linear resistance matrix that is a function of the particle positions, $\boldsymbol{X}$. $\mathcal{R}$ is almost the inverse of the mobility matrix, $\mathcal{M}$, and is a close approximation, though they are not complete inverses of each other [21]. The mobility matrix is more computationally efficient to construct but does not capture the physics of the problem as well as the resistance matrix and can allow the spherical objects to overlap in an unrealistic way. Stokesian Dynamics takes advantage of these properties to reduce computation time. Stokesian Dynamics uses the inverse of the grand mobility matrix, $(\mathcal{M}^\infty)^{-1}$, as an approximation to the grand resistance matrix, $\mathcal{R}^\infty$, which captures the long-range effects. In FLD, the grand mobility matrix is replaced with an isotropic resistance tensor, $\mathcal{R}_{Iso}$, to further improve computational efficiency [19,20]. Stokesian dynamics and FLD both explicitly include the close-range two-body lubrication interactions, $\mathcal{R}_{2B}$, using the lubrication equations developed by Kim and Karrila [21], Jeffrey and Onishi [24] and given succinctly and used in the form developed in Ball and Melrose [25].

The resistance matrix used in FLD is described by:

$$\mathcal{R}_{FLD} = \mathcal{R}_{Iso} + \mathcal{R}_{2B} - \mathcal{R}_{2B}^\infty \tag{3}$$

Since the construction of the mobility matrix and inversion (in Stokesian Dynamics) or the construction of the isotropic resistance matrix approximation include all the pair interactions, the two-body close-range interactions, $\mathcal{R}_{2B}^\infty$, need to be subtracted to not include them twice. Though, part of the $(\mathcal{R})_{FLD}$ is an approximation, the dominant forces are the lubrication pairwise forces.

Ball and Melrose [25] showed that accurate simulations of spherical particles can be performed with only the close-range lubrication interactions and achieve results comparable to more accurate methods for volume fractions of particles of 0.20 and above.

2.4. Bubble Interaction Modeling

Our approach in simulating bubbles is to consider three properties of bubbles: (1) slip surface conditions (F^{Bubble}); (2) the ability of the surface to deflect locally in the interacting region ($F^{Elastic}$); and (3) the repulsive and attractive quality of the surface due to a surfactant (F^{Direct}). The total of these three forces serve to include the predominant properties of realistic bubbles. A direct force that represents the surfactant also serves to prevent nonphysical overlaps of the bubbles during the simulations. A depiction of the three forces is shown in Figure 1. The total hydrodynamic force for the bubbles is then defined by [16]:

$$F^{Total} = F^{Bubble} + F^{Elastic} + F^{Direct} \tag{4}$$

Real bubbles have a mobile interface. Theoretically, the slip on the bubble surface can be considered very high with no shearing resistance. If there is a high amount of slip on the bubble surface then there is no shear traction at the surface boundary condition. Under these conditions, in a pair interaction where the bubbles interact with a relative shearing motion, the bubbles will just slip by each other [21,26]. It can then be assumed that the bubbles have no shearing resistance so the shearing resistance terms all become zero. To simulate surfactant stabilized bubbles though, we need to have a mobile surface boundary and an additional realistic normal direct force to prevent bubble overlap as the surfactant does in real bubble suspensions. The force on approaching bubbles is weaker and will allow bubbles to touch in finite time so we must impose a direct force to avoid unreal overlaps. The direct force was chosen so that it enabled the bubble pairs to approach very close to each other to allow a small surface deflection but did not allow complete bubble overlap.

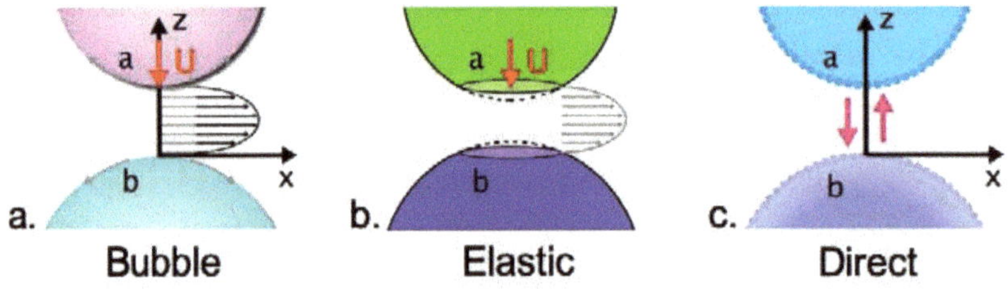

Figure 1. The bubble interaction is a combination of the Bubble Force, F^{Bubble}, the Elastic Force, $F^{Elastic}$, and the Direct Force F^{Direct}.

The hydrodynamic interaction for a pair of bubbles, with the same size diameter, has been derived by [21] for the case of two bubbles approaching each other and is given by [16,21]:

$$F^{Bubble} = a_{sq}^{Bubble} U \tag{5}$$

where a_{sq}^{Bubble} is [21]:

$$a_{sq}^{Bubble} = \pi \mu_f a \left(2 \log \frac{a}{h} + 4 \left(\gamma_{Euler} + \log 2 \right) \right) \tag{6}$$

μ_f is the viscosity of the suspending fluid, a is the radius of the bubble, and h is the gap between the bubbles. γ_{Euler} is Euler's constant (γ_{Euler} = 0.577216...).

The elastic force used in the bubble interaction is taken directly from [27]:

$$F^{Elastic} = a_{Elas} E \delta \tag{7}$$

$$a_{Elas} = \sqrt{a(2h + |\delta|)} \tag{8}$$

$$\delta = \frac{(2a + h - r)}{2} \tag{9}$$

In the region of interaction between the bubble pair, the surface is considered to become flattened. a_{Elas} is the region of the bubble surface that is interacting in the lubrication force after deflection has caused the surface to become flattened. δ is the deflection of the bubble surface. Once the bubbles move away from each other, the deflection of the local surface region is zero and the spherical shape of the bubble is restored. To limit the amount that the bubble surface can deflect, a maximum deflection value was set. E represents an *effective* modulus to deform the interface and bubble and was set to 10^2 for the results shown here. The value of E could be adjusted to match experimental data.

The direct force that was chosen and used for the simulation results reported here is the Lennard-Jones potential force. Surfactants can be attractive and repulsive in nature [28], which is why the Lennard-Jones force was chosen as the direct force. The surfactant forces may include van der Waals and hydrostatic attraction, electrostatic repulsion, steric repulsion, etc. [28], however, it is generally not possible to obtain the specific details about the surfactant due to its proprietary properties. So, no specific surfactant was considered. Using a generalized repulsive force that increases steeply when the particles come in close relative positions to each other has been done in Soft Dynamics [27,29] and the direct force used here is deployed in a similar manor. The Lennard-Jones potential included with LAMMPS [30] was used as the direct force. The Lennard-Jones force in LAMMPS is:

$$F^{Direct} = \frac{4\epsilon}{\sigma} \left[9 \left(\frac{\sigma}{r} \right)^{10} - 6 \left(\frac{\sigma}{r} \right)^{7} \right] \tag{10}$$

This represents a Lennard-Jones potential that is less repulsive than the standard Lennard-Jones potential (The standard Lennard-Jones potential is $\phi_{LJ} = 4\epsilon \left[\left(\frac{\sigma}{r}\right)^{12} - \left(\frac{\sigma}{r}\right)^{6} \right]$). ϵ is the energy scale and σ is the length scale. The influence of the values of ϵ and σ on system dynamics are described in detail in [16], so we do not present them again here. A value of 0.008 was used for ϵ and a value of 0.9 was used for σ. $\left(\frac{\sigma}{r}\right)^{10}$ is the repulsive part and $\left(\frac{\sigma}{r}\right)^{7}$ is the attractive portion.

To simulate bubbles of different sizes, a reduced radius [21,31] was used. If the radius of particle a is a and the radius of particle b is b, then the reduced radius is:

$$a_{reduced} = \frac{a \times b}{a + b} \tag{11}$$

To simulate polydisperse bubbles, $a_{reduced}$ is used in the equations instead of a.

2.5. Simulation Framework

LAMMPS (lammps.sandia.gov Open-source classical (non-quantum) molecular dynamics code developed and maintained at the Sandia National Labs.) (Large-scale Atomic/Molecular Massively Parallel Simulator) was modified with the interactions described in this paper and was used to run the simulations.

2.6. Simulation Inputs

Simulation inputs (i.e., the bubble size and each bubble's $x-, y-, z-$position) were created by randomly placing bubbles with a diameter of 1 in a simulation box of size $10 \times 10 \times 10$. Volume fractions of 10%, 20%, 30%, 40%, 45%, and 50% were created with increasing numbers of bubbles in the simulation box. To remove any initial overlapping bubbles, a soft potential was used to push bubbles apart and remove overlaps. Through a series of test simulations, it was determined that replicating the simulation box four times in each direction eliminated system size effects on the relative viscosity. However, due to replicating the same simulation box, the bubble positions were repeated in each box. To add back the random placement of the bubbles, a Brownian pair interaction was used to randomly move the bubbles a small amount and re-introduce the random nature back into the initial bubble positions. These steps were not part of the dynamics of the simulations but served to create the initial inputs to the simulations.

To create polydisperse bubble systems, the monodisperse bubble positions were used and the bubble size was increased and decreased randomly on the monodisperse bubbles to add slight polydispersity to the distribution of bubble sizes. The system of polydisperse bubbles were then replicated and the positions randomized as described for monodisperse bubbles.

2.7. Implementation of the Shearing Flow

The shearing flow was implemented as described in [13,16,32]. Lees-Edwards boundary conditions were imposed (See the LAMMPS User Manual [30] for the implementation of the Lees-Edwards like boundary conditions, which uses a function called "fix deform" to apply a strain rate to the simulation box in the specified orthogonal box directions.). The strain was imposed on the simulation box in the xy-direction so that the velocity of each bubble becomes a function of its position in the y-direction. The periodic boundary conditions were imposed such that when a bubble crossed the simulation box boundary in any direction, the velocity of the bubble was remapped to correspond to the new position in the simulation box (See e.g., Dayal and James [33] for a discussion of this.). The simulations were all sheared until the stress reached a constant value and the value of $\dot{\gamma} \times t_{total}$ reached 200 so that the simulation results could all be compared. The length of time that the simulation was run is t_{total} and $\dot{\gamma}$ is the strain rate.

An explicit time-integration scheme was used for these interactions and therefore required a small time step. Testing was performed to optimize the dimensionless time step, Δt. It was determined that the time step should be kept below a value of 0.002 for convergence. For the results contained herein, a time step of $\Delta t = 0.001$ was used. Smaller time steps would require the simulation to run for a higher total time to reach the same value of total strain.

2.8. Calculation of Viscosity Ratio of Simulated Suspensions

The relative viscosity can be calculated from the summation of the stress on each bubble. The stress on each bubble is a function of the force, $\boldsymbol{F}$, on each bubble due to its N_p neighbors. The stress is calculated in LAMMPS at each time step and is given by [30,34]:

$$\sigma_{i,\alpha\theta} = \frac{1}{V_i}[-\sum_{n=1}^{N_p}(r_{1_\alpha}F_{1_\theta} + r_{2_\alpha}F_{2_\theta})] \tag{12}$$

α and θ run over the coordinate directions to compute the 6 components of the symmetric stress tensor. $\boldsymbol{r}_1$ and $\boldsymbol{r}_2$ are the positions of bubble 1 and bubble 2, respectively in the pair that has pairwise interactions. And, $\boldsymbol{V}_i$ is the volume of each bubble. In the typical calculation, the value does not consider the bubble volume so to properly calculate the relative viscosity, the total volume, $\boldsymbol{V}$, of the simulation box must be included in the relative viscosity calculation. The relative viscosity or stated another way, the viscosity ratio (i.e., μ over the viscosity of the suspending fluid, μ_f), can be calculated from the average of the total stress once the system has reached steady state:

$$\mu_{\text{relative}} = \frac{\mu}{\mu_f} = \frac{\sum_i \sigma_{i,xy}}{\dot{\gamma}\mu_f V} \tag{13}$$

3. Results

3.1. *Foamed Cement Viscosity from Prior Experimental Results*

3.1.1. Cement Rheology

Cement slurry rheology is an important parameter to understand for designing and implementing a cement job [35,36]. Cement slurries are a suspension of small particles undergoing chemical reactions and are known to have Non-Newtonian rheology, generally assumed to follow a Bingham model [35,37]. Bingham fluids have a yield stress below which no flow occurs and after the yield stress is attained, the fluid maintains a constant viscosity. Cement slurry rheology, however, can be time dependent and shear dependent and depend on shearing history and resting time [35,37]. Other models that have been used to describe cement slurry rheology are the power law model, with limitations, and the Herschel-Bulkley model [35]. Both of these models account for the shear thinning behavior of cement and the Herschel-Bulkley model includes a yield stress. The time dependent behavior of rheology—thixotropy: time-dependent shear thinning or rheoplexy: time-dependent shear thickening—is considered to be reversible. With cement, however, the hydration process will also change the rheology as time progresses [35,37] so it is not a truly reversible process. Cement slurry rheology is also dependent on shearing and resting history. Particles within the suspension can cluster together due to random movements under static or no flow conditions [35,37,38]. During shearing, the particle clusters can be broken apart and this may be the mechanism of shear thinning that occurs with time [35,37,38]. At low shear rates, cement slurries have been shown to be thixotropic, which means that as they are sheared, the viscosity decreases with time, though thickening can occur due to the hydration reaction. Foamed cement is a complex system and inherently dynamic, which makes it a challenge to study rheology with reliable results. Cement rheology is a function of several factors including the cement slurry design and the cement slurry production methods, and can be affected by the measurement techniques themselves [35].

3.1.2. Rheology of Foamed Cement

Foamed cement rheology is further complicated by the addition of bubbles suspended in the cement. The changing downhole conditions in the well as the cement is pumped can affect the rheology as the quality changes. Few studies have actually been done to explore the rheological properties of foamed cement, even at ambient conditions. In the book Well Cementing [10], Chapter 14: Foamed Cement [39] a comment is made about the lack of rheological information and the importance of the data. At the time only one study had been referenced [40] and only two other studies [41,42] have been done since the publishing of that book. Some of the conclusions of the studies contradict each other. The earliest study [40] uses two different capillary flow tubes (0.091 inch I.D. × 8 ft long, 0.130 inch I.D. × 8 ft long) to measure the rheology. A description of the cement, even the type used, was not included in this thesis. The cement mixer is not described fully and these specifics can influence the rheology. The cement was mixed with a "cement mixer" and injected into the system with a "variable flow rate, positive displacement plunger pump", rated to 3000 psi. The foaming agent was injected into the base slurry with a "variable flow rate positive displacement gear pump" and mixed into the cement as the slurry flowed through a small tube coil. After air injection, the cement slurry was foamed by passing it through three stainless steel foam generator tubes (each 12 inch long × 0.33 inch I.D.) that were filled with stainless steel shavings to cause mixing and foaming.

Results span foamed cement qualities from 30% up to 65% and shear rate values from 1000 1/s to 10,000 1/s [40]. It was shown that the viscosity increased with increasing quality for a given shear rate and the viscosity decreased with increasing shear rate. At high shear rates (7000 1/s and higher) and qualities lower than 50%, the viscosity seems to become constant, however, typical shear rates during placement are below 1000 1/s [39]. The viscosity as a function of quality varied between the two diameter flow tubes, with the larger tube having higher reported values. They conclude that the Power

Law model or the Bingham plastic model describe the data. While a first of its kind study, pertinent details were left out of this thesis and the shear rates explored are well above those encountered during a cementing job.

A later study [41] uses an inline flow-through rotational viscometer to measure the rheology. The base slurry consisted of Class G cement, water, and a few additives. The base slurry was made by first mixing the water, dispersant, and a fluid loss additive at 200 rpm for about 1 min. Then the cement was added while mixing at 200 rpm for 50 s. The cement slurry (without foamer) was agitated at 1500 rpm for 5 min. Rheology measurements were taken using a regular rotational bob and sleeve viscometer. The original base slurry was removed from the viscometer, placed back in the original mixing container, foamer was added and mixed gently without foaming it. To foam the cement, the cement mixture with foamer was injected into a mixing chamber where pressurized nitrogen was also injected. The cement, additives, and foamer were agitated at 1600 rpm for 5 min with the mixer. To measure viscosity, the foamed cement flowed through the inline viscometer at a constant rate and measured at various shear rates.

The study of [41] is the first and only known attempt to assess the influence of pressure on the rheology. They also are the first to report viscosity values within relevant shear rates (1000 1/s or less). In conclusion, they find that a Power Law model provides the best fit to the data but at lower shear rates. However, the cement slurry has a yield stress, which is not captured with a Power Law model. They also show that the viscosity increases with quality for 10%, 20%, and 30% foam qualities. The discrepancy with previous reports/experience comes in their conclusion that the base slurry viscosity is higher than both the 10% and 20% foam qualities measured and, rather, is close to that of the 30% foam quality and is almost the same as the 30% foam quality for shear rates of 50 1/s and above. They question the claims that foamed cement displaces drilling muds better than neat cement as experience from industry suggests. It is unclear if these results are real or due to experimental issues. It is also possible that in measuring the foam viscosity, the foam was destabilized and the foamer reduced the slurry viscosity as compared to the 'base' slurry, which had no foamer. The previous study from 1977 does not report data for qualities below 30% or for the neat cement.

The most recent study, [42], uses a uniquely shaped paddle rather than a bob and sleeve viscometer to measure the viscosity and yield point. The paddle devise, a Fann Yield Stress Adapter (FYSA), was used to minimize wall slippage that can be an issue with bob and sleeve viscometers [42]. The FYSA is designed to minimize wall slippage, measures higher volumes of material, and keeps the bubbles and cement particles better dispersed. The base slurry was prepared with Class H cement and water. After initial mixing at high speed in a blender, the foaming agent was added and stirred by hand without producing any foaming and then mixed in a blender to foam as recommended according to API Recommended Practices [43]. The viscosity measurements were then conducted in the viscometer unlike the former recorded measurements, which were done with flow through methods. The results can be considered the most accurate viscosity measurements of foamed cement due to the advantages and reduction in errors due to the design of the FYSA. Measurements were reported for a range of volume fractions of gas: 10%, 20%, 30%, and 40%.

It is challenging to measure the rheology of a foamed cement as the methods used could affect the rheology [39]. Foamed cements used in well cementing are pressurized and to replicate pressurized foamed cement in the laboratory requires specialized equipment to maintain the foam under pressure and perform the measurements under pressure.

3.2. Viscosity Ratio of Suspensions

An important result of adding bubbles to a fluid is to increase the effective viscosity. In this section, we examine the effect of the bubble volume fraction in the suspending fluid on the viscosity, calculated from the simulated results. We provide equations to describe the viscosity ratio as a function of the bubble volume fraction. The equations represent a valuable contribution to determine the influence of the bubbles on the viscosity compared to the base fluid alone.

The definition of the viscosity ratio for the simulated results is given by Equation (13). This equation was used to compute the viscosity ratio determined from each simulation that was run. The values of the computed viscosity ratio can then be fitted to equations and the coefficients determined to define the relationship between bubble volume fraction and the subsequent increase in viscosity.

The relationship between the volume fraction (ϕ) of objects in a Newtonian fluid (with a viscosity of μ_f) and the corresponding change in viscosity can be described by several equations with varying levels of complexity. The effects of object concentration on the suspension viscosity is summarized in [44]. As the bubbles, in the foamed cement, act like particles, we examine typical equations used to describe suspensions.

Einstein's classical equation accounts for the increase in viscosity due to the object volume fraction and is linear in form. The Einstein equation is therefore valid for low object volume fractions. The form of the Einstein Equation [45] is:

$$\mu = \mu_f(k + l\phi) \tag{14}$$

Krieger and Dougherty account for the limits of the amount of objects that can be in the suspension with the maximum packing fraction (ϕ_{max}). The form of the Krieger and Dougherty Equation [46] is:

$$\mu = \mu_f \left(m - \frac{\phi}{\phi_{max}} \right)^{-n\phi_{max}} \tag{15}$$

Batchelor and Green's equation is a second order polynomial and fits the simulation and experimental data the best of all the model equations. The form of the Batchelor and Green Equation [47] is:

$$\mu = \mu_f(o + p\phi + q\phi^2) \tag{16}$$

The coefficients for the above equations for both particles and bubbles are summarized in Table 1.

Taylor's equation allows for fluid spheres of a different fluid from the base suspending fluid. However, the Taylor Equation did not fit the simulation results well and so the coefficients are not given. Here, for bubbles, the fluid is air and the viscosity (μ_{bubble}) is much lower than the suspending fluid. The form of Taylor Equation [48] is:

$$\mu = \mu_f \left[kk + ll\phi \left(\frac{\mu_{bubble} + mm\mu_f}{\mu_{bubble} + \mu_f} \right) \right] \tag{17}$$

Table 1. Coefficients for fits of Equations (14)–(17) for the simulated results with a 95% confidence interval in the fit. The equation fit was determined with Matlab.

Coefficient	Monodisperse Particle	Polydisperse Particle	Monodisperse Bubble	Polydisperse Bubble
k	7.463×10^{-7}	3.578×10^{-6}	2.479×10^{-8}	0.4661
l	0.1605	0.1609	0.1665	0.09995
m	1.057	1.014	1.048	0.9535
n	0.02799	0.02725	0.0281	0.0198
o	0.9777	0.9	0.943	1.006
p	0.007433	1.777×10^{-10}	3.793×10^{-14}	0.01958
q	0.003042	0.003415	0.003393	0.001581

The fit of Equations (14)–(16) to the simulated data results are shown in Figures 2 and 3. It is obvious that the second order polynomials, including the Batchelor form of the viscosity equation, fit the data the best. Figure 2 shows the equation fits to the simulated data of the monodisperse spherical rigid particles and monodisperse bubbles. Figure 3 shows the equation fits to the simulated data of the polydisperse spherical rigid particles and polydisperse bubbles. In both figures, the Einstein equation is only fit through the point $[0, 1]$ and the first two data points, since it is valid at low volume fractions.

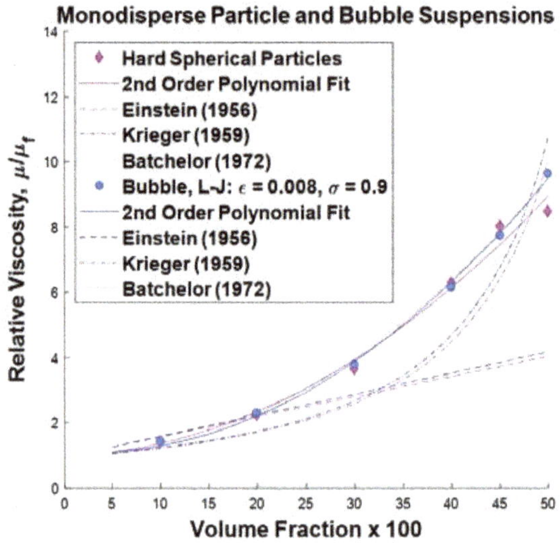

Figure 2. The relative viscosity determined from the simulated results is indicated by points and the lines are the equations fit to the data.

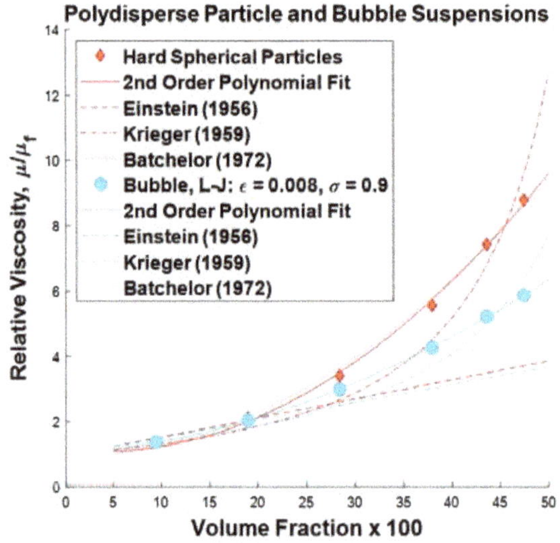

Figure 3. The relative viscosity determined from the simulated results is indicated by points and the lines are the equations fit to the data.

4. Discussion

When comparing the data between the particle suspensions and bubble suspensions, for the case of the same monodisperse particles and bubbles with different surface conditions, the results on the effects to the viscosity are not significant. However, when comparing polydisperse suspensions, which are more realistic, the bubbles do not increase the relative viscosity as much as particles. The viscosity is a function of the stress on each particle or bubble and since the particles are rigid and have no slip surfaces, the overall system stress can increase more at higher volume fractions. The bubbles are able to slide by each other and therefore the overall stress does not increase as much.

The simulated relative viscosity data can be compared with some of the known experimental data (Figure 4). The most comprehensive collection of the viscosity of foamed cement is contained in Al-Mashat's thesis [40]. However, Al-Mashat's method for measuring the viscosity is dependent on the size of the pipe used in the setup. So it can be expected that the measurements would be different with another size of pipe. The data does show the increase in the viscosity due to the addition of bubbles. Due to the discrepancy of the results of Ahmed [41] compared to the other experimental results, they are not displayed here along with the simulated results. Olowolagba [42] reports more accurate experimental foamed cement viscosity results and are used to verify the simulated results. The line fitted through the experimental data of Olowolagba [42] is a second order polynomial fit through the data at the lowest shear rate. The simulated data of bubbles suspended in a fluid match the most accurate experimental measurements [42].

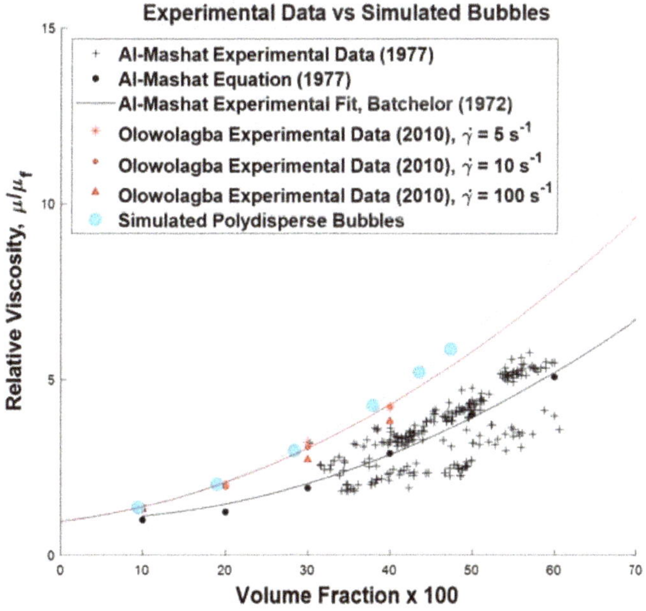

Figure 4. The relative viscosity determined from the simulated results is shown with the experimental data of Al-Mashat [40] and Olowolagba [42].

5. Conclusions

We have shown that simulated bubble suspensions match with experimental measurements. To determine what effects bubbles will have on a suspension viscosity, the equations with the coefficients determined here can be used to predict the effects of the addition of particles and bubbles to the suspension fluid. Experiment setups can influence or change the properties trying to be measured and require specific and careful consideration and relevant simulations can take significant computing resources. However, the relationships shown herein can be used to predict the influence of bubbles and particles on the viscosity.

Author Contributions: The conceptualization, all authors; methodology, all authors; simulations, E.R.; validation, all authors; formal analysis, E.R.; investigation, all authors; resources, all authors; writing—original draft preparation, all authors; writing—review and editing, all authors; visualization, E.R.; supervision, M.M, K.D.; project administration, E.R.; funding acquisition, E.R., K.D.

Funding: This project was funded by the Department of Energy, National Energy Technology Laboratory an agency of the United States Government. Neither the United States Government nor any agency thereof, nor any of their employees, makes any warranty, express or implied, or assumes any legal liability or responsibility for the accuracy, completeness, or usefulness of any information, apparatus, product, or process disclosed, or represents that its use would not infringe privately owned rights. Reference herein to any specific commercial product, process, or service by trade name, trademark, manufacturer, or otherwise does not necessarily constitute or imply its endorsement, recommendation, or favoring by the United States Government or any agency thereof. The views and opinions of authors expressed herein do not necessarily state or reflect those of the United States Government or any agency thereof; Computing resources were provided by the National Science Foundation through the Extreme Science and Engineering Discovery Environment (XSEDE) program; specifically the XSEDE Bridges Computing Resources at the Pittsburgh Supercomputing Center [grant number TG-DMR160018]; Kaushik Dayal was supported through an appointment to the National Energy Technology Laboratory Faculty Research Participation Program sponsored by the U.S. Department of Energy and administered by the Oak Ridge Institute for Science and Education.

Acknowledgments: Kaushik Dayal acknowledges an appointment to the National Energy Technology Laboratory Faculty Research Participation Program sponsored by the U.S. Department of Energy and administered by the Oak Ridge Institute for Science and Education; Eilis Rosenbaum acknowledges the award of the Bradford and Diane Smith Graduate Fellowship that was created through the generosity of Bradford Smith, a 1969 graduate of the Carnegie Mellon University (CMU) College of Engineering, and his wife Diane, a 1971 CMU graduate of the Dietrich College of Humanities and Social Sciences to help support the graduate studies of highly deserving students in the College of Engineering at CMU; This paper draws from the Ph.D. Thesis of Eilis Rosenbaum at Carnegie Mellon University [32].

Conflicts of Interest: The authors declare no conflict of interest.

Abbreviations

The following abbreviations are used in this manuscript:

MDPI	Multidisciplinary Digital Publishing Institute
FLD	Fast Lubrication Dynamics
LAMMPS	Large-scale Atomic/Molecular Massively Parallel Simulator
FYSA	Fann Yield Stress Adapter
API	American Petroleum Institute

References

1. Frisch, G.J.; Services, H.E.; Graham, W.L. SPE 55649 Assessment of Foamed—Cement Slurries Using Conventional Cement Evaluation Logs and Improved Interpretation Methods. In Proceedings of the SPE Rocky Mountain Regional Meeting, Gillette, Wyoming, 15–18 May 1999.
2. Rae, P. Cement Job Design. In *Well Cementing;* Nelson, E.B., Ed.; Schlumberger Educational Services: Sugar Land, TX, USA, 1990; Chapter 11, pp. 11.1–11.17. doi:10.1016/S0376-7361(09)70309-7.
3. American Petroleum Institute. *Isolating Potential Flow Zones During Well Construction—API Standard 65 Part 2;* American Petroleum Institute: Washington, DC, USA, 2010.
4. Benge, O.G.; McDermott, J.R.; Langlinais, J.C.; Griffith, J.E. Foamed cement job successful in deep HTHP offshore well. *Oil Gas J.* **1996**, *94*, 58–63.
5. Bour, D.; Rickard, B. Application of Foamed Cement on Hawaiian Geothermal Well. *Geotherm. Resour. Counc. Trans.* **2000**, *24*, 55–60.
6. O'Rourke, T.J.N.S.C.; Crombie, D.N.S.C. A Unique Solution to Zonal Isolation Utilizing Foam-Cement and Coiled-Tubing Technologies. In Proceedings of the SPE/ICoTA Coiled Tubing Roundtable, Houston, TX, USA, 25–26 May 1999.
7. White, J.; Moore, S.; Miller, M.; Faul, R.; Services, H.E. IADC/SPE 59136 Foaming Cement as a Deterrent to Compaction Damage in Deepwater Production. In Proceedings of the 2000 IADC/SPE Drilling Conference, New Orleans, LA, USA, 23–25 February 2000. doi:10.2523/59136-MS.
8. Dusterhoft, D.M. Foamed & Lightweight Cements. In Proceedings of the Canadian International Petroleum Conference, Calgary, AB, Canada, 10–12 June 2003.

9. Economides, M.J. 1. Implications of Cementing on Well Performance. In *Well Cementing*; Nelson, E.B., Ed.; Schlumberger Educational Services: Sugar Land, TX, USA, 1990; Chapter 1, pp. 1.1–1.6. [CrossRef]
10. Nelson, E.B. *Well Cementing*; Elsevier: Oxford, UK, 1990; Volume 28.
11. Loeffler, N. Foamed Cement: A Second Generation. In Proceedings of the Permian Basin Oil and Gas Recovery Conference, Midland, TX, USA, 8–9 March 1984. doi:10.2523/12592-MS.
12. Griffith, J.E.; Lende, G.; Ravi, K.; Saasen, A.; Nø dland, N.E.; Jordal, O.H. Foam Cement Engineering and Implementation for Cement Sheath Integrity at High Temperature and High Pressure. In Proceedings of the IADC/SPE Drilling Conference, Dallas, TX, USA, 2–4 March 2004.
13. Rosenbaum, E.; Massoudi, M.; Dayal, K. Effects of Polydispersity on Structuring and Rheology in Flowing Suspensions. *J. Appl. Mech.* **2019**, *86*, 081001. doi:10.1115/1.4043094. [CrossRef]
14. Kutchko, B.; Crandall, D.; Gill, M.; McIntyre, D.; Spaulding, R.; Strazisar, B.; Rosenbaum, E.; Haljasmaa, I.; Benge, G.; Cunningham, E.; et al. *Computed Tomography and Statistical Analysis of Bubble Size Distributions in Atmospheric-Generated Foamed Cement*; Technical Report August; U.S. Department of Energy: Pittsburgh, PA, USA, 2013.
15. Dalton, L.E.; Brown, S.; Moore, J.; Crandall, D.; Gill, M. *Evolution Using CT Scanning: Insights from Elevated-Pressure Generation*; SPE: Richardson, TX, USA, 2018; pp. 1–11.
16. Rosenbaum, E.; Massoudi, M.; Dayal, K. Surfactant Stabilized Bubbles Flowing in a Newtonian Fluid. *Math. Mech. Solids* **2019**. [CrossRef]
17. Kutchko, B.; Crandall, D.; Moore, J.; Gill, M.; McIntyre, D.; Rosenbaum, E.; Haljasmaa, I.; Strazisar, B.; Spaulding, R.; Harbert, W.; et al. *Field-Generated Foamed Cement: Initial Collection, Computed Tomography, and Analysis*; Technical Report July; U.S. Department of Energy, National Energy Technology Laboratory: Pittsburgh, PA, USA, 2015.
18. Brady, J. Stokesian Dynamics. *Annu. Rev. Fluid Mech.* **1988**, *20*, 111–157. doi:10.1146/annurev.fluid.20.1.111. [CrossRef]
19. Bybee, M.D. Hydrodynamic Simulations of Colloidal Gels: Microstructure, Dynamics, and Rheology. Ph.D. Thesis, University of Illinois, Champaign County, IL, USA, 2009.
20. Kumar, A. Microscale Dynamics in Suspensions of Non-Spherical Particles. Ph.D. Thesis, University of Illinois, Champaign County, IL, USA, 2010.
21. Kim, S.; Karrila, S.J. *Microhydrodynamics: Principles and Selected Applications*, 2nd ed.; Dover Publications, Inc.: Mineola, NY, USA, 2005.
22. Kumar, A.; Higdon, J.J.L. Origins of the anomalous stress behavior in charged colloidal suspensions under shear. *Phys. Rev. Stat. Nonlinear Soft Matter Phys.* **2010**, *82*, 1–7. doi:10.1103/PhysRevE.82.051401. [CrossRef]
23. Bossis, G.; Brady, J.F. Dynamic simulations of sheared suspensions. {I.} General method. *J. Chem. Phys.* **1984**, *80*, 5141–5154. [CrossRef]
24. Jeffrey, D.J.; Onishi, Y. Calculation of the resistance and mobility functions for two unequal rigid spheres in low-Reynolds-number flow. *J. Fluid Mech.* **1984**, *139*, 261. doi:10.1017/S0022112084000355. [CrossRef]
25. Ball, R.; Melrose, J.R. A simulation technique for many spheres in quasi-static motion under frame-invariant pair drag and Brownian forces. *Phys. Stat. Mech. Appl.* **1997**, *247*, 444–472. [CrossRef]
26. Bechinger, C.; Sciortino, F.; Ziherl, P. *Physics of Complex Colloids*; IOS Press: Washington, DC, USA, 2013.
27. Rognon, P.G.; Einav, I.; Gay, C. Internal relaxation time in immersed particulate materials. *Phys. Rev. Stat. Nonlinear Soft Matter Phys.* **2010**, *81*, 1–9, doi:10.1103/PhysRevE.81.061304. [CrossRef]
28. Tabakova, S.S.; Danov, K.D. Effect of disjoining pressure on the drainage and relaxation dynamics of liquid films with mobile interfaces. *J. Colloid Interface Sci.* **2009**, *336*, 273–284. doi:10.1016/j.jcis.2009.03.084. [CrossRef] [PubMed]
29. Rognon, P.G.; Einav, I.; Gay, C. Flowing resistance and dilatancy of dense suspensions: lubrication and repulsion. *J. Fluid Mech.* **2011**, *689*, 75–96. doi:10.1017/jfm.2011.397. [CrossRef]
30. *LAMMPS Users Manual*; Sandia National Laboratories: Washington, DC, USA, 2003.
31. Davis, R.H.; Schonberg, J.; Rallison, J.M. The lubrication force between two viscous drops. *Phys. Fluids Fluid Dyn.* **1989**, *77*. doi:10.1063/1.857525. [CrossRef]
32. Rosenbaum, E. A Computational Study of Bubble Suspensions and Foamed Cement using Extended Stokesian Dynamics. Ph.D. Thesis, Carnegie Mellon University, Pittsburgh, PA, USA, 2019. [CrossRef]

33. Dayal, K.; James, R.D. Nonequilibrium molecular dynamics for bulk materials and nanostructures. *J. Mech. Phys. Solids* **2010**, *58*, 145–163. [CrossRef]
34. Tadmor, E.B.; Miller, R.E. *Modeling Materials: Continuum, Atomistic and Multiscale Techniques*; Cambridge University Press: Cambridge, UK, 2011.
35. Guillot, D. Rheology of Well Cement Slurries. *Dev. Pet. Sci.* **1990**, *28*, 4-1–4-37. [CrossRef]
36. Tao, C.; Kutchko, B.G.; Rosenbaum, E.; Wu, W.T.; Massoudi, M. Steady Flow of a Cement Slurry. *Energies* **2019**, *12*, 2604. doi:10.3390/en12132604. [CrossRef]
37. Roussel, N. Steady and transient flow behaviour of fresh cement pastes. *Cem. Concr. Res.* **2005**, *35*, 1656–1664. doi:10.1016/j.cemconres.2004.08.001. [CrossRef]
38. Banfill, P.F.G. Rheology of Fresh Cement and Concrete. *Rheol. Rev.* **2006**, *2006*, 61–130. [CrossRef]
39. de Rozieres, J.; Griffin, T.J. Chapter 14 Foamed Cements. In *Well Cementing*; Nelson, E.B., Ed.; 300 Schlumberger Drive; Schlumberger Educational Services: Sugar Land, TX, USA, 1990; pp. 14.1–14.19.
40. Al-Mashat, A.M. Rheology of Foam Cement. Ph.D. Thesis, Colorado School of Mines, Golden, CO, USA, 1977.
41. Ahmed, R.M.; Takach, N.E.; Khan, U.M.; Taoutaou, S.; James, S.; Saasen, A.; Godø y, R. Rheology of foamed cement. *Cem. Concr. Res.* **2009**, *39*, 353–361. doi:10.1016/j.cemconres.2008.12.004. [CrossRef]
42. Olowolagba, K.O.; Brenneis, C. Techniques for the study of foamed cement rheology. In Proceedings of the SPE Production and Operations Conference and Exhibition. Society of Petroleum Engineers, Tunis, Tunisia, 8–10 June 2010.
43. API 10B-4. *Recommended Practice on Preparation and Testing of Foamed Cement Slurries at Atmospheric Pressure ANSI / API Recommended Practice 10B-4*; Technical Report July; ISO: London, UK, 2004.
44. Massoudi, M.; Wang, P. Slag behavior in gasifiers. Part II: Constitutive modeling of slag. *Energies* **2013**, *6*, 807–838. [CrossRef]
45. Einstein, A. *Investigations on the Theory of the Brownian Movement*; Dover Publications: New York, NY, USA, 1956; Volume 58.
46. Krieger, I.M.; Dougherty, T.J. A mechanism for non-Newtonian flow in suspensions of rigid spheres. *Trans. Soc. Rheol.* **1959**, *3*, 137–152. [CrossRef]
47. Batchelor, G.; Green, J. The determination of the bulk stress in a suspension of spherical particles to order c 2. *J. Fluid Mech.* **1972**, *56*, 401–427. [CrossRef]
48. Taylor, G.I. The viscosity of a fluid containing small drops of another fluid. *Math. Phys.* **1932**, *138*, 41–48. [CrossRef]

fluids

MDPI

Article

Slug Translational Velocity for Highly Viscous Oil and Gas Flows in Horizontal Pipes

Yahaya D. Baba [1,*], Archibong Archibong-Eso [2], Aliyu M. Aliyu [3], Olawale T. Fajemidupe [4], Joseph X. F. Ribeiro [5], Liyun Lao [6] and Hoi Yeung [6]

[1] Department of Chemical and Biological Engineering, University of Sheffield, Sheffield S1 3JD, UK
[2] Department of Mechanical Engineering, University of Birmingham, Dubai International Academic City, P.O. Box 341799 Dubai, UAE; archibong.eso@gmail.com
[3] Faculty of Engineering, University of Nottingham, Nottingham NG7 2RD, UK; Ali.Aliyu@nottingham.ac.uk
[4] Mewbourne School of Petroleum and Geological Engineering, University of Oklahoma, OK 73019, USA; tayefajem2000@gmail.com
[5] Department of Mechanical Engineering, Kumasi Technical University, P.O. Box 854, Kumasi, Ghana; joxaro@gmail.com
[6] Oil and Gas Engineering Centre, Cranfield University, Bedfordshire MK43 0AL, UK; l.lao@cranfield.ac.uk (L.L.); h.yeung@cranfield.ac.uk (H.Y.)
* Correspondence: y.baba@sheffield.ac.uk or y.baba9550@gmail.com

Received: 27 June 2019; Accepted: 27 August 2019; Published: 12 September 2019

Abstract: Slug translational velocity, described as the velocity of slug units, is the summation of the maximum mixture velocity in the slug body and the drift velocity. Existing prediction models in literature were developed based on observation from low viscosity liquids, neglecting the effects of fluid properties (i.e., viscosity). However, slug translational velocity is expected to be affected by the fluid viscosity. Here, we investigate the influence of high liquid viscosity on slug translational velocity in a horizontal pipeline of 76.2-mm internal diameter. Air and mineral oil with viscosities within the range of 1.0–5.5 Pa·s were used in this investigation. Measurement was by means of a pair of gamma densitometer with fast sampling frequencies (up to 250 Hz). The results obtained show that slug translational velocity increases with increase in liquid viscosity. Existing slug translational velocity prediction models in literature were assessed based on the present high viscosity data for which statistical analysis revealed discrepancies. In view of this, a new empirical correlation for the calculation of slug translational velocity in highly viscous two-phase flow is proposed. A comparison study and validation of the new correlation showed an improved prediction performance.

Keywords: Gamma densitometer; high viscosity oil; slug translational velocity; closure relationship

1. Introduction

Recently, high viscosity oils (i.e., unconventional oil resources) have been acknowledged as one of the most important future energy sources. This is credited to the increasing world energy demand amidst depletion of lighter of hydrocarbon resources (i.e., conventional oil resources). Recent investigations, as illustrated in Figure 1, have shown that unconventional oil resources constitute the largest available world oil reserves. A good understanding and accurate prediction of high-viscous multiphase flow has become imperative since the behaviour of high viscous liquids in two-phase flowing conditions differs significantly from those of low-viscous liquids. This is needed to ensure efficient production and transportation of high-viscosity oils in pipelines.

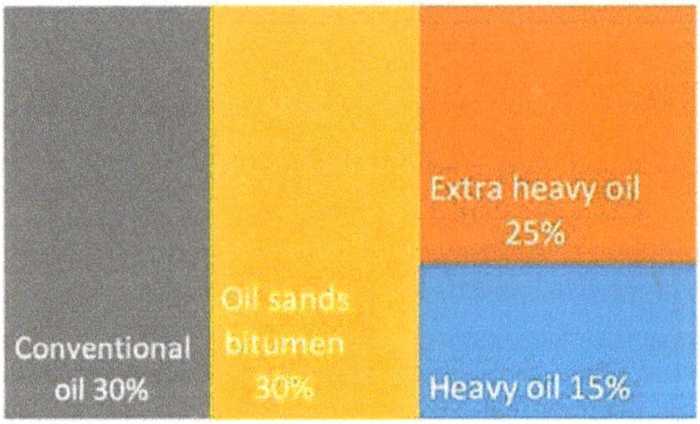

Figure 1. Percentage of world oil reserves showing conventional and unconventional resources [1].

The simultaneous flow of gases and liquids in pipelines is characterised by patterns in which the phases are spatially distributed depending on individual flow rates, pipe size and inclination. This occurs in many industrial applications such as chemical, petroleum and nuclear industries. Slug flow pattern has been the dominant flow regime for highly viscous oils occurs over a wide range of superficial velocities, thus making the knowledge of slug flow a major significance for the oil and gas industries.

The most relevant characteristic of slug flow is its intermittency. This is characterised by the intermittent flow of two distinct flow regions (i.e., slug body region and slug film region) making up the slug unit. Intermittent flows have been thoroughly studied theoretically and experimentally owing to their relevance in the oil and gas industry with much more effort dedicated to low viscous liquids. However, in recent times, there has been increasing interest in its understanding for highly viscous multiphase flows.

The equation that describes the velocity of a slug unit is a crucial closure relationship in slug flow modelling. It is required as an input parameter for the estimation of mean liquid holdup and pressure gradient [2,3]. It is also required as closure equation for slug flow models e.g., Wang et al. [4], Zhang et al. [5] and others needed for the prediction of slug flow features and pressure drop in pipes.

Recent high viscosity work by Archibong-Eso et al. [6,7] have shown significant effects on slug flow parameters (such as slug frequency, mean liquid holdup, slug body holdup) with changes in pipe internal diameter. It would therefore not be far-fetched to consider that similar changes in pipe diameter may also affect slug translational velocities.

Air-medium viscous oil ($\mu = 0.1 - 0.584$ Pa·s) experiments were performed using horizontal pipe with internal diameter of 0.0508-m [8,9] The viscosity oil was varied by changing the temperature of oil used. His investigation noted significant increase in two phase flow parameters such pressure gradient and liquid holdup as oil viscosity increases. New correlations were proposed for drift velocity and the slug frequency for high viscosity liquid.

Furthermore, the current authors [10–13] conducted experiments using high viscosity oils ranging from 0.6 to 7.0 Pa·s and gas in both horizontal and inclined experimental test facility with internal diameter of 0.025-m and 0.0762-m located within the Oil and Gas Engineering Centre at Cranfield University. Their investigations noted very significant effects of high liquid viscosity on two phase flows on slug frequency, holdup, length and translational velocities.

2. Literature Review

Nicklin et al. [14] were the first researchers to investigate elongated bubble motion in flowing liquids. They noted that translational velocity can be estimated by the superimposition of the velocity in stagnant liquid and the influence and the influence of the moving liquid. The expression below was then proposed for the estimation of bubble translational velocity in vertical flow.

$$V_T = C_0 V_m + C_1 \sqrt{gD} \tag{1}$$

Since the second term in Equation (1) is meaningless in horizontal flow, the translational velocity in horizontal flow is denoted as;

$$V_T = C_0 V_m \tag{2}$$

where V_T = Translational velocity, V_D = Drift velocity, C_0 = a constant that quantifies the influence of the mixture on bubble velocity (called the distribution parameter) given as 1.2, V_m = Mixture velocity, C_1 = Constant (Froude number) to evaluate drift velocity, and g = gravitational acceleration. Several other investigators have proposed different values and Equations for C_0 summarised in Table 1. We note that such an expression (Equation (2)) can be used to calculate the velocity of the liquid slug if C_0 is appropriately defined.

Quite a number of correlations have been developed in the literature from diverse experimental data sources for the determination of slug translational velocity in literature. Slug translational velocity estimation by these models (shown in Table 1) only accounts for a limited number of variables such as the distribution parameter and mixture velocity. As can be seen, the studies of Gregory and Scott [15], Mattar and Gregory [16], Dukler and Hubbard [17], Nicholson et al. [18], Dukler et al. [19], and Kouba and Jepson [20], all proposed correlations of the Nicklin [14] type, mainly based on superficial velocities, and do not have a liquid viscosity term embedded. However, only few existing [9,12,21–23] have indicated that the fluid properties such as the liquid viscosity have significant effects on slug translational velocity. It was noted by these studies that slug translational velocity increases as liquid viscosity increases. Furthermore, most of these investigations were carried out using liquid viscosity limited to less than 1.0 Pa·s in addition to using relatively smaller pipe internal diameter test facilities.

Other researchers have investigated the effects of fluid properties on drift velocity which forms an integral part of translational velocity as indicated in Equation (2). Most recent and notable amongst these investigators are the works of Gokcal et al. [8] and Jeyachandra et al. [24] which were conducted using an inclinable facility of internal diameter 0.0508-m for liquid viscosities that are within the range 0.0154–0.574 Pa·s. Brito et al. [22] using same test facility noted that slug translational velocity increases as liquid viscosity increases. Additionally, it is worth noting that most of these investigations were carried out using liquid viscosity oil limited to less than 1.0 Pa·s and also using relatively smaller pipe internal diameter test facilities.

Other researchers have investigated the effects of fluid properties on drift velocity which forms an integral part of translational velocity as indicated in Equation (2). Most recent and notable amongst these investigators are the works of Gokcal et al. [9] and Jeyachandra et al. [24] which were conducted using an inclinable facility of internal diameter 0.0508-m for liquid viscosities that are within the range 0.0154–0.574 Pa·s.

In recent studies, Al-Kaiyem et al. [25] and Bendiksen et al. [26] experimentally investigated slug flow in a horizontal pipeline using medium viscosities. While Al-kayiem et al. [25] carried out statistical assessment of slug body length and translational velocity using water as the liquid phase, they noted that for a fixed water velocity, the slug length and translational velocity increases with an increase in the superficial air velocity while the slug frequency decreased. Bendiksen et al. [26] utilised oil of viscosity ranging from 0.240–0.730 Pa·s in a 0.057-m ID horizontal pipe to study slug bubble velocity. The authors noted liquid viscosity has a strong effect on bubble shape and velocity.

In the present investigation, oils of varying oil viscosity in the range of 0.07–5.5 Pa·s were used to conduct an experiment which differed from the majority of experiments presented in the literature where low to medium viscosity is generally used. In this work, we also propose a new correlation that integrates the effects of liquid viscosity for the prediction of slug translational velocity has been proposed.

Table 1. A summary of current models in the open literature for slug translational velocity.

Authors/Year	Experimental Conditions	C_0	Model/Correlations Developed for Slug Translational Velocity
Nicklin et al. [14]	Theoretical model	1.2 at high Reynolds numbers and 2.0 at low Reynolds numbers	$V_T = C_0 V_m + C_1 \sqrt{gD}$
Moissis and Griffith [27]	Theoretical model	-	$V_T = V_{T,\infty}\left[1 + 8exp\left(-1.06\frac{L_s}{D}\right)\right]$
Gregory and Scott [15]	Theoretical model	1.35	$V_T = C_0 V_m$
Mattar and Gregory [16]	Air-oil(9 cP) horizontal to 10º incline	1.32	-
Dukler and Hubbard [17]	Air-water; 3.75 m horizontal pipe	$1.022 + 0.021ln(Re_s)$.	$V_T = (1 + C_0)V_m$
Nicholson et al. [18]	Theoretical model validated with air-oil horizontal flow data	{1.19; $D = 0.0258$ *m* 1.128; $D = 0.0512$ *m*	$V_T = \begin{cases} 1.19V_m \ +0:27;\ D = 0.0258m \\ 1.128V_m \ +0:28;\ D = 0.0512m \end{cases}$
Dukler et al. [19]		1.225	$V_T = 1.225V_m; Horizontal$
Kouba and Jepson [20]	Air-water, Diameter= 0.15 m,	-	$V_T = 1.21\left(0.1134 + 0.94V_{sl} + V_{sg}\right)$.
Fabre and Line [28]	-	$C_o = \frac{2.27}{1+(Re/1000)^2} + \frac{1.2}{1+(1000/Re)^2}$	-
Manolis [29]	-	{1.033; Fr_m < 2.86 1.216; Fr_m ≥ 2.86	-
Woods and Hanratty [30]	Horizontal air-water in a 0.0953-m pipe at atmospheric pressure	1.1; Fr_m < 3.1 1.2; Fr_m ≥ 3.1	$V_T = \left(C_o - \frac{V_m}{1+(s-1)\epsilon}\right)(1-\epsilon)$
Petalas and Aziz [31]	-	$1.64Re_{mL}^{-0.031}$	-
Choi et al. [32]	Air-water/light oil inclinations of -10º to 10º	$C_o = \frac{2.27}{1+(Re/1000)^2} + \frac{1.2-0.2\sqrt{\left(\frac{\rho_G}{\rho_L}\right)}(1-exp(18\alpha_G))}{1+(1000/Re)^2}$	-
Archibong [11]	Air-oil in two horizontal pipes of 0.0508, 0.0762 m internal diameter	$C_0 = \frac{\Psi_1}{\Psi_2+\left(\frac{Re}{1000}\right)^2} + \frac{\Psi_3+\Psi_4\sqrt{\frac{\rho_g}{\rho_l}}(1-exp(1-exp\ exp\ (-18\alpha_G)\ )}{\left(\frac{Re}{1000}\right)^2}$ Ψ_1,Ψ_2 Ψ_3 and Ψ_4 were respectively obtained as 0.272, 0.236, 0.471 and 17.143	-
Kim et al. [33]	Air-oil in a horizontal pipe of internal diameters: 0.0508 and 0.0762 m	$C_o = 1.13 * \frac{V_{Lmaxi}}{V_{SLi}}$ Where are V_{Lmaxi} and V_{SLi} are the in-situ maximum liquid and superficial liquid velocities at each PIV x-grid.	-

$Re_{mL} = \frac{\rho_L V_m D}{\mu_L}$; $Fr_m = \frac{V_m}{\sqrt{gD}}$; $\Gamma = 1 + \left(\frac{Fr_m}{Fr_{crit.}} Cos(\theta)\right)$; $Fr_{crit.} = 3.5$; *β_{TB} and Ψ_{TB} are emperical constant whose value* 5.5 & 0.6; $n =$ 7 *corresponding to the* 1/7*th power law* $Re_s \frac{\alpha_{GS}\rho_G+(1+\alpha_{GS})\rho_L}{\alpha_{GS}\mu_G+(1+\alpha_{GS})\mu_L}$; $V_{T,\infty}$*is the translational vel. of a long, stable slug at the same mixture velocity;* L_s*, stable was stated to be between* 10 *and* 15 *pipe diameters*

3. Experimental Setup

3.1. Description of Test Facility

The test facility we used for this study is situated at Cranfield University's Oil and Gas Engineering Centre. The facility has previously been used for other slug flow studies [7,11,13,34]. It is made up of a 0.0762-m internal diameter (ID) horizontal pipe constructed from a transparent pipe with an L/D = 223. A schematic representation of the flow facility is presented in Figure 2a. It comprises of three main sections: the fluid handling section and that for the instrumentation and data acquisition. The measurement and observation section is 14 m from the pipe inlet. A separator is at the end of the pipe for phase separation.

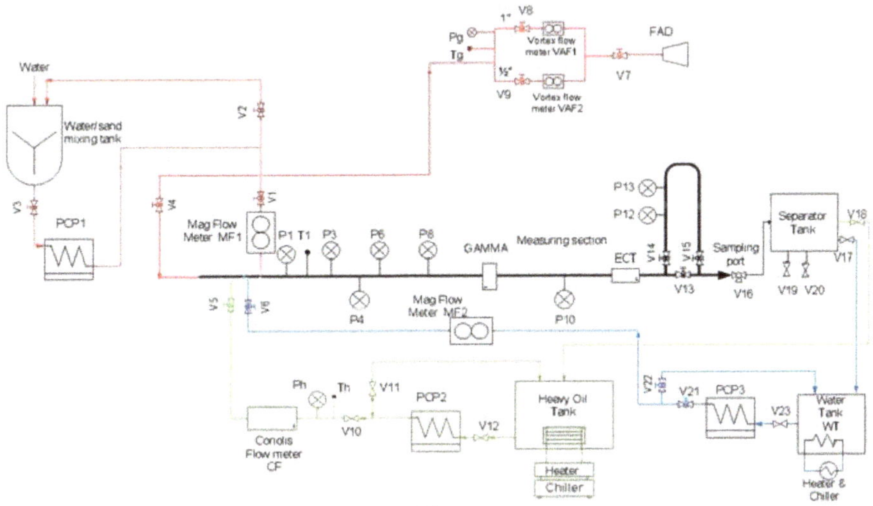

(a)

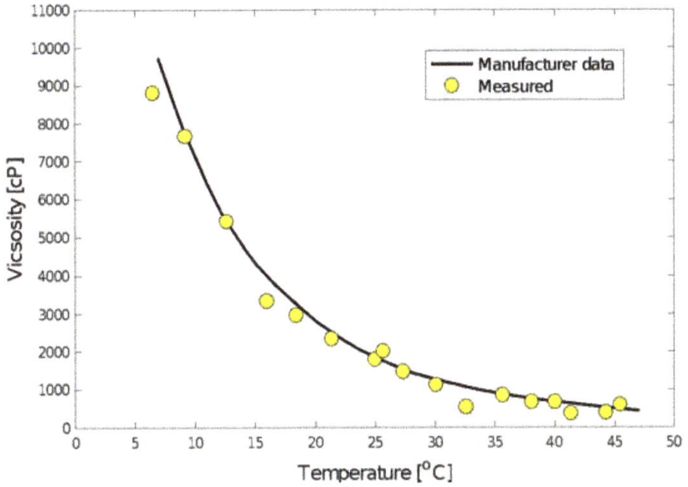

(b)

Figure 2. Experimental test facility (**a**) Schematic (**b**) A comparison of measured with manufacturer-supplied temperature response of viscosity.

The mineral oil used for the experiments is CYL680, manufactured by Total Limited UK, is stored in a steel metal tank of 2-m^3 capacity and introduced to the test section via a T-junction by means of a progressive cavity pump manufactured by Fluid Pumps Limited, United Kingdom. The oil flow rate is metered using a Coriolis flow meter (Endress + Hauser, Promass 83F80 DN80) with an accuracy of ±0.035%. Oil recirculation to the tank is done using a bypass line before the commencement of experimental runs in order to achieve a uniform oil viscosity. The temperature of the oil is controlled using a bath circulator (Thermal Fisher) with a temperature range of 0–50 ± 0.01 °C. The oil stored in the tank is heated or cooled to a desired temperature resulting to change in oil viscosity in the tank. In the laboratory, Brookfield DV-I™ prime viscometer was used to measure oil viscosity and the obtained result compared well with the manufacturer-supplied values given in Figure 2b.

Air is pumped by a screw compressor (Anglian Compressors, Peterborough, UK) is monitored using two vortex flow meters (Prowirl 72F15 DN15, Endress + Hauser, Reinach, Switzerland) and a vortex flow meter (1.5-inch Prowirl 72F40 DN40, Endress + Hauser, Reinach, Switzerland) with different measuring ranges (0–20 and 10–130 m^3/h). To avoid fluctuating air supply to the test section, the air is first charged into a 2.5-m^3 and delivered at a pressure of 7 barg. Afterwards, filtered air from the compressor is afterwards injected into the test line with the aid of 2-inch steel pipe which is at L/D = 150 upstream of the observation and measurement section. The collected two-phase mixture is allowed to stay 48 h or more in the separator to reach full separation and all air bubbles were removed by agitation and buoyancy. This left the oil in the tank which was pumped for reuse.

3.2. *Flow Facility Instrumentation, Data Acquisition and Processing*

3.2.1. Thermocouples and Pressure Transducers

For the temperature of the experimental test fluids, J-type thermocouples manufactured by Reotemps Instruments Limited (Reotemps, San Diego, CA, USA) are used for its measurement. The thermocouples which have an accuracy of ±0.1 °C are installed at regular intervals along the test line. Pressure measurement is achieved by means of differential pressure transducers manufactured by GE Druck (Leicester, UK) and installed at 4 m and 13 m downstream of the main test line and connected. Pressure readings were obtained from differential pressure transducers, flowmeters and temperature sensors. These are saved to a PC using LabVIEW 8.6.1 (National Instruments, Austin, TX, USA). This system is comprised of a National Instruments (NI) USB-6210 (National Instruments, Austin, TX, USA) connector board interface that outputs signals from the instrumentation via BNC coaxial cables.

3.2.2. Gamma Densitometers

Gamma densitometers are devices used for measuring the density of fluids flowing in a conduit which uses a radioactive source of gamma rays and a detector. Standard gamma densitometers use 137Cs (cesium) or 241Am (americium) to induce Compton scattering. The count rate at the detector depends mainly on the density of the flowing fluids. Advantages of gamma densitometers include non-intrusiveness, and they work for non-transparent pipe sections. Moreover, they are not affected by friction or kinetic effects.

The single-beam gamma densitometers we used were manufactured by Neftemer Limited, Russia. Its major components are highlighted in Figure 3. These gamma densitometers consist of a single energy source that emits gamma rays at 662 keV known as the high-energy level (hard spectrum or hard count) and a soft spectrum that emits lower-energy gamma rays, specifically 100–300 keV. They contain Caesium-137 acquired via a sodium iodide scintillator. In this work, a proprietary data acquisition (DAQ) system was used for voltage signal acquisition. In addition, an ICP i-7188 (ICP DAS, Fresno, CA, USA) programmable logic controller was used for the conversion of the raw voltage to gamma counts (counts refer to the remaining attenuated signals after absorption through the fluid).

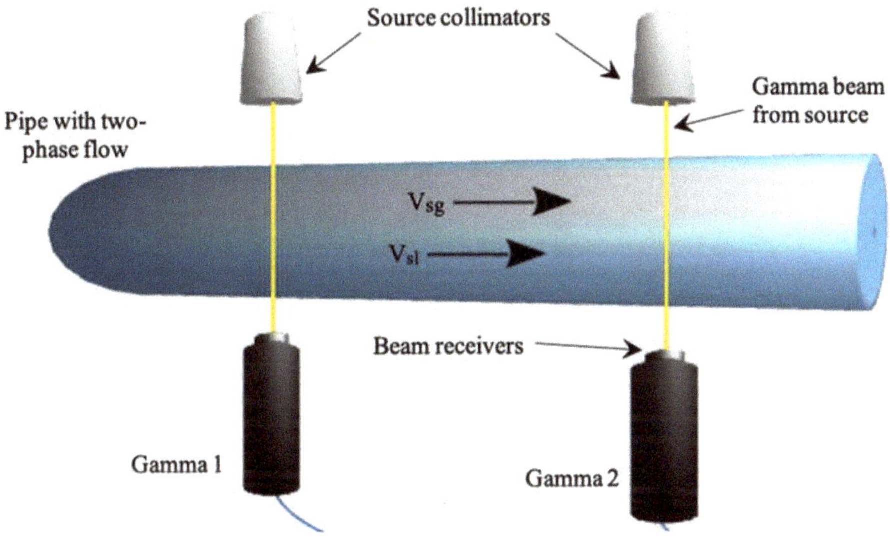

Figure 3. Representation of the installed gamma radiation source on the test section.

The Beer-Lambert Equation given in Equation (3) is used for the calculation of the liquid holdup from the linear attenuation coefficients, and is as follows:

$$H_L = \left[\frac{ln\left(\frac{I_M}{I_A}\right)}{ln\left(\frac{I_L}{I_A}\right)} \right] \tag{3}$$

where I_M = mean gamma count from gas-liquid mixture in the pipe; I_A = mean calibrated gamma count for an empty pipe (i.e., 100% Air); I_L = mean calibrated gamma count for the pipe containing pure liquid; and H_l = Liquid holdup. A characteristic plot of the time series of the liquid holdup obtained by the gamma densitometers are presented in Figure 4a, which exhibits an intermittent slug flow behaviour. This is characterised by troughs and crests, while the crests suggest the movement of liquid slugs, the troughs regions indicate the slug film. Two gamma densitometers that we positioned at L/D = 103 and 124 were used to measure holdup. A comprehensive statistical uncertainty analysis for the densitometer measurements have earlier been presented in Baba et al. [34]. The time series data were analysed and used to calculate the slug translational velocity using Equation (4). This was achieved through cross-correlation by utilising the "xcorr" function of MATLAB's signal processing toolbox.

$$V_T = \frac{\Delta l_{Gamma}}{T_1 - T_2} = \frac{\Delta l_{Gamma}}{\tau} \tag{4}$$

where τ is the time lag between the two signals determined by cross-correlation, the procedure of which is explained in the next paragraph. For the purpose of this investigation, noisy inputs associated with the acquired signals were minimised by using signal filters in MATLAB (i.e. the "smooth" function). Sample raw as well as filtered signal outputs from the gamma densitometers are presented respectively in Figure 4.

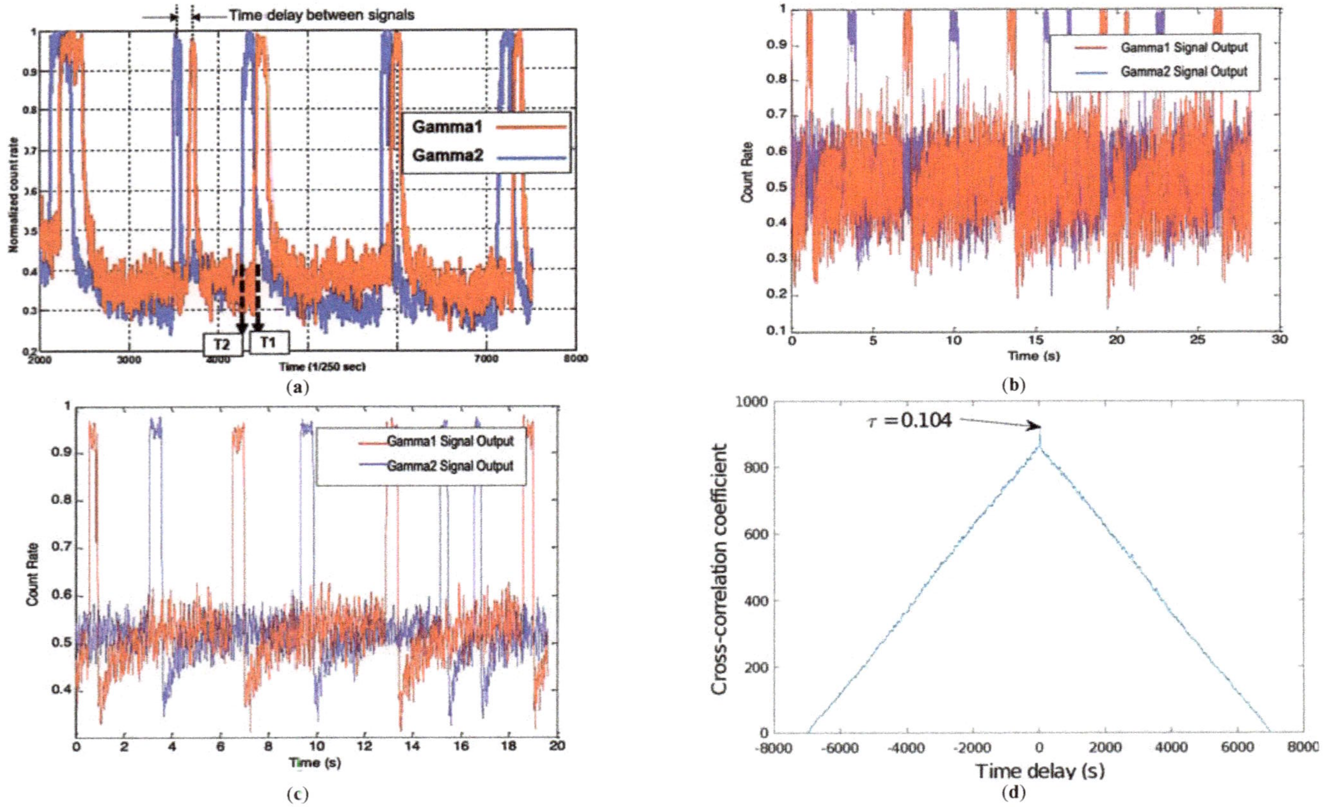

Figure 4. (**a**) Depiction of the determination of time delay for slug translational velocity from gamma densitometer time series (**a**) Typical liquid holdup plot showing the concept of signal time delay (**b**) an example of a raw signal output from gamma photon counts (**c**) an example of a filtered gamma signal output (**d**) Cross-correlation coefficient and time delay plot.

Cross-correlation which is a standard method that is used to measure the degree of similarity between two signals and for the determination of the time difference that exists between them was implemented on the two gamma densitometer signal time series as presented in Equations (5) and (6). Assuming two time series are $X(t_n)$ and $Y(t_n)$, where $n = 0, 1, 2, 3 \ldots$. $N-1$, then the cross-correlation coefficient is given as:

$$R_{xy}(\tau) = \frac{C_{xy}(\tau)}{\sqrt{C_y(0)C_y(0)}} \tag{5}$$

$$C_{xy}(\tau) = \frac{1}{N-\tau}\sum_{n-1}^{N-\tau} X(t_n)\, Y(\tau + t_n) \tag{6}$$

where τ is the temporal lag.

3.3. *Fluid Properties*

Table 2 summarises the properties of the fluids used in the experiments and the test matrix. The measurement uncertainties of the listed parameters were obtained from manufacturers' specification of flow meters, viscometer, and gamma sensor. These agreed with values obtained from repeated tests to ascertain accuracy.

Table 2. Test matrix for experimental run and measurement uncertainties.

Density (kg/m^3)	Test Fluids	Viscosity (cP)	Interfacial Tension (25 °C, N/m)	Test Matrix (m/s)	API Gravity
1.293	Air	0.017	0.033	0.3–9.0	-
≈918	CYL680	1000~6000	0.033	0.06–0.3	22.67

Measurement	Uncertainty (%)
Superficial liquid velocity	±0.5
Superficial gas velocity	±2.1
Liquid viscosity	±1
Pressure drop	±2
Liquid holdup	±10

4. Results

4.1. *Flow Regime Map*

Flow pattern maps are a means of representing local flow patterns as a function of gas and liquid velocities. Generally, these are plots of two-dimensional graphs showing separate areas corresponding to different flow patterns defined transition criteria. Undoubtedly, no universally accepted flow pattern map has been developed however, a number of flow patterns maps have been proposed by early researchers and widely used in the oil and gas industry [35,36].

In this study, flow pattern maps were constructed based on experimental observations in this study using superficial velocities of oil and gas as ordinate and abscissa respectively. Figure 5 shows the flow regime maps which highlight the effects of liquid viscosity on oil-gas two phase flow.

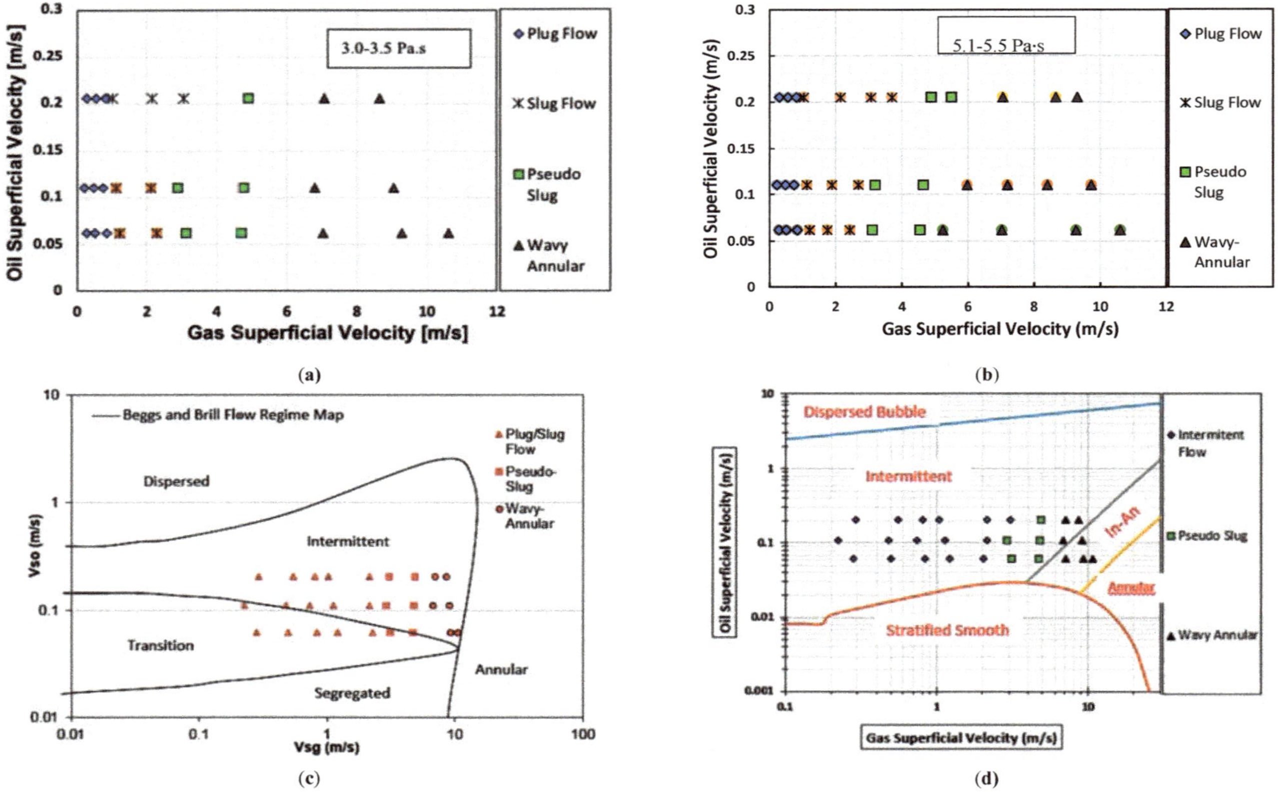

Figure 5. Experimental data points for (**a**) 3–3.5 Pa·s and (**b**) 5.5–5.5 Pa·s (**c**) plotted against the Beggs and Brill [35] flow pattern pap for gas-Liquid horizontal flow, shown in (**c**). A comparison of the observed flow patterns with the flow pattern map of Taitel and Dukler [36] is given in (**d**).

As can be seen in Figure 5, the flow pattern changes from intermittent region (i.e., plug and slug) to transition region (i.e., pseudo-slug) and then to separated flow region (i.e., annular flow pattern). Plug flow is observed from experimental observations within the range of superficial velocity of oil and superficial gas velocities of 0.3–1.0 m/s. Slug flow pattern is then observed as the gas phase gains more kinetic energy owing to increase in superficial gas velocity. Entrainment of droplets from the elongated liquid body occurs with increasing turbulence and this eventually leads to breaking up the liquid body into shorter ones. It is worth noting here that, the liquid viscosity range investigated showed that the intermittent region (i.e., plug flow and slug flow) dominates the flow map and even become enlarged as the viscosity of liquid increases conforming previous findings [9–11,37,38] and this can be attributed to the increase in shear in the pipe walls owing to viscosity effects. The intermittent flow region is also amplified as the superficial velocity of liquid increases credited to increased liquid height which enhances the formation of slug. With further increase in the gas phase resulting in reduction of the liquid fraction translates into insufficient liquid height to aid slug formation thereby initiating the transition from slug flow to annular flow (i.e., pseudo-slug) and occurs generally at 3–5 m/s superficial velocity of gas with the appearance of rolling waves at the interphase. In Figure 5c, a comparison of observed flow against the prediction of flow regime map proposed by Beggs and Brill [35] is presented. The Beggs and Brill [35] flow regime map shows some differences in the prediction of the flow regimes for the current experiments. As can be seen, the prediction of regime transition from intermittent to annular flow is over predicted by the map. This could be due to diameter difference as Beggs and Brill [35] used 1-inch and 1.5-inch diameter pipes and viscosity effects. In Figure 5d, a comparison of the flow pattern map for this study with that of Taitel and Dukler [36] shows an agreement in terms of non-existence of the stratified pattern region, the map however under predicted the annular flow region.

4.2. Mechanism of Slug Flow Formation and Liquid Viscosity on Translational Velocity

The continuous growth of disturbance waves in stratified gas-liquid flows in a test channel result in the formation of slug flow. If the velocity of the gas phase is continually increased resulting to increased momentum of the gas phase and at a suitable liquid height in the test flow line. Due to increase in momentum of the gas phase, there is a resulting decrease in interfacial pressure. The decrease in pressure results in suction forces (commonly referred to as Bernoulli Effect) that acts on the liquid phase. A combination of these forces thereby overcome the forces of gravitational as well as surface tension forces. This mechanism is known as the Kelvin-Helmholtz (K-H) instability which governs transition from stratified to slug flow. Wave peaks continue to grow eventually bridging the pipe cross-section.

Figure 6 shows a typical slug flow pattern observed in our experiments. The flow pattern is characterised by two distinct regions; the film region the slug body. As depicted, the slug front had high momentum and is highly energetic, traveling at a velocity higher than that of that the test fluid mixture velocity. As a consequence of this high velocity, the slug body accelerates the film region causing an entrainment of gas in the slug body. Furthermore, entrained gas bubbles in the liquid film around the pipe wall causes aeration in the slug body. Researchers [39,40] have noted that entrained gas bubbles cause losses from the slug front to the leading film tail, on account of recirculation.

Figure 7a shows the measured translational velocity plotted as a function of mixture velocity. The result, as can be seen, illustrates a linear relationship between the experimental translational velocity and mixture velocity for the different viscosities investigated. The error bars shown represent measurement error of ±8% being the maximum deviation from the mean slug translational velocity values determined using the gamma densitometers. Expectedly, the measured translational velocity grows with increase in mixture velocity with the slope of the graph found to be 2.1–2.3. The obtained slope represents the flow coefficient C_o as expressed in the translational velocity in Equation (1). The result also shows that increased oil viscosity slightly affects the flow distribution coefficient C_o. It can be concluded that the experiments conducted are in the laminar flow region as widely reported the literature.

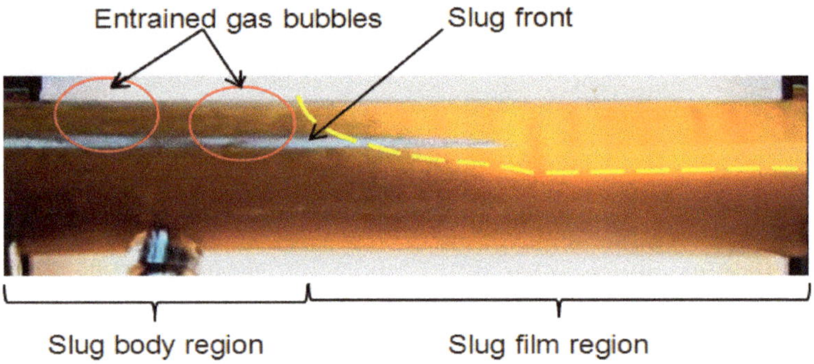

Figure 6. A typical Slug flow for gas and high viscosity oil flow observed for this study (figure from Archibong-Eso et al. [7]).

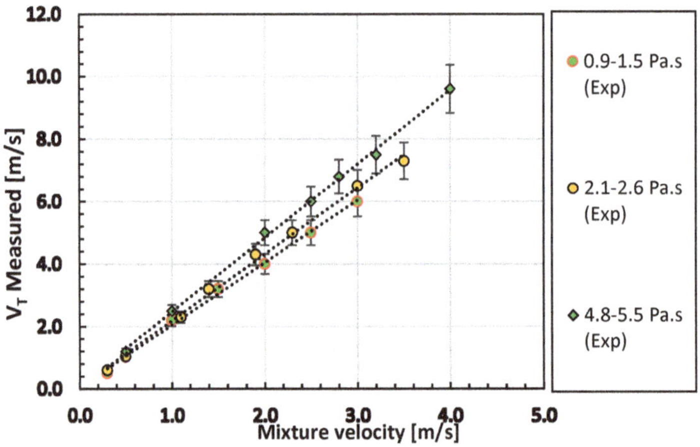

(**a**)

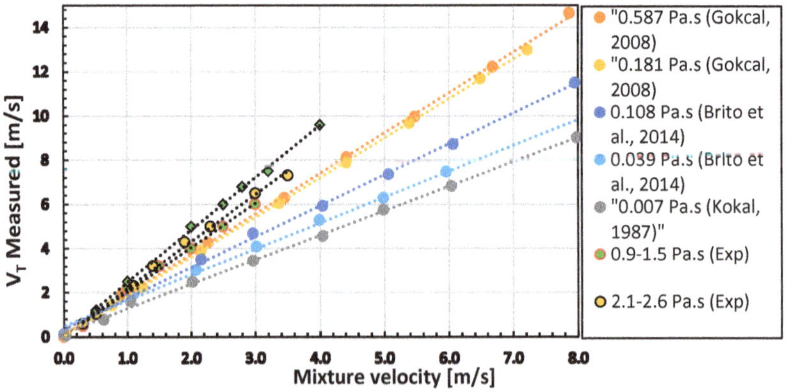

(**b**)

Figure 7. (**a**) Data from this work and (**b**) slug translational velocity vs mixture velocity for data from this work compared with previously reported at different flow conditions (Vsg = 0.3–7.0 m/s and Vsl = 0.2–0.4 m/s).

Correspondingly, translational velocity—as can be seen in Figure 7b—increased with an increase in viscosity as can be seen from the plot. We attribute this to the fact that the length of the slug was reduced with an increase in liquid viscosity (as confirmed by many researchers [34,37–41] thereby reducing the mass of liquid body travelling per unit time which translated to higher momentum. In other words, slug translational velocity is proportional with liquid viscosity as a result of reduced slug length and increased frequency. Though the laminar sublayer increases with increasing liquid viscosity, this increase is dominated by the effects associated with the reduction in the slug body (as a result of the decrease in the slug length. This trend conforms to those reported for medium liquid viscosities previously reported [9,22,34].

5. Correlation of Slug Translational Velocity Data

As stated earlier, translational velocity is a key parameter for slug flow modelling. Existing correlations found in the literatures show significant performance deficiencies for application in high viscosity usage within specified limits. A correlation based on experiments was first proffered by Nicklin et al. [14] to determine the velocity of a Taylor bubble in slug flow in vertical pipes but have largely been applied to different pipe inclinations by many researchers in recent times as given in Equation (2). The value of the distribution parameter C_o was found to be nearer 1.2 for fully-developed turbulent flow and nearing 2 for laminar flow. Previous studies have shown that for low viscosity liquids, the distribution parameter C_o ranges between $1.0 < C_o < 1.2$. However, Wallis [42] noted that the value of C_o can even be higher than 2 for fully developed laminar flow though it was stated in his work that the exact behaviour was to be determined. This has been confirmed by the works of Gokcal [9] who suggested a larger distribution parameter. Choi et al. [32] proposed $C_o = 2.27$ for relatively medium viscosity oils from Equation (7). Lacy [43] also suggested 2.3 for C_o and most recently 2.26 by Archibong-Eso [11].

$$C_o = \frac{2}{1+\left(\frac{Re}{1000}\right)^2} + \frac{1.2-0.2\sqrt{\frac{\rho_G}{\rho_L}}\left(1-exp(-18\alpha_G)\right)}{1+\left(\frac{1000}{Re}\right)^2} \tag{7}$$

Thus the proposed translational velocity V_T was correlated from the experimental dataset for this study ranging from 1.5 to 5.5 Pa·s and Gokcal's [9] dataset ranging from 0.108 to 0.587 Pa·s. By utilising the least squares regression method, the error margin in prediction between the proposed correlation and that of the experimental data is obtained, afterwards minimised by way of fine-tuning C_1 to obtain an optimum local solution. The new optimum solution obtained for C_1 based on the present data set for high viscosity oil in the range 1.5–5.5 Pa·s was calculated as 0.79.

Seven correlations were used for a comparative study [14,24,32,44–47]. Figure 8 shows a comparison of the current high viscosity translational velocity and predictions of four selected correlations on the basis of performance from the aforementioned [14,24,44,46]. It may be seen that there are large deviations from the error bands of ± 15% and are mostly under predictions.

A statistical quantification of the deviations between the experimental data and the seven correlations is given in Table 3. The statistical parameters used ε_1–ε_6 defined in the Appendix A and vary from standard deviation of relative error to absolute mean actual error, have previously been used by other authors [6,13,33]. It therefore becomes imperative to extend the validity of existing correlations as their accurate prediction is crucial in the design of pipelines and downstream unit operation equipment such as separators and slug catchers. To do this, we introduce the so-called viscosity number by adding it to the distribution parameter C_o so as to account for the effect of liquid viscosity and its effect on the velocity of the liquid slug in highly viscous oil flows. The viscosity number has been successfully demonstrated in the past to adequately characterize highly viscous flows in multiphase pipelines [34,40,41]. Its addition yields an expression for the slug translational velocity as follows:

$$V_T = \left(C_o + N_\mu\right)V_m + +C_1\sqrt{gD} \tag{8}$$

where N_μ is the viscosity number given by:

$$N_\mu = \frac{V_m \mu_L}{g D^2 (\rho_l - \rho_g)} \tag{9}$$

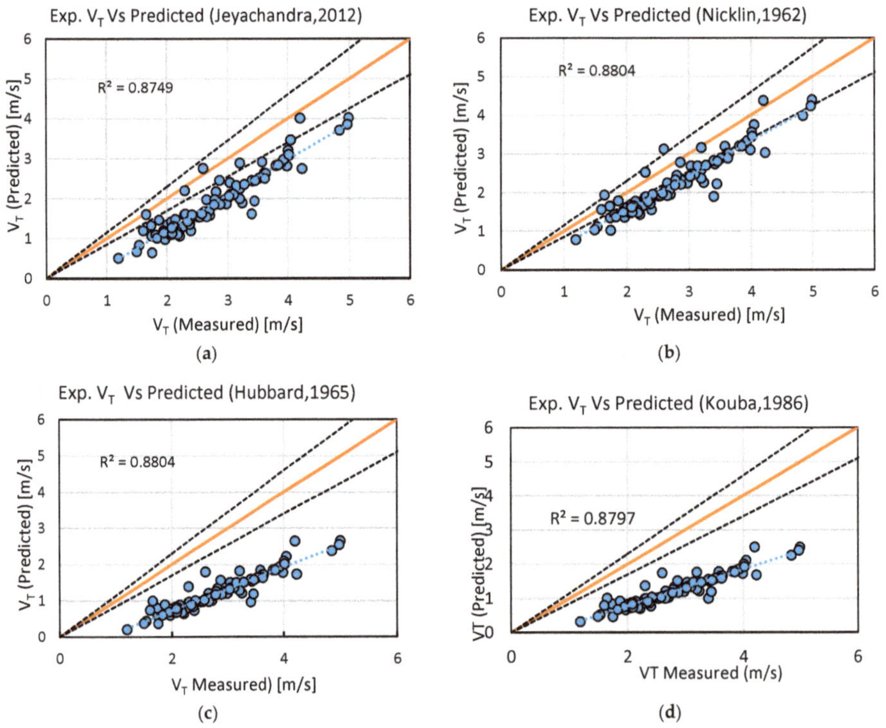

Figure 8. Comparison of the experimental data from the present study with several model predictions and experimental data for the present study. Model predictions are by (**a**) Jeyachandra et al. [24], (**b**) Nicklin [14], (**c**) Hubbard [44], and (**d**) Kouba [46]. The error bands represent a ± 15% deviation of the model from the present experimental slug translational velocities.

Table 3. A comparison between our proposed and existing correlations.

	Nicklin [14]	Hubbard [44]	Benjamin, [47]	Kouba [46]	Jepson [45]	Jeyachandra et al. [24]	Choi et al. [32]	Eqn. 8
ε_1	−19.85	−58.20	−19.85	−57.63	−54.32	−33.22	−7.67	−0.31
ε_2	20.96	58.20	20.96	57.63	54.32	33.35	17.79	8.05
ε_3	10.61	8.64	10.61	6.95	14.96	12.16	31.31	6.55
ε_4	−0.54	−1.59	−1.59	−1.59	−1.44	−0.88	−0.11	0.00
ε_5	0.56	1.59	1.59	1.59	1.44	0.89	0.54	0.21
ε_6	0.29	0.38	0.38	0.42	0.35	0.29	1.43	0.26

The results of our comparison are presented in Figures 8 and 9. Figure 9 shows a comparison between the predictions of Equation (8) and the current experimental data on slug translational velocity. Compared with Figure 8, it represents a significant increase in predictive performance of slug translational velocity. Only 9 of over 80 data points were outside the ±15% error bands. Statistical comparison using the parameters ε_1–ε_6 also shows improved predictions when compared to the previous correlations with their performance is presented in Table 3.

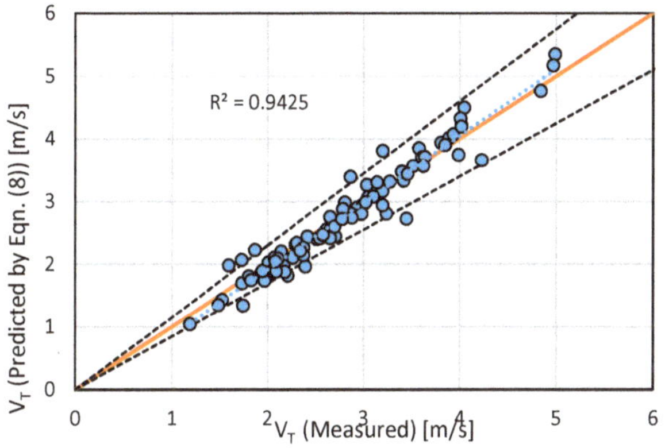

(**a**)

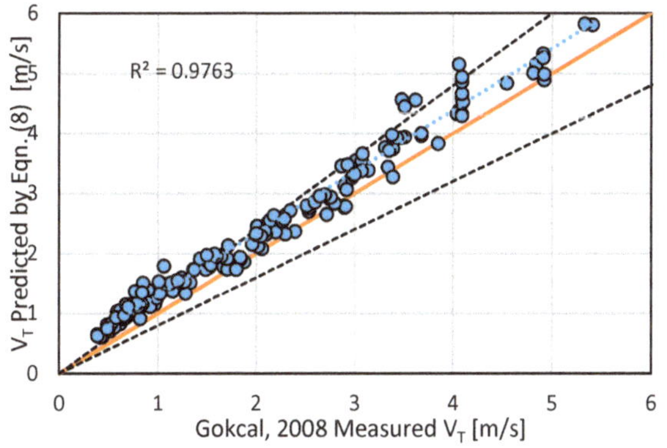

(**b**)

Figure 9. (**a**) Cross plot of the new correlation's prediction against the experimentally measured slug body translational velocity (Equation (8)). The error bands are ± 15% limits of deviation between experiments and predicted points. (**b**) Cross-plot of the model prediction (Equation 8) vs. the experimental measurements of Gokcal [9]. Error bands represent ± 15% deviation of the model from the experimental slug translational velocities.

A further demonstration of the predictive capabilities of the proposed correlation is carried out. We show this by comparing its predictions with the data of Gokcal [9]. Gokcal's dataset is for liquid viscosity of 0.181–0.587 Pa·s, a range lower than that investigated in this study (0.9–5.5 Pa·s). Notwithstanding this, Figure 9b shows that the correlation compares favourably, illustrating its consistency against independent data. While there are over-predictions for most data points, these are well within +20% of Gokcal's data. We note that the developed slug translational velocity correlation needs to be further tested against a lower viscosity dataset as well as those with a wider range of fluid properties for robustness.

6. Conclusions

Experimental measurement of slug translational velocity in high viscosity oil-gas flow conducted in this study by using cross correlated data from two gamma densitometers with the same sampling frequency of 250 Hz. The results show that slug translational velocity increased with rise in oil viscosity. It also increased with increase in mixture velocity of flow. This is similar to observations made by previous authors [9,22]. Comparison of seven existing models/correlations on for slug translational velocity against data obtained in this study showed that the measured slug translational velocity was under predicted by several degrees of magnitude. The proposed slug translational velocity developed from data with oil viscosity from 0.108–5.5 Pa·s. A statistical performance evaluation of the proposed correlation against independent data obtained from Ref. [9] showed that the proposed correlation had better predictive capabilities for the high liquid viscosity range. We note that the slug translational velocity correlation developed in this study needs to be further tested against a more a wider range dataset to make it more robust. The reason for this is that, for any correlation, the more the data used to derive it, the more its predictive power across a wide range of conditions, fluid properties, and pipe sizes. It is impossible to cover all those in an experimental campaign as experiments are expensive to conduct. As a result, our measurements had to be limited, and were obtained with one pipe diameter (0.0762m), and one oil which we varied the viscosity between 1 and 6 Pa·s. To increase the range of applicability of the derived correlation at high oil viscosities, we further added the data of Gokcal obtained at viscosities between 0.2 and 0.6 Pa·s. Consequently, the correlation can only be used with confidence at these conditions.

Author Contributions: Conceptualization, Y.D.B. and A.M.A.; methodology, O.T.F. and A.A.E.; data analysis: J.X.F.R., Y.D.B. and A.M.A.; writing—original draft preparation, Y.D.B.; writing—review and editing, Y.D.B., A.M.A., A.A.E., J.X.F.R.; proofreading, A.M.A., O.T.F.; supervision, L.L. and H.Y.

Funding: This research received no external funding.

Acknowledgments: The authors (Y.D.B. and A.M.A.) are grateful to the Petroleum Technology Development Fund (PTDF), which forms part of the Nigerian Government for funding their PhD studies in Cranfield University. We also acknowledge the support and kind assistance of laboratory manager Stan Collins, Laboratory technicians, David Withington, Kelvin White and Sheridan Cross. The authors are also grateful to research colleagues at Cranfield University's Oil and Gas Engineering Centre.

Conflicts of Interest: All authors declare no conflict of interest.

Nomenclature

Symbols	*Denotes*	*Units*
A	*Area*	m^2
C	*Constant*	[-]
D	*Pipe diameter*	m
Fr	*Froude number*	[-]
fs	*Slug Frequency*	s^{-1}
g	*Acceleration due to gravity*	$m.\ s^{-2}$
L	*length*	m
$N\mu$	*Viscosity number*	[-]
HL	*Holdup*	[-]
HF	*Average film holdup*	[-]
HS	*Average slug holdup*	[-]
Nf	*Inverse viscosity number*	[-]
Re	*Reynolds number*	[-]
St	*Strouhal number*	[-]
Vm	*Mixture Velocity*	m/s
Ls	*Liquid slug length*	m
VT	*Translational velocity*	m/s

Greek letters

μ	*Viscosity*	*cP*
ρ	*Density*	*kg/m*^{3}
τ	*Shear stress*	*Pa*
ε	*Relative error*	*[-]*
Subscripts		
f	*Film zone*	
g	*Gas phase*	
l	*Liquid phase*	
m	*Mixture phase*	
s	*Superficial*	
t	*Translational*	

Appendix A

Statistical parameters for evaluating the performance of predictive correlations relative to the experimental data acquired. These parameters, six of them, have also been used by numerous researchers [6,13,34,48–50] and are evaluated based on two types of errors; actual and relative error respectively. Results are given in Table 3 and the best performing correlations are those with the least magnitude of the statistical parameter concerned. They are:

$$\varepsilon_i = \frac{y_{predicted} - y_{measured}}{y_{measured}} * 100 \tag{A1}$$

$$\varepsilon_j = y_{predicted} - y_{measured} \tag{A2}$$

The error margins from estimated actual error and relative error above, six other statistical parameters are defined from Equations (A3) to (A8)

The average relative error is given as:

$$\varepsilon_1 = \frac{1}{N}\sum_{i=1}^{N} y_i \tag{A3}$$

The absolute of average relative error is given as:

$$\varepsilon_2 = \frac{1}{N}\sum_{i=1}^{N} |y_i| \tag{A4}$$

The standard deviation of the relative error is defined as follows:

$$\varepsilon_3 = \sqrt{\frac{\sum_{i=1}^{N} (y_i - Y_1)^2}{N-1}} \tag{A5}$$

The average actual error

$$\varepsilon_4 = \frac{1}{N}\sum_{i=1}^{N} y_j \tag{A6}$$

The absolute of the mean actual error is calculated using:

$$\varepsilon_5 = \frac{1}{N}\sum_{i=1}^{N} |y_j| \tag{A7}$$

The standard deviation of actual errors is calculated as follows:

$$\varepsilon_6 = \sqrt{\frac{\sum_{j=1}^{N} (y_j - Y_4)^2}{N-1}} \tag{A8}$$

The mean relative error ε_1 and the mean actual error ε_4 measure the disparity between the predicted and measured parameters. If they are positive, this indicates over-estimation of the parameter and vice versa. Individual errors can be either positive or negative, and can cancel each other out, masking the true performance.

The mean absolute percentage relative error ε_2 and the mean absolute actual error ε_5 do not mask the true performance. Nevertheless, they signify the magnitude of the error on average. The standard deviations, ε_3 and ε_6, indicate the degree of scattering compared to their corresponding mean errors, ε_1 and ε_4.

References

1. Baba, Y.D.; Aliyu, A.M.; Archibong, A.E.; Almabrok, A.A.; Igbafe, A.I. Study of high viscous multiphase phase flow in a horizontal pipe. *Heat Mass Transf.* **2018**, *54*, 651–699. [CrossRef]
2. Cook, M.; Behnia, M. Pressure drop calculation and modelling of inclined intermittent gas–liquid flow. *Chem. Eng. Sci.* **2000**, *55*, 4699–4708. [CrossRef]
3. Xiao, J.J.; Shonham, O.; Brill, J.P. A Comprehensive Mechanistic Model for Two-Phase Flow in Pipelines. In Proceedings of the SPE Annual Technical Conference and Exhibition, New Orleans, LA, USA, 23–26 September 1990.
4. Wang, S.; Zhang, H.; Sarica, C.; Pereyra, E. A Mechanistic Slug Liquid Holdup Model for Wide Ranges of Liquid Viscosity and Pipe Inclination Angle. In Proceedings of the Offshore Technology Conference, Houston, TX, USA, 6–9 May 2013; pp. 1–11.
5. Zhang, H.-Q.; Wang, Q.; Sarica, C.; Brill, J.P. Unified Model for Gas-Liquid Pipe Flow via Slug Dynamics−Part 1: Model Development. *J. Energy Resour. Technol.* **2003**, *125*, 266–273. [CrossRef]
6. Archibong-Eso, A.; Okeke, N.; Baba, Y.; Aliyu, A.; Lao, L.; Yeung, H. Estimating slug liquid holdup in high viscosity oil-gas two-phase flow. *Flow Meas. Instrum.* **2019**, *65*, 22–32. [CrossRef]
7. Archibong-Eso, A.; Baba, Y.; Aliyu, A.; Zhao, Y.; Yan, W.; Yeung, H. On slug frequency in concurrent high viscosity liquid and gas flow. *J. Pet. Sci. Eng.* **2018**, *163*, 600–610. [CrossRef]
8. Gokcal, B. Effects of High Oil Viscosity on Oil/Gas Flow Behavior in Horizontal Pipes. Master's Thesis, University of Tulsa, Tulsa, OK, USA, 2006.
9. Gokcal, B. An Experimental and Theoretical Investigation of Slug Flow for High Oil Viscosity in Horizontal Pipes. Ph.D. Thesis, The University Tulsa, Tulsa, OK, USA, 2008.
10. Zhao, Y. High Viscosity Liquid Two-Phase Flow. Ph.D Thesis, Cranfield University, Cranfield, UK, 2014.
11. Archibong, A. Viscous Multiphase Flow Characteristics in Pipelines. Ph.D. Thesis, Cranfield University, Cranfield, UK, 2015.
12. Baba, Y.D. Experimental Investigation of High Viscous Multiphase Flow in Horizontal Pipelines. Ph.D. Thesis, Cranfield University, Cranfield, UK, 2016.
13. Baba, Y.D.; Archibong, A.E.; Aliyu, A.M.; Ameen, A.I. Slug frequency in high viscosity oil-gas two-phase flow: Experiment and prediction. *Flow Meas. Instrum.* **2017**, *54*, 109–123. [CrossRef]
14. Nicklin, D.; Wilkes, J.; Davidson, J. Two Phase Flow in Vertical tubes. *Trans. Inst. Chem. Eng.* **1962**, *40*, 61–68.
15. Gregory, G. a.; Scott, D.S. Correlation of liquid slug velocity and frequency in horizontal cocurrent gas-liquid slug flow. *AIChE J.* **1969**, *15*, 933–935. [CrossRef]
16. Mattar, L.; Gregory, G.A. Air-Oil slug flow in an upward-inclined pipe-I: Slug velocity, holdup and pressure gradient. *J. Can. Pet. Technol.* **1974**, *13*, 69–76. [CrossRef]
17. Dukler, A.E.; Hubbard, M.G. A Model for Gas-Liquid Slug Flow in Horizontal and Near Horizontal Tubes. *Ind. Eng. Chem. Fundam.* **1975**, *14*, 337–347. [CrossRef]
18. Nicholson, M.K.; Aziz, K.; Gregory, G.A. Intermittent Two Phase Flow in Horizontal Pipes: Predictive Models. *Can. J. Chem. Eng.* **1978**, *56*, 653–663. [CrossRef]
19. Dukler, A.E.; Maron, D.M.; Brauner, N. A Physical Model For Predicting The Minimum Stable Slug Length. *Chem. Eng. Sci.* **1985**, *40*, 1379–1385. [CrossRef]
20. Kouba, G.E.; Jepson, W.P. The Flow of Slugs in Horizontal, Two-Phase Pipelines. *J. Energy Resour. Technol.* **1990**, *112*, 20–24. [CrossRef]
21. Pan, J. Gas Entrainment in Two-Phase Gas-Liquid Slug Flow. Ph.D. Thesis, Imperial College London, London, UK, 2010.
22. Brito, R.; Pereyra, E.; Sarica, C. Experimental study to characterize slug flow for medium oil viscosities in horizontal pipes. In Proceedings of the 9th North American Conference on Multiphase Technology, Banff, AB, Canada, 11–13 June 2014; pp. 403–417.
23. Al Awadi, H. Multiphase Characteristics of High Viscosity Oil. Ph.D. Thesis, Cranfield University, Cranfield, UK, 2011.

24. Jeyachandra, B.C.; Gokcal, B.; Al-Sarkhi, A.; Serica, C.; Sharma, A.K. Drift-Velocity Closure Relationships for Slug Two-Phase High-Viscosity Oil Flow in Pipes. *Soc. Pet. Eng.* **2012**, *17*, 593–601. [CrossRef]
25. Al-kayiem, H.H.; Mohmmed, A.O.; Al-hashimy, Z.I.; Time, R.W. Statistical assessment of experimental observation on the slug body length and slug translational velocity in a horizontal pipe. *Int. J. Heat Mass Transf.* **2017**, *105*, 252–260. [CrossRef]
26. Bendiksen, K.H.; Langsholt, M.; Liu, L. An experimental investigation of the motion of long bubbles in high viscosity slug flow in horizontal pipes. *Int. J. Multiph. Flow* **2018**, *104*, 60–73. [CrossRef]
27. Moissis, R.; Griffith, P. Entrance Effects in Slug Flow. In Proceedings of the Transaction of American Society of Mechanical Engineers (A.S.M.E.), New York, NY, USA, 1 February 1962; pp. 29–39.
28. Fabre, J.; Line, A. Modeling of Two-Phase Slug Flow. *Annu. Rev. Fluid Mech.* **1992**, *24*, 21–46. [CrossRef]
29. Manolis, I.G. High Pressure Gas-Liquid Slug Flow. Ph.D. Thesis, Imperial College London, London, UK, 1995.
30. Woods, B.D.; Hanratty, T.J. Relation of slug stability to shedding rate. *Int. J. Multiph. Flow* **1996**, *22*, 809–828. [CrossRef]
31. Petalas, N.; Aziz, K. A Mechanistic Model for Multiphase Flow in Pipes. *J. Can. Pet. Technol.* **2000**, *39*, 43–55. [CrossRef]
32. Choi, J.; Pereyra, E.; Sarica, C.; Park, C.; Kang, J. An Efficient Drift-Flux Closure Relationship to Estimate Liquid Holdups of Gas-Liquid Two-Phase Flow in Pipes. *Energies* **2012**, *5*, 5294–5306. [CrossRef]
33. Kim, T.W.; Aydin, T.B.; Pereyra, E.; Sarica, C. Detailed flow field measurements and analysis in highly viscous slug flow in horizontal pipes. *Int. J. Multiph. Flow* **2018**, *106*, 75–94. [CrossRef]
34. Baba, Y.D.; Aliyu, A.M.; Archibong, A.E.; Abdulkadir, M.; Lao, L.; Yeung, H. Slug length for high viscosity oil-gas flow in horizontal pipes: Experiments and prediction. *J. Pet. Sci. Eng.* **2018**, *165*, 397–411. [CrossRef]
35. Beggs, D.H.; Brill, J.P. A Study of Two-Phase Flow in Inclined Pipes. *J. Pet. Technol.* **1973**, *25*, 607–617. [CrossRef]
36. Taitel, Y.; Dukler, A.E. A model for predicting flow regime transitions in horizontal and near horizontal gas-liquid flow. *AIChE J.* **1976**, *22*, 47–55. [CrossRef]
37. Brito, R.; Pereyra, E.; Sarica, C. Effect of Medium Oil Viscosity on Two-Phase Oil-Gas Flow Behavior in Horizontal Pipes. In Proceedings of the Offshore Technology Conference, Offshore Technology Conference, Houston, TX, USA, 6–9 May 2013; p. 285.
38. Zhao, Y.; Lao, L.; Yeung, H. Investigation and prediction of slug flow characteristics in highly viscous liquid and gas flows in horizontal pipes. *Chem. Eng. Res. Des.* **2015**, *102*, 124–137. [CrossRef]
39. Al-safran, E.; Gokcal, B.; Sarica, C. High Viscosity Liquid Effect on Two-Phase Slug Length in Horizontal Pipes. In Proceedings of the 15th International Conference on Multiphase Production Technology, Cannes, France, 15–17 June 2011; pp. 257–276.
40. Al-safran, E.M.; Gokcal, B.; Sarica, C. Investigation and Prediction of High-Viscosity Liquid Effect on Two-Phase Slug Length in Horizontal Pipelines. *SPE Prod. Oper.* **2013**, *28*, 12–14. [CrossRef]
41. Gokcal, B.; Sarica, C. Analysis and Prediction of Heavy Oil Two-Phase Slug Length in Horizontal Pipelines. In Proceedings of the SPE Heavy Oil Conference and Exhibition, Kuwait City, Kuwait, 12–14 December 2011.
42. Wallis, G.B. *One-Dimensional Two-Phase Flow*; American Institute of Chemical Engineers: New York, NY, USA, 1969.
43. Lacy, C.E. Applicability of Slug Flow Models to Heavy Oils. In Proceedings of the SPE Heavy Oil Conference Canada, Calgary, AB, Canada, 12–14 June 2012; pp. 266–275.
44. Hubbard, M.G. An Analysis of Horizontal Gas-liquid Slug Flow. Ph.D. Thesis, University of Houston, Houston, TX, USA, 1965.
45. Jepson, W.P. Modelling the transition to slug flow in horizontal conduit. *Can. J. Chem. Eng.* **1989**, *67*, 731–740. [CrossRef]
46. Kouba, G.E. *Horizontal Slug Flow Modeling and Metering*; University of Tulsa: Tulsa, OK, USA, 1986.
47. Benjamin, B. Gravity currents and related phenomena. *J. Fluid Mech* **1968**, *31*, 209–248. [CrossRef]
48. Gokcal, B.; Al-Sarkhi, A.; Sarica, C.; Alsafran, E.M. Prediction of Slug Frequency for High Viscosity Oils in Horizontal Pipes. In Proceedings of the SPE Annual Technical Conference and Exhibition, New Orleans, LA, USA, 4–7 October 2009.

49. Al-Safran, E. Investigation and prediction of slug frequency in gas/liquid horizontal pipe flow. *J. Pet. Sci. Eng.* **2009**, *69*, 143–155. [CrossRef]
50. Kora, C.; Sarica, C.; Zhang, H.; Al-Sarkhi, A.; Al-Safran, E. Effects of High Oil Viscosity on Slug Liquid Holdup in Horizontal Pipes. In Proceedings of the Canadian Unconventional Resources Conference, Calgary, AB, Canada, 15–17 November 2011.

fluids

MDPI

Review

Wormlike Micellar Solutions, Beyond the Chemical Enhanced Oil Recovery Restrictions

Emad Jafari Nodoushan [1], Taeil Yi [2], Young Ju Lee [1,3] and Namwon Kim [1,4,*]

1 Materials Science, Engineering, and Commercialization, Texas State University, San Marcos, TX 78666, USA; e_j107@txstate.edu (E.J.N.); y_l39@txstate.edu (Y.J.L.)
2 School of Mechanical Engineering, Kyungnam University, Changwon 51767, Korea; yti0811@kyungnam.ac.kr
3 Department of Mathematics, Texas State University, San Marcos, TX 78666, USA
4 Ingram School of Engineering, Texas State University, San Marcos, TX 78666, USA
* Correspondence: n_k43@txstate.edu

Received: 10 June 2019; Accepted: 10 September 2019; Published: 17 September 2019

Abstract: While traditional oil recovery methods are limited in terms of meeting the overall oil demands, enhanced oil recovery (EOR) techniques are being continually developed to provide a principal portion of our energy demands. Chemical EOR (cEOR) is one of the EOR techniques that shows an efficient oil recovery factor in a number of oilfields with low salinity and temperature ranges. However, the application of cEOR under the harsh conditions of reservoirs where most of today's crude oils come from remains a challenge. High temperatures, the presence of ions, divalent ions, and heterogeneous rock structures in such reservoirs restrict the application of cEOR. Polymer solutions, surfactants, alkaline-based solutions, and complex multi-components of them are common chemical displacing fluids that failed to show successful recovery results in hostile conditions for various reasons. Wormlike micellar solutions (WMS) are viscoelastic surfactants that possess advantageous characteristics for overcoming current cEOR challenges. In this study, we first review the major approaches and challenges of commonly used chemical agents for cEOR applications. Subsequently, we review special characteristics of WMS that make them promising materials for the future of cEOR.

Keywords: wormlike micellar solutions (WMS); enhanced oil recovery (EOR); chemical EOR (cEOR); viscoelastic surfactants (VES)

1. Introduction

In primary oil extraction methods, the natural pressure of the reservoir and artificial lift oil devices are in use for oil extraction, which results in an average recovery factor of less than 30% of oils in a reservoir [1]. Secondary oil recovery methods, such as water or gas injection, are then applied to force the oil resting in the reservoir and push them to the surface of the earth. However, more than 50% of oil in a reservoir remains trapped. Tertiary or enhanced oil recovery (EOR) methods are techniques seeking to increase reservoir pressure or alter immobile oils properties to make trapped oils more conducive for extraction [1,2]. A general classification of EOR methods had been provided by Thomas, as represented in Figure 1 [3]. Capillary Number (N_c) and mobility ratio (M) are two major parameters that determine the mobilization of the residual oil in the reservoirs. N_c and M are defined, as the following:

$$N_c = \frac{v\eta}{\sigma} \quad (1)$$

$$M = \frac{\lambda ing}{\lambda ed} \quad (2)$$

where v is the Darcy velocity (m/s), η is the dynamic viscosity of a displacing fluid (Pa·s), σ is the interfacial tension (N/m), λ_{ing} is the mobility of the displacing fluid (e.g., water), and λ_{ed} is the mobility of the displaced fluid (e.g., oil). The mobility factor of each fluid is defined as:

$$\lambda = \frac{k}{\eta} \tag{3}$$

where k is the effective permeability (m^2) [3,4]. In chemical EOR methods (cEOR), a chemical formulation is added to displacing fluid with the aim of decreasing the mobility ratio and/or enhancing the capillary number. Two major approaches in cEOR are (1) increasing the viscosity of the displacing fluid, which results in an enhancement in the mobility and improving the volumetric sweep efficiency and (2) adding chemicals to decrease the interfacial tension (IFT, σ in Equation (1)) between the oil and displacement fluid that allows trapped oil to flow through low-permeability zones [5].

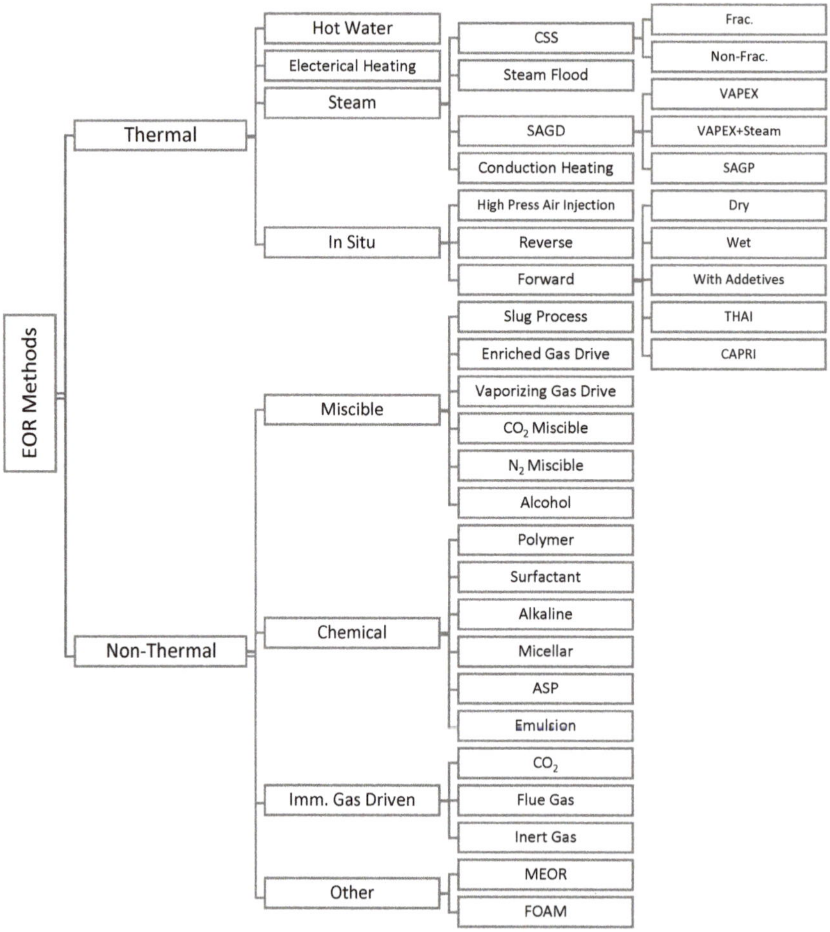

Figure 1. Classification of enhanced oil recovery (EOR) methods. Reproduced from reference [3] under a Creative Commons license.

Water-based polymer solutions, surfactants, low-salinity water, and alkalizes were reported as useful displacing fluids for cEOR [6–9]. Commercial EOR chemicals were applied for years in reservoirs with low temperatures and low-salinity brines [10]. However, these chemical agents generally suffer from poor salt and temperature tolerances, in addition to problems, such as loss of viscosity due to shear degradation when injecting them into reservoirs under harsh conditions [10,11]. The challenge lies in developing EOR chemicals for high-temperature and high-salinity reservoirs [12,13]. In next sections, we survey the most important polymeric and surfactant-based chemical agents for cEOR and address their major problems restricting efficient EOR. Afterwards, we review the wormlike micellar solutions (WMS) and their characteristics that make them promising candidates for the future of cEOR applications.

2. Non-Newtonian Displacing Fluids

Displacing fluids, such as polymer solutions and viscoelastic surfactant solutions, are Non-Newtonian fluids [14–17]. These fluids do not follow the Newtonian law of viscosity ($\tau = \eta\dot{\gamma}$), where viscosity (η) is independent of the applied shear rate ($\dot{\gamma}$). Instead, the majority of displacing fluids show shear-thinning behavior that shows decreased viscosity (η ($\dot{\gamma}$) $\neq$ Constant) with higher shear rates [18]. Shear-thinning behavior is of importance in cEOR, because it facilitates the injectivity of displacing fluids by decreasing the viscosity of fluid under high stresses presented in oil reservoirs [19]. However, the shear-thinning characteristic of the fluid is unfavorable for sweeping oil, because the loss of viscosity would lower the mobility ratio (see Equations (2) and (3)). In addition, shear-thinning displacing fluids tend to flow through high permeability zones of oil reservoir, while low permeability zones remain un-swept [20,21]. Therefore, displacing fluids showing shear-thickening characteristics attracted interest [15,22,23]. Shear-thickening, an enhancement of fluid viscosity when fluid is subjected to incremental stresses, is observed in wormlike micellar solutions (WMS) when the solutions are sheared beyond a critical shear rate [24,25]. The shear-thickening characteristic is attributed to the formation of shear-induced structures (SIS) (see Section 6.3 for more details) and it is advantageous for retarding the flow of displacing fluids that are injected in high-permeability zones of oil reservoirs [20,26]. A further discussion on the characteristics of displacing fluids caused by their non-Newtonian rheology is presented in Sections 6.1–6.3.

3. Polymeric Solutions Restrictions as cEOR Agents

Various polymers, such as hydrolyzed polyacrylamide (HPAM) [27], xanthan gum [28], salinity-tolerant polymer KYPAM [29], hydrophobically associating polymers, and 2-acrylamide-2-methyl propane-sulfonate co-polymers (AMPS) [30], have been examined for applications in cEOR. Even though HPAM or xanthan gum are used as one of main components for polymer displacing fluids, xanthan gum is different from HPAM, as it is a biopolymer. Since biocides are required to avoid the biological degeneration of the biopolymers, the process is quite expensive and economically unfeasible. When considering this, HPAM is the most widely used polymer for mobility control [7,31]. The backbone of HPAM consists of monomer units of acrylamide and acrylic acids. In aqueous solutions with low salinity, such as deionized water, the charge repulsion of the carboxylic groups on the HPAM backbone elongates the chain structure, which results in a highly viscous HPAM solution. However, the carboxylic group's charges are neutralized or shielded by salt ions of high-salinity brine solutions and, in turn, the HPAM chains lose their stretched states and the viscosity of solution decreases [31].

The temperature of the reservoir is also another critical factor that affects HPAM hydrolysis in cEOR applications [32–34]. The reservoirs, which are dormant candidates for cEOR methods, in certain areas, such as in Kuwait, Malaysia, Canada, Germany, etc., have issues that are related to this [34–36]. The salinity of reservoir brine is usually more than 200,000 ppm and the temperatures vary from 70 °C to 120 °C in those regions [7,35,37]. However, the working temperature of HPAM polymers is below 60 °C and the salinity resistance of them is below 35,000 ppm of NaCl, including 1000 ppm of divalent salts. Copolymers of acrylamide and sodium salt of acrylamide propyl sulfonated (AMPS) acid were introduced to overcome those shortcomings, which are more tolerant against the salts and high temperature when compared to HPAM polymers. Nevertheless, the AMPS polymers have lower viscosities than HPAMs due to the low solubility of AMPS in water [35,37–39]. Low-permeability layers in a reservoir will be blocked after improper employment of AMPS fluids due to the presence of the bulky polymer chains and surfactant foams. This event will cause a drop in the amount of oil production from these regions. Another problem that is associated with polymer solutions is the rise in the effect of preferential flow in high-permeability zones because of the decreased viscosity that is caused by the shear-thinning nature of the polymers [12,40,41].

In many well site treatments, the operational complexity of fluid delivery significantly influences the total treatment cost, speed, and reliability. Traditional oilfield fluid thickeners consist of high-molecular-weight biopolymers, such as guar and derivatized cellulose and, sometimes, synthetic polymers such as polyacrylamides [42]. Although these polymers have effective viscosities, they often require specific mixing and hydration procedures, as well as many field-specific chemical additives. If a polymer solution is to be crosslinked for enhanced viscosity, additional crosslinking agents, such as borate or zirconium ions, must be introduced along with appropriate pH buffers. If a crosslinked polymer solution cannot be reliably delivered through a tube due to shear degradation or excessive frictional pressure, then additional treatments that allow for temporal suspension of the crosslinking must be consolidated into the fluid design [42,43]. The treatments that can be considered include using competing ligands to temporarily bind the metal crosslinking agents, adding pH buffers to control the crosslinking reaction kinetics, or using two or more organometallic complexes as crosslinking agents, in which one can limit the crosslinking for initial transportation in the tube and another can initiate the crosslinking in the fracture site. However, organometallic crosslinked fluids require complex formulations that have to be modulated for each treatment operation [43]. Various oil field conditions, including mix-water composition, tubular shear rate, well depth, and temperature, require corresponding performance-adjusting additives and make it difficult to routinely achieve maximum efficiency [8]. While considering the discussion, it is important to review alternative materials that are technically applicable for cEOR.

4. Surfactant-Based Fluids

Surfactants are added to displacing fluid to decrease IFT between oil and injected recovery fluid to enhance trapped oil recovery through tight pore necks and low-permeability zones of reservoirs [3,35,44]. However, these solutions are not capable of withstanding the harsh conditions of reservoirs. The mineral components of rock formations play a key role in reaction with and absorption of surfactants. Moreover, cleaning the surfactants emulsions and disposal treatment of them remain a challenge [44,45]. Materials containing alkaline components can react with the acid portion of the crude oil. The reaction results in the production of surfactants in situ. The major purpose of using alkaline chemicals is to reduce the IFT, because it can cause wettability alterations [9,46,47]. The complexity of the process design because of the various undesired reactions in the reservoirs that can occur and the low recovery factor of 0–3% of oil in place (OIP) are discouragements of using these chemical agents [46].

Alkaline-surfactant-polymer displacing fluids have been evaluated through laboratory experiments and reported during recent decades [48]. However, multi-component displacing fluids are exposed to chromatographic separation in the oil reservoirs. This was demonstrated to occur in the Daqing reservoirs in China [49,50].

5. Viscoelastic Surfactants

Viscoelastic surfactants (VES) were initially introduced by Dow Chemicals as wormlike micelles and were used as thickeners in consumer products [51]. VES are also promising chemical agents for cEOR that provide a combination of viscosifying agents (i.e., water-based polymer solutions) and a lowering of IFT chemicals (i.e., surfactants), which results in advantages in the recovery of oil. Surfactants that are added to displacing fluids form spherical micelles of oil in water, while VES are self-assemblies with longer and complex structures that maintain a high viscosity in displacing fluid, even at a low concentration [52–54]. At the same time, their surface-active molecules decrease the IFT [21,52]. Morvan et al. investigated the effect of the addition of wormlike micelles to displacing fluids. In core flood experiments with low-salinity brine, WMS injection resulted in a significantly higher recovery of oil than with a water injection alone [54]. There is a complex interplay of properties and characteristics that make VES/WMS powerful candidates for cEOR. In the next sections, WMS and unique characteristics of wormlike micelles are reviewed in more detail.

6. Wormlike Micellar Solutions (WMS)

WMS have emerged as a suitable alternative material to polymer solutions for cEOR. The surfactants begin to form micelles through self-assembly above the critical micelle concentration (CMC) [12,55]. Amongst the several types of supramolecular micelles, the very long cylindrical and flexible wormlike micelles have attained considerable interest from both theorists and experimentalists in virtue of their unique micellar constructions and remarkable rheological responses. The rheological characteristics of these WMS are comparable to polymer solutions [56]. The ability of wormlike micelles in increasing the viscosity of water-based solutions (water thickening) are identified as being similar to that of water-soluble polymers. Although the long cylindrical chains of wormlike micelles show similarity to the long chain of high-molecular-weight polymers, the rheological behavior of WMS is more complex than that of polymer solutions. Typically, WMS systems are formed from a mixture of surfactants and salts, and any change in the concentration of each component can contribute to a change in the rheological behavior of WMS. Moreover, the solubility of surfactant and salts in aqueous solvents is a function of temperature [57]. The driving force in the formation of micellar aggregations is hydrophobic interactions, which are physical interactions that are weaker than the covalent bonds that form the polymers' backbone structures. Thus, unlike polymer chains that are difficult to break once they are formed, micellar structures break and reform continuously. The kinetics of formation and scission of wormlike micellar structures is also a function of temperature [58]. Therefore, WMS are highly sensitive to temperature, because any change in temperature results in altering the solubility of micellar components in the solution and, also, the change in kinetics of micellar reversible scission. In this way, WMS chains are more sensitive to conditions, such as temperature. Wang et al. [59] reported the rheological comparison between the WMS and ultrahigh molecular weight (UHMW) polymer aqueous solution. They chose erucyl dimethyl amidoprpyl betaine (EDAB) to form long WMS chains that are independent of salt and nonionic polyacrylamide (PAM) for a 12 MDa UHMW polymer solution, which demonstrated a strong thickening capability in saline water at room temperature. It has been reported that both WMS and UHMW polymer solution are effective water thickeners and are able to retain considerable viscosity up to 85 °C. However, the dynamic viscosity of WMS follows the Maxwellian model with a single relaxation time, while the UHMW PAM solution demonstrates a spectrum of relaxation times. Figure 2 shows the zero-shear viscosity (η_0) of both WMS and UHMW PAM in an Arrhenius plot. While both show an exponential decrease in their zero-shear viscosity at higher temperatures, the higher slope of WMS confirms that they are more sensitive to temperature than the UHMW polymer solution [59].

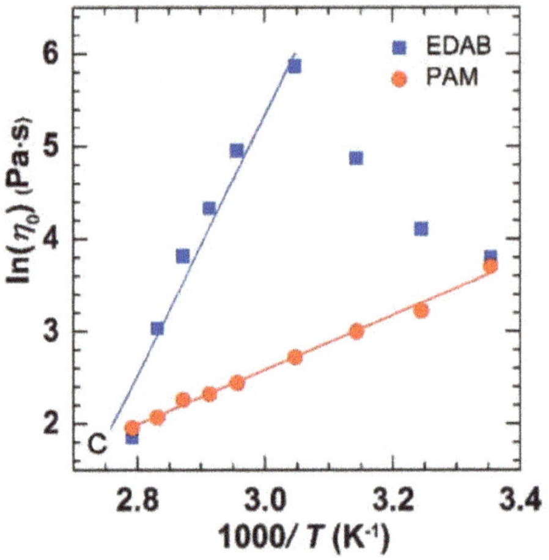

Figure 2. Arrhenius plot of zero-shear viscosity of erucyl dimethyl amidoprpyl betaine (EDAB) and ultrahigh molecular weight (UHMW) polyacrylamide (PAM) verses 1/T. Reprinted from reference [59] with permission from the Royal Society of Chemistry.

6.1. Shear-Thickening Behavior

Several methods have been developed to induce wormlike micelles from spherical micelles. One way is to increase the concentration of a surfactant with extra counter ions, which leads to the growth of the micelles in one dimension, resulting in the wormlike configuration. The nature of viscosity depends on both the size and shape of the solutes in a solution at the same concentration. Even for similarly sized solutes, a solution with solutes of a relatively higher aspect ratio, such as wormlike micelles, has a higher viscosity than a solution with lower aspect ratio solutes. As aforementioned, WMS behave in a similar way as many polymer solutions in terms of mechanical disturbance and viscoelasticity [60]. The majority of viscoelastic fluids that are being used for cEOR show shear-thinning behavior [14,15]. However, Liu and Pine reported unusual behavior in viscosity of WMS that they prepared while using aqueous solutions of hexadecyltrimethyl ammonium bromide (CTAB) and sodium salicylate (NaSal) [61]. Figure 3 shows the viscosities of WMS as a function of shear rate for three different concentrations of CTAB/NaSal in water. When the solutions are sheared beyond a critical shear rate, they show a sudden increase in viscosity, so-called shear-thickening behavior [62–64].

The power-law model given by Equation (4) is a simple mathematical model that describes the behavior of fluids:

$$\eta(\dot{\Upsilon}) = K\dot{\Upsilon}^{n-1} \tag{4}$$

where K is the flow consistency index and n is the flow behavior index. Based on the value of n, the fluid behavior is either shear-thinning ($n < 1$), Newtonian ($n = 1$), or shear-thickening ($n > 1$) [65]. However, viscosity curve of a shear-thickening fluid is typically constituted of three regions: a slight shear-thinning at low shear rates, a subsequent viscosity increases over a critical shear rate, and a shear-thinning region at high shear rates [66]. Therefore, a power law model fails to describe the fluid behavior in the entire range of applied shear rates. For example, while a power-law model (with $n > 1$) can describe the shear-thickening behavior of fluid in intermediate shear rates, it fails to fit the regions of low and high shear rates where the fluid shows the shear-thinning (Figure 3b). Galindo-Rosales et al. [67] suggested a composite function that is given by Equation (5), taking all regions of viscosity curve into account:

$$\eta(\dot{\gamma}) = \begin{cases} \eta_{I}(\dot{\gamma}) & \text{for } \dot{\gamma} \le \dot{\gamma}_c \\ \eta_{II}(\dot{\gamma}) & \text{for } \dot{\gamma}_c < \dot{\gamma} \le \dot{\gamma}_{max} \\ \eta_{III}(\dot{\gamma}) & \text{for } \dot{\gamma}_{max} < \dot{\gamma} \end{cases} \quad (5)$$

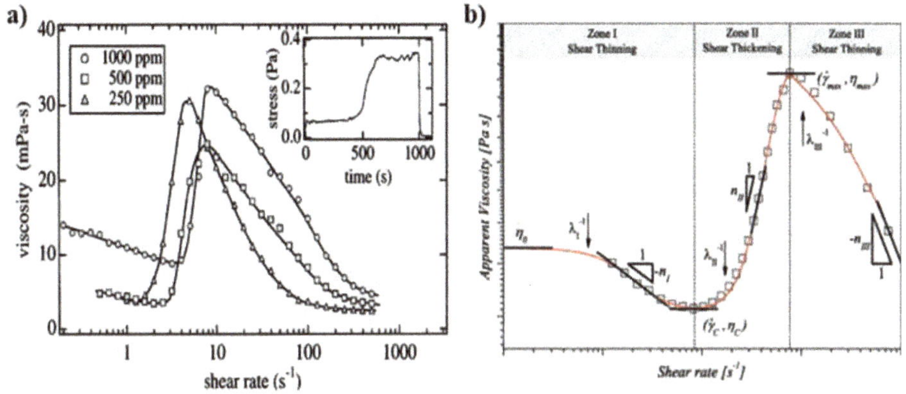

Figure 3. (**a**) Variation in the viscosity as a function of shear rate for hexadecyltrimethyl ammonium bromide/sodium salicylate wormlike micellar solutions (CTAB/NaSal WMS). The inset illustrates the shear stress over time after applying a constant shear rate. The stress increases after an induction period of 400–500 s and relaxes in a few seconds upon cessation of shear flow. Reprinted from reference [61] with permission from the American Physical Society. (**b**) Three typical zones of I: shear-thinning, II: shear-thickening, and III: shear-thinning in a log–log viscosity plot of a shear-thickening fluid. Reprinted from reference [66] with permission from Elsevier.

In this equation, $\dot{\gamma}_c$ is the critical shear rate and $\dot{\gamma}_{max}$ is the shear rate in which fluid show the highest viscosity. The viscosity functions of η_i (for i = I, II, III) are branched equations based on the Cross model [68]:

$$\eta = \eta_\infty + \frac{\eta_0 - \eta_\infty}{1 + (\lambda\dot{\gamma})^n} \quad (6)$$

where η_∞ is the infinite shear viscosity, η_0 is the zero-shear viscosity, and λ is the relaxation time of the fluid. The final form of suggested model is as follows:

$$\eta(\dot{\gamma}) = \begin{cases} \eta_{I}(\dot{\gamma}) = \eta_c + \frac{\eta_0 - \eta_c}{1+\left[K_I\left(\frac{\dot{\gamma}^2}{\dot{\gamma}-\dot{\gamma}_c}\right)\right]^{n_I}} & \text{for } \dot{\gamma} \le \dot{\gamma}_c \\ \eta_{II}(\dot{\gamma}) = \eta_{max} + \frac{\eta_c - \eta_{max}}{1+\left[K_{II}\left(\frac{\dot{\gamma}-\dot{\gamma}_c}{\dot{\gamma}-\dot{\gamma}_{max}}\right)\dot{\gamma}\right]^{n_{II}}} & \text{for } \dot{\gamma}_c < \dot{\gamma} \le \dot{\gamma}_{max} \\ \eta_{III}(\dot{\gamma}) = \frac{\eta_{max}}{1+\left[K_{III}(\dot{\gamma}-\dot{\gamma}_{max})\right]^{n_{III}}} & \text{for } \dot{\gamma}_{max} < \dot{\gamma} \end{cases} \quad (7)$$

where K_i (for i = I, II, III) are time constant parameters, n_i (for i = I, II, III) are dimensionless exponents that are related to the slope of the each power-law regime, η_c is the viscosity of fluid at the critical shear rate, and η_{max} is the viscosity of fluid at $\dot{\gamma}_{max}$ (Figure 3b). The proposed model is useful for describing both the shear-thinning (either for $\dot{\gamma} \le \dot{\gamma}_c$ or $\dot{\gamma}_{max} < \dot{\gamma}$) and shear-thickening ($\dot{\gamma}_c < \dot{\gamma} \le \dot{\gamma}_{max}$) regimes that were observed in the viscosity curve of a shear-thickening fluid [66,67]. However, a drawback of this model is that it possesses a high number of independent constants.

Shear-thickening behavior in WMS is attributed to shear-induced gelation that formed by long micellar chains [26]. In this approach, gelation occurs when the shear rate is high enough, i.e., above a critical shear rate $\dot{\gamma}_c$, to overcome the rotational Brownian diffusion (D_r) of the rod-like micelles [69]:

$$\dot{\gamma}_c \sim D_r \sim \frac{k_B T}{\pi L^3 \eta_s} \tag{8}$$

where k_B is the Boltzmann constant, T is the absolute temperature, L is the length of the rods, and η_s is the viscosity of the solvent. Barentin and Liu [70] presented a new mechanism for the formation of shear-induced structure, in which the micelles collide under the shear flow and aggregate into bundles. Unlike the approach explaining the shear-induced gelation due to overcoming the Brownian diffusion, they concluded that shear-induced gelation is attributed to micellar–micellar interactions. Section 6.3 presents further discussion on shear-induced structures (SIS).

6.2. Thermal-Thickening Response

A major challenge in cEOR is that displacing fluids lose their viscosity at high temperatures of oil reservoirs. One major factor in determining the viscosity of WMS is the contour length (L) of micellar chains [71]. In the mean-field theory of Cates and Candau [72], the average contour length ($\bar{L}$) is associated with temperature, as follows:

$$\bar{L} \sim \phi^{\frac{1}{2}} \exp\left(\frac{E_c}{2k_B T}\right) \tag{9}$$

where ϕ is the volume fraction of surfactants and Ec is the excess free energy that is associated with hemispherical end caps of micelles as compared to their cylindrical body [72,73]. The decay in contour length with temperature leads to the decrease of viscosity upon increasing the temperature following an Arrhenius law:

$$\eta_0 \sim G_0 A \exp\left(\frac{E_a}{RT}\right) \tag{10}$$

where E_a is the flow activation energy, G_0 is the plateau modulus of WMS, R is the gas constant, and A is the pre-exponential factor [74]. Even though decreasing viscosity with temperature is a common characteristic of WMS, some of WMS have been reported to show a thermal-thickening response where the viscosity increases with temperature. In the following, we focus on studies that reported the WMS with the thermal-thickening behavior that can be beneficial for displacing fluids under the elevated temperature of an oil reservoir.

Sharma et al. [75] studied mixed nonionic WMS whileusing long polyoxyethylene chain phytosterol (Ph_yEO_{30}) and polyoxyethylene dodecyl ether ($C_{12}EO_n$, n = 3 and 4) surfactants in aqueous solutions. They reported the variation of η_0 as a function of temperature for 5 wt% of Ph_yEO_{30} + $C_{12}EO_3$ system at X = 0.36—the weight fraction of $C_{12}EO_3$ in total surfactant. As presented in Figure 4, increasing the temperature from 15 °C to 30 °C led to the increase of η_0. Temperature affects the micellar growth and in turn, the viscosity of WMS. These nonionic WMS, including polyoxyethylene head groups, are significantly influenced by temperature because the hydration of oxyethylene is sensitive to temperature. In wormlike micellar structures, the free energy density at the end-caps of wormlike micelles is higher than that of its cylindrical body. The micelles grow in length by reducing the number of end-caps in order to minimize the excess energy at the end-caps. Increasing the temperature enhances the dehydration of the EO-chain of lipophilic cosurfactant, $C_{12}EO_3$, and decreases the average cross-sectional area of the surfactant. These changes increase the end-cap energy and result in enhanced one-dimensional micellar growth.

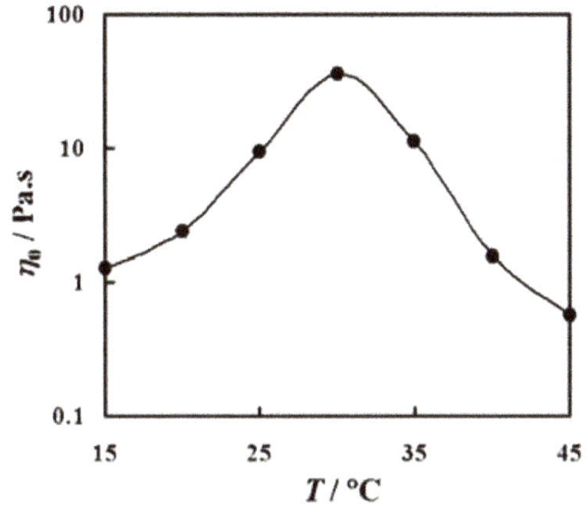

Figure 4. Variation of zero-shear viscosity (η_0) as a function of temperature (T) for 5 wt % Ph_yEO_{30} + $C_{12}EO_3$ WMS at X = 0.36. Reprinted from reference [75] with permission from the American Chemical Society.

Varade et al. [76] reported the thermal-thickening effect of nonionic WMS that was prepared by $C_{14}EO_3$ and Tween 80 in aqueous media. Figure 5a shows the typical variation of steady-state viscosity as a function of shear rate for the water/Tween-80/$C_{14}EO_3$ system at the weight fraction of $C_{14}EO_3$ in total amphiphiles, X = 0.17 at different temperatures (20, 25, and 30 °C). Increasing the temperature from 20 °C to 30 °C results in an increased viscosity of the Newtonian plateau. In addition, the shear-thinning behavior begins at lower shear rates when the temperature is elevated, which means that more micellar structures are formed at higher temperatures. Figure 5b represents the zero-shear viscosity of the water/Tween-80/$C_{14}EO_3$ solution as a function of X at different temperatures. Increasing the temperature shifts the viscosity curve to the left, which means that a lower X is required to obtain a comparable viscosity for higher temperatures. The thermal-thickening response is a result of dehydration of polyoxyethylene chains at elevated temperatures. The cross-sectional area at the hydrophile–lipophile interface and, in turn, the curvature of the cylindrical body of micelles decrease upon dehydration of the polyoxyethylene chains. Thus, the increasing length of micelles is more favorable than the formation of end-caps. Micellar growth as a function of temperature results in higher viscosity and a thermal-thickening response.

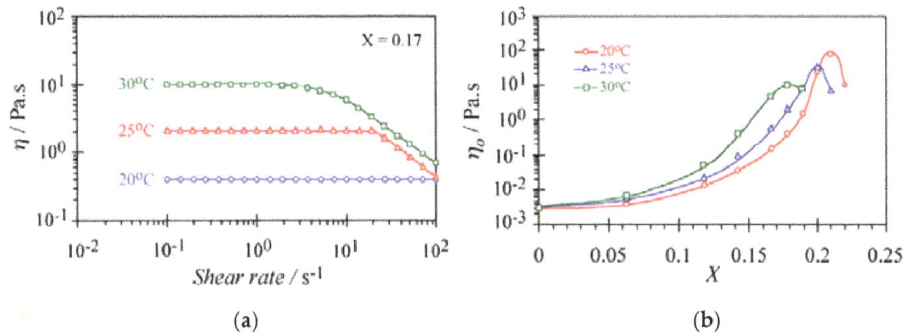

Figure 5. (**a**) Steady shear rate of the water/Tween-80/$C_{14}EO_3$ WMS at X = 0.17 at different temperatures. (**b**) Variation of zero-shear viscosity as a function of X at different temperatures in water/Tween-80/$C_{14}EO_3$ WMS. The concentration of Tween-80 in water is fixed at 15 wt%. Reprinted from reference [76] with permission from Elsevier.

The important feature of Sharma [75] and Varade [76] studies is that, even by a slight increase in temperature (15 °C and 10 °C, respectively), the zero-shear viscosity is enhanced by an order of magnitude, which is in contrast with typical thermal-thinning behavior of viscoelastic solutions described by Equation (10) (see Figure 2) [59,77]. However, in the case of polyoxyethylene solutions, the dehydration of polyoxyethylene leads to a thermal-thickening response.

Chu and Feng [78] reported the thermo-switchable behavior of WMS, in which the solution shows a reversible transition in its viscosity in response to a thermal stimuli. They used a synthesized sulfobetaine amphiphile, 3-(N-palmitamido propyl-N, N-dimethyl ammonium) propane sulfonate (PDAS) as a zwitterionic surfactant. The viscosity of 1.0 M PDAS micellar solution in the presence of 0.5 M NaCl is shown in Figure 6 as a function of shear rate. The shear viscosity of the solution at 30 °C endures a constant shear rate that matches conventional Newtonian fluid behavior. The micellar structures in the lower viscous PDAS/NaCl solution at 30 °C are associated with short rod-like micelles. However, by increasing the temperature to 40 °C, the measured viscosity of the solution is higher than the constant viscosity at a lower temperature through the entire range of shear rates. It also should be noted that the solution displays a non-Newtonian fluid behavior—shear-thinning—which is well acknowledged as evidence of the alignment of wormlike micelles in the direction of the flow field. The viscosity of this solution could be turned on and off by changing the temperature (see Figure 7).

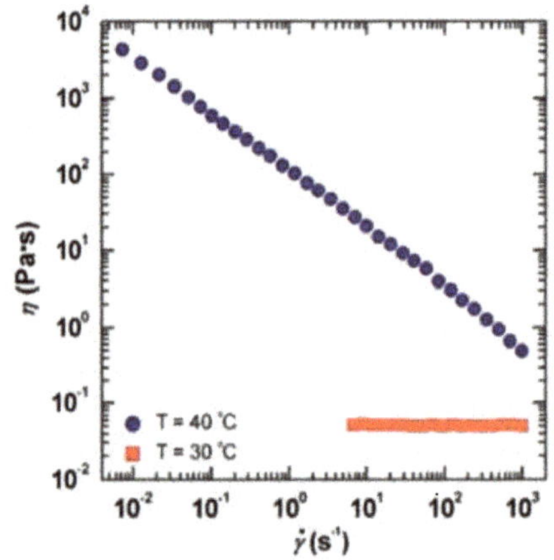

Figure 6. The thermal-thickening behavior of propane sulfonate (PDAS)/NaCl WMS. Reprinted from reference [78] with permission from the Royal Society of Chemistry.

The response of PDAS/NaCl solution to thermal stimulus is attributed to the decreased solubility of the hydrophobic parts that are influenced by heating. This behavior is similar to what occurs within polymer gels in terms of their weakening gelation upon heating [59,60].

A salting-out effect is a term that explains the decreased solubility of nonelectrolytes in water caused by the addition of electrolytes. Increasing the temperature to 40 °C leads to an enhanced salting-out effect, thus the solubility of the hydrophobic moieties of PDAS tails decreases. This results in micellization and the growth of the micelles from short rods into entangled long wormlike micelles (see Figure 8). These micellar structures immobilize the water and form hydrogels. In contrast, decreasing the temperature to 30 °C dissipates entangled long wormlike micelles due to the transfer of amphiphiles from the micellar phase to the bulk. The long self-assembled micelles revert to the short rods and the overall viscosity of the solution significantly decreases [78].

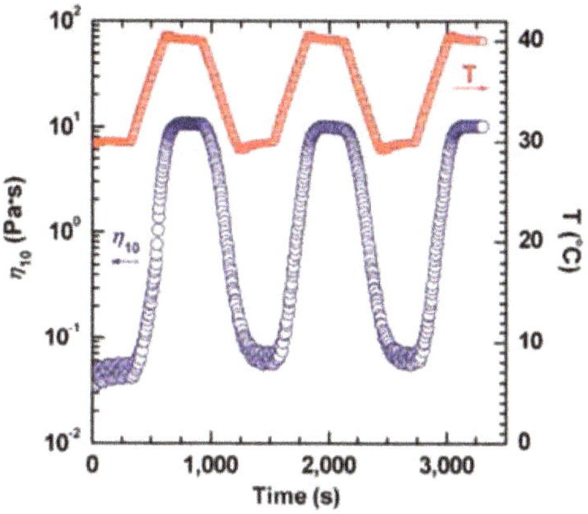

Figure 7. The thermo-reversible rheological response of PDAS/NaCl WMS at shear rate of 10 s^{-1}. Reprinted from reference [78] with permission from the Royal Society of Chemistry.

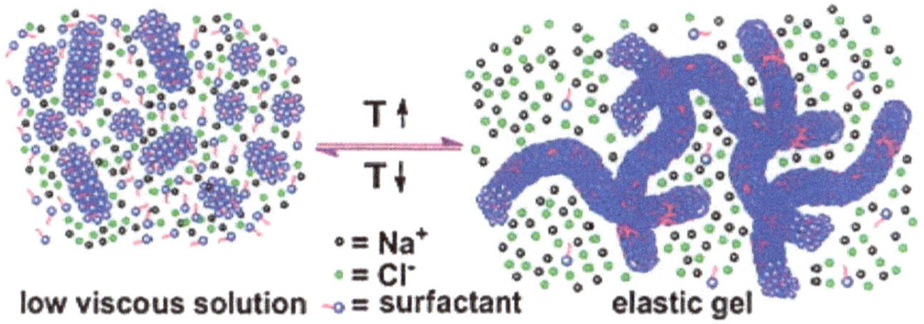

Figure 8. A schematic illustration of the switchable gelation mechanism. Reprinted from reference [78] with permission from the Royal Society of Chemistry.

6.3. Formation of Shear-Induced Structure (SIS)

In WMS systems, micellar chains continuously break and reform due to their weak physical bonds (see Section 6.5 for more details). This results in a broad and dynamic distribution of micellar chain lengths that change along with a shear or an extensional flow regime. The resulting flow behavior is even more complex when compared to that of polymeric fluids. Above a critical shear rate in the steady-state shear flow, the shear viscosity increases as a new and more viscous phase develops. This shear-induced state is called shear-induced phase (SIP) [79] or shear-induced structure (SIS) [80]. Rehage and Hoffmann [81,82] observed that an aqueous WMS of CPyCl/NaSal forms a gel-like structure along the direction of shear flow above a critical shear rate. This gel-like structure is referred to as an SIS, a structure that will disintegrate with the cessation of the shear flow [82,83]. The general rheological response of the CPyCl/NaSal solution under the Couette flow condition confirms that the formation and disbanding of SIS increases and decreases the viscosity of the solution, respectively. The formation of SIS occurs when a flow regime causes a stretching and tumbling of wormlike micelles, which invoke micelle–micelle collisions that produce a complex viscous network. Rojas et al. [83] revealed that, while it is not possible for short micelles to form long-range connections under static conditions, these connections can be elongated by imposing an appropriate level of shear stress and, in turn, increase the viscosity. Jones [84] hypothesized that it is possible to employ shear-thickening characteristics to overcome the challenge of preferential flow, which tends to only flow through the high-permeability zones of reservoirs rather than the low-permeability zones, where trapped oils remain upswept. This hypothesis also led the work of Golombok et al. [20,85], who carried out experimental investigations on pressure-driven flows within cylindrical glass core samples with different permeability classes. The shear-thickening response that is caused by an associated flow through the high-permeability zones will result in a higher local fluid viscosity when compared to the viscosity of fluid flowing through the low-permeability zones under the same pressure drop. The fluid viscosity decreases in the low-permeability zone as the shear rate decreases. Different frictional losses will be developed according to the variation of the viscosity in different permeability zones. The formation of SIS in WMS is advantageous for the selective retardation of flow in the most permeable zone of reservoirs. SIS form in higher permeability zones because of higher shear rates and increase the fluid viscosity in those regions. Thus, the fluid velocity at higher permeability zones slows down relative to the fluid in the low-permeability zones. Golombok et al. [20] reported that the formation of SIS led to a 36% increase in the volume of oil that can be swept by WMS displacing fluid when compared to the brine displacing fluid made of 3 wt% of NaCl in demineralized water.

6.4. Salinity Resistance

Conventional polymer solutions that are used for EOR are not able to sustain their viscosity under high-temperature or high-salinity conditions of oil reservoirs. Viscosity loss and precipitation of most water-soluble polymers can occur due to the presence of divalent cations (i.e., Ca^{2+}, Mg^{2+}), which lead to the hydrolysis of polymers [86]. Polymer solutions, in particular HPAM, are widely used as oil-displacing agents [27]; however, their poor salt tolerance and thermal instability impeded their applications in high-salinity and high-temperature oil reservoirs. More salt-tolerant and thermostable monomer or groups, such as SO_3^-, onto the HPAM backbone have been introduced to enhance the characteristics of HPAM in a high-salinity and high-temperature environment [87]. However, the main portion of the polymer backbone is still the acrylamide segment, which demonstrates poor long-term thermal stability and salt tolerance. Zhu et al. [12] introduced a complex of wormlike micellar EDAB and HPAM to overcome the deficiencies. The HPAM/EDAB hybrid samples were studied in comparison with sole HPAM and EDAB in a synthetic brine regarding their rheological behaviors. They performed experiments under simulated high-temperature and high-salinity oil reservoir conditions (T: 85 °C; total dissolved solids, TDS: 32,868 mg/L; $[Ca^{2+}] + [Mg^{2+}]$: 873 mg/L). A major factor for the use of any displacing fluid in high-temperature and high-salinity reservoirs is that they enable a solution to retain a high viscosity after exposure to harsh environments. Figure 9 shows the viscosity variations of sole HPAM, EDAB, and two hybrid solutions of them at 85 °C.

While the HPAM solution shows a significant decay in the viscosity as a function of time, the hybrid solutions sustain their viscosity much more effectively than HPAM solution after aging. Most importantly, the EDAB solutions containing the ultra-long-chain zwitterionic surfactant possess long-term stability when compared to other solutions containing HPAM. Even with its long-term stability, the oil recovery factor when using EDAB has been reported to be lower than that when using EDAB/HPAM hybrid solutions. The recovery factor for EDAB was only 1.9% as compared to the factors of 10.20% and 7.10% for P15E15 and P10E20, respectively. The authors suggested that the collapse of micellar assemblies upon contacting with the oil and disassembly of micelles at shear and elongation forces imposed to WMS at pore throats might describe the lower oil recovery factor of EDAB. However, further studies are required for verification of their hypothesis.

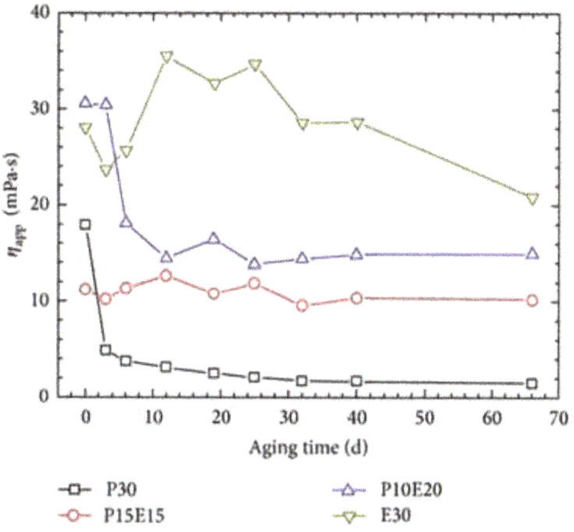

Figure 9. Long-term thermal stability of HPAM/EDAB hybrid samples in comparison with EDAB and HPAM (TDS = 32,868 mg/L, $[Ca^{2+}] + [Mg^{2+}]$ = 873 mg/L, $\dot{\gamma} = 7.34\ s^{-1}$). Both the aging and measuring temperature is 85 °C. P30: 0.3 wt% of HPAM and 0 wt% of EDAB; P10E20: 0.1 wt% of HPAM and 0.2 wt% of EDAB; P15E15: 0.15 wt% of HPAM and 0.15 wt% of EDAB; and E30: 0 wt% of HPAM and 0.3 wt% of EDAB. Reprinted from reference [12] under a Creative Commons license.

Cao et al. [88] studied the effects of adding NaCl, Na_2CO_3 and NaCl/NaOH in NaOA solutions and demonstrated that the zero-shear viscosity of the solution increased with the micellar growth when the concentrations of Na_2CO_3 and NaCl increase (see Figure 10). The reason behind the increased zero-shear viscosity is that, by increasing salt concentrations (electrolytes), the oleic acid salting-out effect leads to longer micellar chains. As shown in Figure 10, both salts induce a significant enhancement of the viscosity. The increase in the length of rod-like micellar chains as a result of the salting-out effect promotes micelles long enough to be wormlike micelles and shows a rapid increase in the zero-shear viscosity.

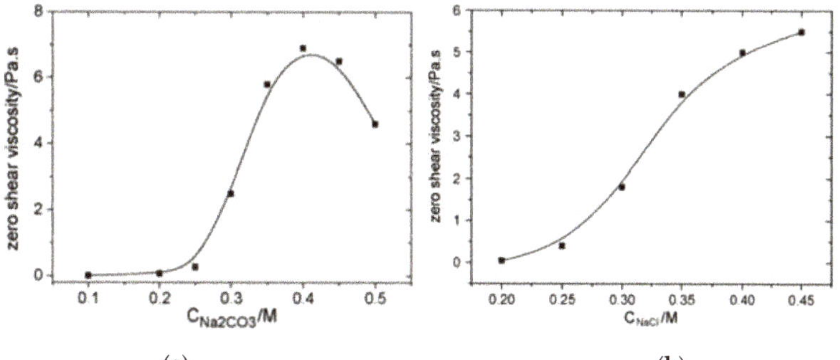

Figure 10. Zero-shear viscosity of NaOA/Na_2CO_3 WMS as a function of (**a**) Na_2CO_3 with a fixed NaOA concentration of 0.060 M and (**b**) NaCl concentrations at fixed NaOA and NaOH concentrations of 0.60 M and 0.10 M, respectively. Reprinted from reference [88] with permission from Elsevier.

However, an excessive concentration of Na_2CO_3 (higher than ~0.4 M) results in a decrease in the zero-shear viscosity. The reason is that excessive electrolyte might further reduce the net charge of the micelles and the linear micellar chains may turn into branched micellar chains. Unlike branching in polymers, it has been reported that branching in micelles assists stress relaxation, which results in a shortened relaxation time and lower zero-shear viscosity. The branched points of wormlike micelles are not chemical connections and the wormlike micelles can slide against one another (i.e., reptation). Reptation can occur along every path of branched wormlike micelles as a mechanism for the stress relaxation, which leads to very low zero-viscosity [89,90]. The difference in the viscosity trend at high salt concentrations between NaCl and Na_2CO_3 is because of hydroxide ions that can suppress the hydrolysis of oleic acid. The hydration of ${CO_3}^{2-}$ is not as prominent as that of Cl^-; therefore, less wormlike micellar chains of oleic acid are subjected to hydrolysis in the presence of Cl^- than ${CO_3}^{2-}$ ions. These studies are evidence that ultra-long chain zwitterionic surfactants, such as EBDA and anionic-based WMS solutions, are promising materials that can enhance the tolerance of displacing fluids against the high-salinity conditions of oil reservoirs.

6.5. Reversible Scission

Even though wormlike micelles show many of the same viscoelastic characteristics of polymers, they are quite physically different from polymers. One of the significant behavioral differences between WMS and polymer solutions stems from the type of bonds in their structures. In polymer solutions, monomers are covalently bonded to each other to form a chain. In this way, once they are formed, the polymer chains are not easily dissociated. In wormlike micelles, surfactants units aggregate together by relatively weak physical attractions. The micellar structures in the equilibrium state break and reform continuously, where the kinetics of breakage and reformation reactions are greatly dependent on the surfactants and salt concentrations, temperature, and flow. The dynamic break and reform of WMS is the reason that they are also called living polymers [59]. Branched wormlike micelles relax from imposed stresses via reptation. According to the living polymer model [91], wormlike micelles can relax from stresses while using an additional mechanism in which the micellar chains breakdown into shorter micellar chains. In wormlike micelles, the micellar breaking time (τ_b) is shorter than reptation time (τ_r), thus the breaking of the micellar structure is the dominant mechanism for stress relaxation time (τ_R). The stress relaxation time of micelles is the geometric mean ($\tau_R \sim \sqrt{\tau_r \tau_b}$) of breaking and reptation times.

To date, various approaches have been taken to model the behavior of WMS by considering their breakage and reformation dynamics. Marques et al. [92] suggested that wormlike micelles can relieve stress by eliminating the entanglement points through a bond interchange mechanism. In this mechanism, two micellar chains that have an entanglement can react and form a transient four-arm star shape of micelles and then break into two new chains of different lengths. Appell et al. [93] suggested ghostlike crossing as another possible mechanism for stress relaxation of wormlike micelles. When a micellar chain collides with another chain, it forms a transient crosslink on the side where it collides, and then separates from the other side of the micellar chain. However, as the authors stated, a theoretical description of such behavior requires further investigation.

Bautista et al. [94] proposed a kinetic equation for WMS by taking the shear rate effect on the breaking and reformation into account. In this model, the breakage rate of micellar chains is considered as a function of the local shear rate. The model describes the appearance of stress plateau in a WMS flow curve at higher shear rates than the first critical shear rate. The model also predicts that WMS with higher micellar breakage rates have stress plateau values at a lower critical shear rate. Vasquez et al. [95] suggested a network scission model in which a micellar chain with the length L can break in the middle to develop two micelles of length L/2. The micellar breakage occurs at a rate that depends on the shear rate and the local elongation rate. The shorter micelles can then recombine to reform a micellar chain back to the initial length.

Several oilfield treatment processes impose different shear stresses on a fluid over the course of its employment. For example, in hydraulic fracturing, a fluid may experience very high shear as it flows through wellbore tubular and perforation orifices, but then subsequently must suspend and transport particles for an extended time (possibly hours) at low shear rates [96]. In the case of using polymeric displacing fluids, these sudden changes in shear rates are not desired, because the viscosity of displacing fluid permanently decreases under the shear rate due to the shear-thinning characteristics of polymer solutions. Thus, the ability of wormlike micellar structures to reform after shear exposure is a fundamental advantage over polymer solutions. Wormlike micelles can display a temporary shear-thinning behavior under injection conditions and then reform their assembly and gain their initial viscosity in the reservoir [97].

6.6. Drag Reduction

High-molecular-weight polymers, including flexible chains, are widely used as drag-reducing agents. They are more effective at lower concentrations, such as a few ppm. However, these materials drop their effectiveness permanently under the extreme shear or extensional stresses due to the breakage of primary bonds, predominantly near the chains' midpoints [98]. A number of studies investigated the influence of wormlike micellar structure and their rheological characteristics on drag reduction [99–101]. The authors reported that there is a correlation between drag reduction and high apparent extensional viscosity. High extensional viscosity enables WMS and polymer solutions to resist against vortex stretching, which results in lower energy dissipation, or so-called drag reduction phenomena [102]. An analysis of drag reduction that is caused by polymer-induced turbulence revealed that there is a maximum drag reduction asymptote (MDRA), which is the maximum extent of drag reduction that can be achieved [103]. Although there are similar mechanisms of drag reduction suggested for polymer and wormlike micelles, MDRAs of WMS drag reducing agents (DRA) have been reported to be 5% to 15% higher than that of polymer DRA, depending on Reynolds number [104]. Brostow [105], Wang et al. [106], and Hong et al. [107] provided findings regarding the drag reduction in transportation of viscoelastic solutions. Despite the fact that the mechanism of drag reduction is still considered puzzling, it has been agreed that injection of macromolecules such as polymer chains and wormlike micelles leads to the reduction of turbulence in a sublayer adjacent to the solid pipe wall and it reduces the friction between the fluid and walls [108]. Drag reduction can be calculated, as

$$\mathrm{DR}\ (\%) = \left(\frac{f_s - f_{DRA}}{f_s}\right) \times 100 = \left(1 - \frac{\Delta P_{DRA} \cdot {Q_s}^2}{\Delta P_s \cdot {Q_{DRA}}^2}\right) \times 100 \tag{11}$$

where f is the friction factor, ΔP is the pressure drop, Q is the volumetric flow rate, and indices of s and DRA refer to the solvent and drag reducing agents, respectively [108]. Kotenko et al. [109] studied the drug reduction while using two WMS prepared by dissolving commercial surfactant-based DRA (Beraid DR-IW 616 and 618, AkzoNobel, Amsterdam, The Netherlands) and sodium nitrite in deionized water. The authors measured the pressure drop in test rigs with main components of a centrifugal pump, a water tank, expansion tanks, and straight sections of pipes, and reported the drug reduction of 60% to 80% when using WMS as compared to pure water. The drag reduction is attributed to the presence of wormlike micellar chains that serve as a buffer between the neighboring turbulent eddies, and consequently reduce turbulence of the flow. Moreover, authors suggested that wormlike micelles form a viscous sublayer adjacent to the wall of the pipes, which reduces the friction between the test fluid and pipe walls (Figure 11a).

In many applications, VES are effective DRA, as well as particle suspension media. The major application of drag reduction is to enhance the flow rate in crude oil pipelines by the addition of polymers and surfactants [110]. When comparing the MDRA of surfactant fluids to that of polymer solutions reveals the superior drag reduction of the surfactant under the same flow conditions [111,112]. Additionally, surfactant DRA have significant advantages, such as the ability to restructure their micellar assemblies and sustained rheological properties, even after being exposed to high shear (Figure 11b). This shear-recovery attribute is a notable advantage over polymer solutions that can be irreversibly degraded during fluid pumping [5,109,113].

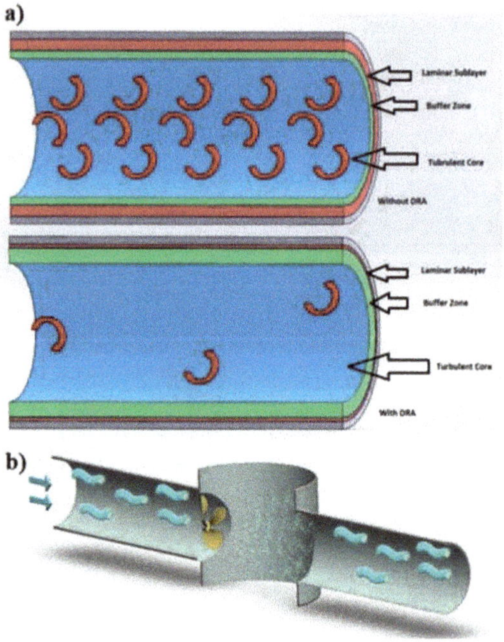

Figure 11. (**a**) The dampening mechanism of the turbulent bursts. (**b**) Rearrangement of broken wormlike micelles after crossing a pump. Reprinted from reference [109] with the permission from Elsevier.

6.7. *Hydrocarbon Responsivity*

The addition of hydrocarbons significantly influences the behavior of WMS systems. Entangled wormlike micellar networks can be eliminated by adding a small quantity of hydrocarbons, and thus the initial viscoelastic WMS is transformed into a Newtonian water-like solution. Surfactant micellar solutions tend to dissolve a certain amount of hydrocarbons by the interaction between the hydrocarbon and the hydrophobic core of micelles. Even though aqueous solutions containing globular micelles are able to trap a small amount of hydrocarbons, the maximum concentration of dispersed hydrocarbons in these solutions is lower by at least one order of magnitude than the concentration of the surfactants. However, solutions containing wormlike micelles dissolve larger amounts of hydrocarbons in which the hydrocarbon/surfactant ratio can reach about one, as reported by Hoffmann and Ebert [81]. Hydrocarbons induce the transformation of entangled wormlike micellar chains into isolated globular aggregates with radii from 20 to 500 Å. Thus, the viscoelastic WMS can be turned into a Newtonian low-viscosity solution.

Shibaev et al. [114] investigated the effect of hydrocarbon addition on the rheological characteristics and the structure of WMS formed by potassium oleate. In their study, the addition of 0.5 wt% n-dodecane results in a drastic decrease in viscosity by up to five orders of magnitude. This is attributed to the complete disruption of wormlike micellar chains and the formation of micro-emulsion droplets. As demonstrated in Figure 12, the transition state of wormlike micellar chains to micro-emulsion droplets can be classified into three regions:

- In region I, the WMS demonstrates high values of viscosity (~10–350 Pa·s), because the micellar chains are still entangled with each other. However, their lengths are reduced by the dissolved hydrocarbons.

- In region II, the micellar chains are transformed to the unentangled regime and the viscosity drastically drops due to a further shortening of micellar chains and the formation of micro-emulsion droplets.
- In region III, the viscosity is as low as ~0.001 Pa·s, in which only microemulsion droplets are present in the solution.

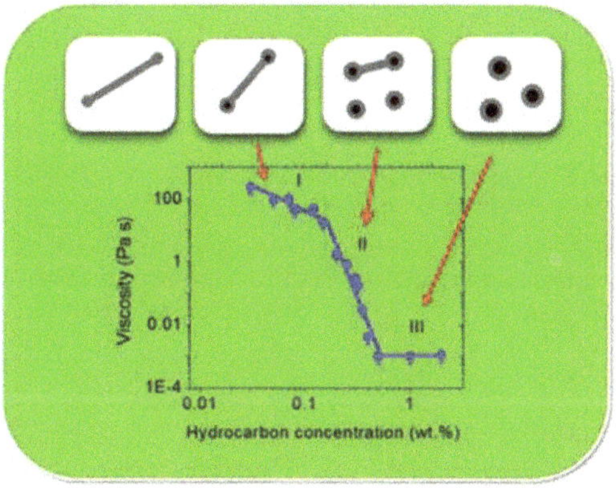

Figure 12. The dependency of zero-shear viscosity on the hydrocarbon (n-dodecane) concentration in 3 wt% potassium oleate solution. Reprinted from reference [114] with permission from the American Chemical Society.

The authors suggested that the shortening mechanism of micellar chains that are involved with the preferential accumulation of the solubilized hydrocarbon at the end-caps of wormlike micellar chains. In the formation of wormlike micelles, the higher energy at the end-caps of micelles when compared to the body leads to the lengthening of micelles and formation of long cylindrical micelles. The drastic decrease in the viscosity is correlated with the crossover from the entangled to the unentangled regimes of the WMS upon the shortening of micelles and the shortened cylindrical micelles coexist along with microemulsion droplets in region II.

In another study, Shibaev et al. [115] investigated the responsiveness of linear and branched anionic/cationic-mixed wormlike surfactant micelles to hydrocarbon. They prepared linear and branched micelles by changing the molar ratios of potassium oleate and n-octyltrimethylammonium bromide (C_8TAB) dissolved in salt-free water. Linear wormlike micelles with the molar ratio of [C_8TAB]/[potassium oleate] = 0.5 show a dramatic drop of viscosity upon the addition of small amounts of n-decane as reported in their previous study [114] (Figure 12). However, in the other case of molar ratio of 1, wormlike micelles are branched, and much higher concentration of n-decane is required for the significant drop of viscosity. The authors suggested that the solubilization of hydrocarbon was proceeded in branching points, not in the end-caps. Thus, micellar chains are prevented from shortening and the rheological properties of WMS are not significantly affected up to the saturation of branching points by the hydrocarbon solubilization [115]. In this way, one can conclude that the rheological stability of branched micellar chains is more resilient to hydrocarbon solubilization when compared to that of linear chains.

Fogang et al. [116] studied the effect of different organic compounds solubilization on the zero-shear viscosity of an aqueous solution of a sulfobetaine viscoelastic surfactant. They prepared the solution by dissolving 3.96 wt% of erucamidopropyl hydroxypropyl sulfobetaine and 6.2 wt% of calcium chloride ($CaCl_2$) in deionized water and investigated the effects of n-decane, crude oil (CO), and extra virgin olive oil (EVOO) on the zero-shear viscosity of WMS. The solubilization of n-decane resulted in three regimes of viscosity change similar to that reported in ref. [114]. However, both EVOO and CO in WMS only resulted in the regime I at 30 °C and regimes of I and II at 60 °C. Authors concluded that higher molecular components of CO and EVOO as compared to that of n-decane resulted in less solubilization in WMS. They also concluded that the viscosity reduction of WMS was dependent on the balance between micellization and oil solubilization.

The interaction between WMS and hydrocarbons can be beneficial for sweeping oil in low-permeability zones. While highly viscous displacing fluids sweep the oil in high-permeability zones, low-permeability zones with high saturated hydrocarbons are retained. Therefore, the interaction between hydrocarbons and WMS in low-permeability zones decreases the viscosity and enhances the permeability of displacing fluids into those areas [90].

7. Conclusions and Outlook

WMS are promising materials for cEOR that combine the advantage of both displacing fluid viscosifiers and lowering IFT of surfactants. The wide application of cEOR methods is restricted due to the degraded functionality of displacing fluids that are influenced by the high-salinity and high-temperature environment of oil reservoirs. The unique characteristics of WMS make them powerful candidates for overcoming cEOR restrictions. One of the major restrictions in using the polymer flooding method is that there is a high probability for the polymer chains to block the low-permeability zones of oil reservoirs. The shear-thinning property of WMS and breakage of wormlike micelles under high shear stress conditions lead to a better injectivity and provide the ability to sweep the trapped oil in low-permeability zones. Permanent degradation is another risk when using polymer solutions in cEOR. The reversible scission of wormlike micelles makes them tolerant against permanent shear degradation. Polymer-based solutions are highly sensitive to the presence of ions in oil reservoir environments. Surfactant- and alkaline-based solutions are also susceptible to a reaction with ions and absorption to the rock structures. Complex solutions of alkaline-surfactant-polymers are shown to have low stability in high-salinity conditions because of the risk of chromatographic separation. In contrast to all of them, ultra-long chain zwitterionic surfactants and anionic surfactant-based WMS display sustained high viscosity and aging stability against harsh conditions. A major challenge of EOR arises from the heterogeneous structures of oil reservoirs that are made up of higher and lower permeability zones in which displacing fluids tend to pass through high-permeability zones and leave the low-permeability zones unswept. The formation of SIS in WMS increases the viscosity and retards the flow in the high-permeability zones relative to the less permeable areas. Thus, WMS displacing fluids divert more flow to the less permeable zones and increase the oil recovery. In addition to these common useful characteristics of WMS, the living characteristic of wormlike micelles provides a broad research field of investigations to find or design smart wormlike structures. The thermal-thickening of WMS based on the salting-out effect is an example of using the living nature of micelles to change their rheological response.

Although WMS, such as EBDA, show a higher tolerance of high-salinity conditions when compared to polymer solutions, their oil recovery factors are still limited. The collapse of micelles due to the hydrocarbon–micellar interaction is suggested as the main reason for the lower recovery factor of EBDA. On the other hand, hybrid displacing fluids of polymer/WMS show higher recovery factors. Investigations on the hydrocarbon-responsive characteristics of WMS point out that the hydrocarbon–WMS interaction is advantageous for diverting the displacing fluid into the less permeable and unswept zones. When considering the heterogeneous structures of oil reservoirs, further studies must be conducted on designing hybrid displacing fluids or multi-step recovery processes that sweep the oil in all reservoir areas. Injecting a shear-thinning displacing fluid for sweeping high-permeability zones and the subsequent injection of WMS hydrocarbon-responsive fluid for sweeping oil in unswept zones is an example of the desired multi-step recovery processes.

Author Contributions: Conceptualization, E.J.N. and N.K.; Investigation and Analysis, E.J.N. and N.K.; Writing—Original Draft, E.J.N.; Writing—Review and Editing, E.J.N., T.Y., Y.J.L. and N.K.; Funding Acquisition, Y.J.L and N.K.

Funding: This research was funded by American Chemical Society Petroleum Research Fund (ACS-PRF 57552-ND9).

Acknowledgments: The authors would like to acknowledge the financial support from the Office of Provost & Vice President for Academic Affairs and the REP program at Texas State University as well as the Donors of the American Chemical Society Petroleum Research Fund.

Conflicts of Interest: The authors declare no conflict of interest.

References

1. Kokal, S.; Al-Kaabi, A. *Enhanced Oil Recovery: Challenges & Opportunities*; World Petroleum Council: London, UK, 2010.
2. Al-Mjeni, R.; Arora, S.; Cherukupalli, P.; van Wunnik, J.; Edwards, J.; Felber, B.J.; Gurpinar, O.; Hirasaki, G.J.; Miller, C.A.; Jackson, C. Has the Time Come for EOR? *Oilfield Rev.* **2010**, *2011*, 4.
3. Thomas, S. Enhanced oil recovery—An overview. *Oil Gas Sci. Technol.-Rev. D'IFP Energ. Nouv.* **2008**, *63*, 9–19. [CrossRef]
4. Terry, R.E. Enhanced oil recovery. *Encycl. Phys. Sci. Technol.* **2001**, *18*, 503–518.
5. Muggeridge, A.; Cockin, A.; Webb, K.; Frampton, H.; Collins, I.; Moulds, T.; Salino, P. Recovery rates, enhanced oil recovery and technological limits. *Phil. Trans. R. Soc. A* **2014**, *372*, 20120320. [CrossRef] [PubMed]
6. Delshad, M.; Kim, D.H.; Magbagbeola, O.A.; Huh, C.; Pope, G.A.; Tarahhom, F. Mechanistic interpretation and utilization of viscoelastic behavior of polymer solutions for improved polymer-flood efficiency. In Proceedings of the SPE Symposium on Improved Oil Recovery, Tulsa, OK, USA, 20–23 April 2008; Society of Petroleum Engineers: Richardson, TX, USA.
7. Sorbie, K.S. *Polymer-Improved Oil Recovery*; Springer Science & Business Media: Berlin, Germany, 2013.
8. Morvan, M.; Degre, G.; Beaumont, J.; Colin, A.; Dupuis, G.; Zaitoun, A.; Al-maamari, R.S.; Al-Hashmi, A.A.R.; Al-Sharji, H.H. Viscosifying surfactant technology for heavy oil reservoirs. In Proceedings of the SPE Heavy Oil Conference Canada, Calgary, AB, Canada, 12–14 June 2012; Society of Petroleum Engineers: Richardson, TX, USA.
9. Mayer, E.H.; Berg, R.L.; Carmichael, J.D.; Weinbrandt, R.M. Alkaline injection for enhanced oil recovery—A status report. *J. Pet. Technol.* **1983**, *35*, 209–221. [CrossRef]
10. Lake, L.W.; Venuto, P.B. A niche for enhanced oil recovery in the 1990s. *Oil Gas J.* **1990**, *88*, 62–67.
11. Austad, T.; Shariatpanahi, S.; Strand, S.; Black, C.; Webb, K. Conditions for a low-salinity enhanced oil recovery (EOR) effect in carbonate oil reservoirs. *Energy Fuels* **2011**, *26*, 569–575. [CrossRef]
12. Zhu, D.; Zhang, J.; Han, Y.; Wang, H.; Feng, Y. Laboratory study on the potential EOR use of HPAM/VES hybrid in high-temperature and high-salinity oil reservoirs. *J. Chem.* **2013**, *2013*, 1–8. [CrossRef]
13. Han, M.; AlSofi, A.; Fuseni, A.; Zhou, X.; Hassan, S. Development of chemical EOR formulations for a high temperature and high salinity carbonate reservoir. In Proceedings of the IPTC 2013: International Petroleum Technology Conference, Beijing, China, 26–28 March 2013.
14. Zhang, Z.; Li, J.; Zhou, J. Microscopic roles of "viscoelasticity" in HPMA polymer flooding for EOR. *Transp. Porous Media* **2011**, *86*, 199–214. [CrossRef]
15. Wei, B.; Romero-Zerón, L.; Rodrigue, D. Oil displacement mechanisms of viscoelastic polymers in enhanced oil recovery (EOR): A review. *J. Pet. Explor. Prod. Technol.* **2014**, *4*, 113–121. [CrossRef]
16. Nasr-El-Din, H.; Hawkins, B.; Green, K. Viscosity behavior of alkaline, surfactant, polyacrylamide solutions used for enhanced oil recovery. In Proceedings of the SPE International Symposium on Oilfield Chemistry, Anaheim, CA, USA, 20–22 February 1991.
17. Ghannam, M.; Esmail, N. Flow behavior of enhanced oil recovery alcoflood polymers. *J. Appl. Polym. Sci.* **2002**, *85*, 2896–2904. [CrossRef]
18. Savins, J. Non-Newtonian flow through porous media. *Ind. Eng. Chem.* **1969**, *61*, 18–47. [CrossRef]
19. Lee, K.S. Performance of a polymer flood with shear-thinning fluid in heterogeneous layered systems with crossflow. *Energies* **2011**, *4*, 1112–1128. [CrossRef]

20. Golombok, M.; van der Wijst, R. Permeability thickening fluids for improved secondary oil recovery. *J. Pet. Sci. Eng.* **2013**, *110*, 22–26. [CrossRef]
21. Lakatos, I.J.; Toth, J.; Bodi, T.; Lakatos-Szabo, J.; Berger, P.D.; Lee, C.H. Application of viscoelastic surfactants as mobility-control agents in low-tension surfactant floods. In Proceedings of the International Symposium on Oilfield Chemistry, Houston, TX, USA, 28 February–2 March 2007.
22. Kjøniksen, A.-L.; Beheshti, N.; Kotlar, H.K.; Zhu, K.; Nyström, B. Modified polysaccharides for use in enhanced oil recovery applications. *Eur. Polym. J.* **2008**, *44*, 959–967. [CrossRef]
23. Taylor, K.C.; Nasr-El-Din, H.A. Water-soluble hydrophobically associating polymers for improved oil recovery: A literature review. *J. Pet. Sci. Eng.* **1998**, *19*, 265–280. [CrossRef]
24. Berret, J.-F.; Gamez-Corrales, R.; Oberdisse, J.; Walker, L.; Lindner, P. Flow-structure relationship of shear-thickening surfactant solutions. *EPL (Europhys. Lett.)* **1998**, *41*, 677. [CrossRef]
25. Gamez-Corrales, R.; Berret, J.-F.; Walker, L.; Oberdisse, J. Shear-thickening dilute surfactant solutions: Equilibrium structure as studied by small-angle neutron scattering. *Langmuir* **1999**, *15*, 6755–6763. [CrossRef]
26. Bruinsma, R.; Gelbart, W.M.; Ben-Shaul, A. Flow-induced gelation of living (micellar) polymers. *J. Chem. Phys.* **1992**, *96*, 7710–7727. [CrossRef]
27. Jung, J.C.; Zhang, K.; Chon, B.H.; Choi, H.J. Rheology and polymer flooding characteristics of partially hydrolyzed polyacrylamide for enhanced heavy oil recovery. *J. Appl. Polym. Sci.* **2013**, *127*, 4833–4839. [CrossRef]
28. Jang, H.Y.; Zhang, K.; Chon, B.H.; Choi, H.J. Enhanced oil recovery performance and viscosity characteristics of polysaccharide xanthan gum solution. *J. Ind. Eng. Chem.* **2015**, *21*, 741–745. [CrossRef]
29. Luo, J.-H.; Pu, R.-Y.; Wang, P.-M.; Bai, F.-L.; Zhang, Y.; Yang, J.-B.; Liu, Y.-Z. Performance Properties of Salt Tolerant Polymer KYPAM for EOR. *Oilfield Chem.* **2002**, *19*, 64–67.
30. Ju, S.; Zhonghua, W.; Ning, H.; Xudong, L. Synthesis and performance evaluation of AMPS/AM tercopolymer for eor. *Chem. Eng. Oil Gas* **2001**, *4*, 013.
31. Sheng, J.J. Chapter 5—Polymer Flooding. In *Modern Chemical Enhanced Oil Recovery*; Gulf Professional Publishing: Boston, MA, USA, 2011; pp. 101–206. [CrossRef]
32. Seright, R.S.; Campbell, A.; Mozley, P.; Han, P. Stability of Partially Hydrolyzed Polyacrylamides at Elevated Temperatures in the Absence of Divalent Cations. In Proceedings of the SPE International Symposium on Oilfield Chemistry, The Woodlands, TX, USA, 20–22 April 2009. [CrossRef]
33. Levitt, D.; Pope, G.A. Selection and Screening of Polymers for Enhanced-Oil Recovery. In Proceedings of the SPE Symposium on Improved Oil Recovery, Tulsa, OK, USA, 20–23 April 2008.
34. Algharaib, M.; Alajmi, A.; Gharbi, R. Investigation of Polymer Flood Performance in High Salinity Oil Reservoirs. In Proceedings of the SPE/DGS Saudi Arabia Section Technical Symposium and Exhibition, Al-Khobar, Saudi Arabia, 15–18 May 2011.
35. Puerto, M.; Miller, C.A.; Hirasaki, G.J.; Barnes, J.R. Surfactant systems for EOR in high-temperature, high-salinity environments. In Proceedings of the SPE Improved Oil Recovery Symposium, Tulsa, OK, USA, 24–28 April 2010.
36. Ibrahim, Z.B.; Manap, A.A.A.; Hamid, P.A.; Hon, V.Y.; Lim, P.H.; Wyatt, K. Laboratory Aspect of Chemical EOR Processes Evaluation for Malaysian Oilfields. In Proceedings of the SPE Asia Pacific Oil & Gas Conference and Exhibition, Adelaide, Australia, 11–13 September 2006.
37. Pope, G.A. Recent developments and remaining challenges of enhanced oil recovery. *J. Pet. Technol.* **2011**, *63*, 65–68. [CrossRef]
38. Qiuzhi, Z.T.W.Z.G.; Xingqi, H. Synthesis and Evaluation of Anionic AM/AMPS Copolymer. *Adv. Fine Petrochem.* **2006**, *12*, 006.
39. Al-Sofi, A.M.; La Force, T.; Blunt, M.J. Sweep impairment due to polymers shear thinning. In Proceedings of the SPE Middle East Oil and Gas Show and Conference, Manama, Bahrain, 15–18 March 2009.
40. Silva, J.A.K.; Smith, M.M.; Munakata-Marr, J.; McCray, J.E. The effect of system variables on in situ sweep-efficiency improvements via viscosity modification. *J. Contam. Hydrol.* **2012**, *136*, 117–130. [CrossRef] [PubMed]
41. Lee, W.; Makarychev-mikhailov, S.M.; Lastre Buelvas, M.J.; Abad, C.; Christiawan, A.B.; Narayan, S.; Singh, M.M.; Sengupta, S. Fast Hydrating Fracturing Fluid for Ultrahigh Temperature Reservoirs. In Proceedings of the Abu Dhabi International Petroleum Exhibition and Conference, Abu Dhabi, UAE, 10–13 November 2014.

42. Lei, C.; Clark, P.E. Crosslinking of guar and guar derivatives. In Proceedings of the SPE Annual Technical Conference and Exhibition, Houston, TX, USA, 26–29 September 2004.
43. Abad, C.; Mirakyan, A.L.; Parris, M.D.; Chen, Y.; Mueller, F.A. Rheological Characterization of Novel Delayed-Transition Metal Crosslinked Fracturing Fluids: Correlation with First Field Applications. In Proceedings of the SPE Annual Technical Conference and Exhibition, New Orleans, LA, USA, 4–7 October 2009.
44. Siggel, L.; Santa, M.; Hansch, M.; Nowak, M.; Ranft, M.; Weiss, H.; Hajnal, D.; Schreiner, E.; Oetter, G.; Tinsley, J. A new class of viscoelastic surfactants for enhanced oil recovery. In Proceedings of the SPE Improved Oil Recovery Symposium, Tulsa, OK, USA, 14–18 April 2012.
45. Somasundaran, P.; Hanna, H. Adsorption of sulfonates on reservoir rocks. *Soc. Pet. Eng. J.* **1979**, *19*, 221–232. [CrossRef]
46. Mohan, K. Alkaline surfactant flooding for tight carbonate reservoirs. In Proceedings of the SPE Annual Technical Conference and Exhibition, New Orleans, LA, USA, 4–7 October 2009; Society of Petroleum Engineers: Richardson, TX, USA, 2009.
47. Li, N.; Zhang, G.; Ge, J.; Luchao, J.; Jianqiang, Z.; Baodong, D.; Pei, H. Adsorption behavior of betaine-type surfactant on quartz sand. *Energy Fuels* **2011**, *25*, 4430–4437. [CrossRef]
48. Olajire, A.A. Review of ASP EOR (alkaline surfactant polymer enhanced oil recovery) technology in the petroleum industry: Prospects and challenges. *Energy* **2014**, *77*, 963–982. [CrossRef]
49. Hou, J.; Liu, Z.; Zhang, S.; Yang, J. The role of viscoelasticity of alkali/surfactant/polymer solutions in enhanced oil recovery. *J. Pet. Sci. Eng.* **2005**, *47*, 219–235. [CrossRef]
50. Li, D.; Shi, M.-Y.; Wang, D.; Li, Z. Chromatographic separation of chemicals in alkaline surfactant polymer flooding in reservoir rocks in the Daqing oil field. In Proceedings of the SPE International Symposium on Oilfield Chemistry, The Woodlands, TX, USA, 20–22 April 2009.
51. Dreiss, C.A. Wormlike micelles: Where do we stand? Recent developments, linear rheology and scattering techniques. *Soft Matter* **2007**, *3*, 956–970. [CrossRef]
52. Kamal, M.S. A review of gemini surfactants: Potential application in enhanced oil recovery. *J. Surfactants Deterg.* **2016**, *19*, 223–236. [CrossRef]
53. Ezrahi, S.; Tuval, E.; Aserin, A. Properties, main applications and perspectives of worm micelles. *Adv. Colloid Interface Sci.* **2006**, *128*, 77–102. [CrossRef] [PubMed]
54. Morvan, M.; Moreau, P.; Degre, G.; Leng, J.; Masselon, C.; Bouillot, J.; Zaitoun, A. New Viscoelastic Fluid for Chemical EOR. In Proceedings of the SPE International Symposium on Oilfield Chemistry, The Woodlands, TX, USA, 20–22 April 2009.
55. Fjelde, I.; Austad, T.; Milter, J. Adsorption VII. Dynamic adsorption of a dual surfactant system onto reservoir cores at seawater salinities. *J. Pet. Sci. Eng.* **1995**, *13*, 193–201. [CrossRef]
56. Graf, K.; Kappl, M. *Physics and Chemistry of Interfaces*; John Wiley & Sons: Hoboken, NJ, USA, 2006.
57. González-Pérez, A.; Ruso, J.M.; Prieto, G.; Sarmiento, F. Temperature-Sensitive Critical Micelle Transition of Sodium Octanoate. *Langmuir* **2004**, *20*, 2512–2514. [CrossRef] [PubMed]
58. Acharya, D.P.; Varade, D.; Aramaki, K. Effect of temperature on the rheology of wormlike micelles in a mixed surfactant system. *J. Colloid Interface Sci.* **2007**, *315*, 330–336. [CrossRef] [PubMed]
59. Wang, J.; Feng, Y.; Agrawal, N.R.; Raghavan, S.R. Wormlike micelles versus water-soluble polymers as rheology-modifiers: Similarities and differences. *Phys. Chem. Chem. Phys.* **2017**, *19*, 24458–24466. [CrossRef]
60. Sharma, M.M.; Yen, T.; Chilingarian, G.; Donaldson, E. Some Chemical and Physical Problems in Enhanced Oil Recovery Opzrations. *Dev. Pet. Sci.* **1985**, *17*, 223–249.
61. Liu, C.-H.; Pine, D. Shear-induced gelation and fracture in micellar solutions. *Phys. Rev. Lett.* **1996**, *77*, 2121. [CrossRef]
62. Hu, Y.; Wang, S.; Jamieson, A. Rheological and flow birefringence studies of a shear-thickening complex fluid—A surfactant model system. *J. Rheol.* **1993**, *37*, 531–546. [CrossRef]
63. Hu, Y.; Rajaram, C.; Wang, S.; Jamieson, A. Shear thickening behavior of a rheopectic micellar solution: Salt effects. *Langmuir* **1994**, *10*, 80–85. [CrossRef]
64. Hu, Y.; Matthys, E.F. Characterization of micellar structure dynamics for a drag-reducing surfactant solution under shear: Normal stress studies and flow geometry effects. *Rheol. Acta* **1995**, *34*, 450–460. [CrossRef]
65. Gabbanelli, S.; Drazer, G.; Koplik, J. Lattice Boltzmann method for non-Newtonian (power-law) fluids. *Phys. Rev. E* **2005**, *72*, 046312. [CrossRef]

66. Galindo-Rosales, F.; Rubio-Hernández, F.; Sevilla, A.; Ewoldt, R.H. How Dr. Malcom, M. Cross may have tackled the development of "An apparent viscosity function for shear thickening fluids". *J. Non-Newton. Fluid Mech.* **2011**, *166*, 1421–1424. [CrossRef]
67. Galindo-Rosales, F.; Rubio-Hernández, F.; Sevilla, A. An apparent viscosity function for shear thickening fluids. *J. Non-Newton. Fluid Mech.* **2011**, *166*, 321–325. [CrossRef]
68. Cross, M.M. Rheology of non-Newtonian fluids: A new flow equation for pseudoplastic systems. *J. Colloid Sci.* **1965**, *20*, 417–437. [CrossRef]
69. Doi, M.; Edwards, S.F. *The Theory of Polymer Dynamics*; Oxford University Press: Oxford, UK, 1988; Volume 73.
70. Barentin, C.; Liu, A. Shear thickening in dilute solutions of wormlike micelles. *EPL (Europhys. Lett.)* **2001**, *55*, 432. [CrossRef]
71. Vogtt, K.; Jiang, H.; Beaucage, G.; Weaver, M. Free Energy of Scission for Sodium Laureth-1-Sulfate Wormlike Micelles. *Langmuir* **2017**, *33*, 1872–1880. [CrossRef]
72. Cates, M.; Candau, S. Statics and dynamics of worm-like surfactant micelles. *J. Phys. Condens. Matter* **1990**, *2*, 6869. [CrossRef]
73. Tung, S.-H.; Huang, Y.-E.; Raghavan, S.R. Contrasting effects of temperature on the rheology of normal and reverse wormlike micelles. *Langmuir* **2007**, *23*, 372–376. [CrossRef]
74. Raghavan, S.R.; Kaler, E.W. Highly Viscoelastic Wormlike Micellar Solutions Formed by Cationic Surfactants with Long Unsaturated Tails. *Langmuir* **2001**, *17*, 300–306. [CrossRef]
75. Sharma, S.C.; Shrestha, L.K.; Tsuchiya, K.; Sakai, K.; Sakai, H.; Abe, M. Viscoelastic wormlike micelles of long polyoxyethylene chain phytosterol with lipophilic nonionic surfactant in aqueous solution. *J. Phys. Chem. B* **2009**, *113*, 3043–3050. [CrossRef]
76. Varade, D.; Ushiyama, K.; Shrestha, L.K.; Aramaki, K. Wormlike micelles in Tween-80/CmEO3 mixed nonionic surfactant systems in aqueous media. *J. Colloid Interface Sci.* **2007**, *312*, 489–497. [CrossRef]
77. Croce, V.; Cosgrove, T.; Maitland, G.; Hughes, T.; Karlsson, G. Rheology, cryogenic transmission electron spectroscopy, and small-angle neutron scattering of highly viscoelastic wormlike micellar solutions. *Langmuir* **2003**, *19*, 8536–8541. [CrossRef]
78. Chu, Z.; Feng, Y. Thermo-switchable surfactant gel. *Chem. Commun.* **2011**, *47*, 7191–7193. [CrossRef]
79. Lindner, P.; Bewersdorff, H.; Heen, R.; Sittart, P.; Thiel, H.; Langowski, J.; Oberthür, R. Drag-reducing surfactant solutions in laminar and turbulent flow investigated by small-angle neutron scattering and light scattering. In *Trends in Colloid and Interface Science IV*; Springer: Berlin/Heidelberg, Germany, 1990; pp. 107–112.
80. Boltenhagen, P.; Hu, Y.; Matthys, E.; Pine, D. Observation of bulk phase separation and coexistence in a sheared micellar solution. *Phys. Rev. Lett.* **1997**, *79*, 2359. [CrossRef]
81. Hoffmann, H.; Ebert, G. Surfactants, micelles and fascinating phenomena. *Angew. Chem. Int. Ed.* **1988**, *27*, 902–912. [CrossRef]
82. Rehage, H.; Hoffmann, H. Rheological properties of viscoelastic surfactant systems. *J. Phys. Chem.* **1988**, *92*, 4712–4719. [CrossRef]
83. Rojas, M.R.; Müller, A.J.; Sáez, A.E. Shear rheology and porous media flow of wormlike micelle solutions formed by mixtures of surfactants of opposite charge. *J. Colloid Interface Sci.* **2008**, *326*, 221–226. [CrossRef]
84. Jones, W. Polymer additives in reservoir flooding for oil recovery: Shear thinning or shear thickening? *J. Phys. D Appl. Phys.* **1980**, *13*, L87. [CrossRef]
85. Reuvers, N.; Golombok, M. Shear rate and permeability in water flooding. *Transp. Porous Media* **2009**, *79*, 249–253. [CrossRef]
86. Vermolen, E.; Van Haasterecht, M.J.; Masalmeh, S.K.; Faber, M.J.; Boersma, D.M.; Gruenenfelder, M.A. Pushing the envelope for polymer flooding towards high-temperature and high-salinity reservoirs with polyacrylamide based ter-polymers. In Proceedings of the SPE Middle East Oil and Gas Show and Conference, Manama, Bahrain, 25–28 September 2011.
87. Dupuis, G.; Rousseau, D.; Tabary, R.; Argillier, J.-F.; Grassl, B. Hydrophobically modified sulfonated polyacrylamides for IOR: Correlations between associative behavior and injectivity in the diluted regime. *Oil Gas Sci. Technol.–Rev. D'IFP Energ. Nouv.* **2012**, *67*, 903–919. [CrossRef]
88. Cao, Q.; Yu, L.; Zheng, L.-Q.; Li, G.-Z.; Ding, Y.-H.; Xiao, J.-H. Rheological properties of wormlike micelles in sodium oleate solution induced by sodium ion. *Colloids Surf. A Physicochem. Eng. Asp.* **2008**, *312*, 32–38. [CrossRef]

89. Rogers, S.A.; Calabrese, M.A.; Wagner, N.J. Rheology of branched wormlike micelles. *Curr. Opin. Colloid Interface Sci.* **2014**, *19*, 530–535. [CrossRef]
90. Lequeux, F. Reptation of connected wormlike micelles. *EPL (Europhys. Lett.)* **1992**, *19*, 675. [CrossRef]
91. Cates, M. Reptation of living polymers: Dynamics of entangled polymers in the presence of reversible chain-scission reactions. *Macromolecules* **1987**, *20*, 2289–2296. [CrossRef]
92. Marques, C.M.; Turner, M.S.; Cates, M.E. Relaxation mechanisms in worm-like micelles. *J. Non-Cryst. Solids* **1994**, *172*, 1168–1172. [CrossRef]
93. Appell, J.; Porte, G.; Khatory, A.; Kern, F.; Candau, S. Static and dynamic properties of a network of wormlike surfactant micelles (cetylpyridinium chlorate in sodium chlorate brine). *J. Phys. II* **1992**, *2*, 1045–1052. [CrossRef]
94. Bautista, F.; Soltero, J.; Pérez-López, J.; Puig, J.; Manero, O. On the shear banding flow of elongated micellar solutions. *J. Non-Newton. Fluid Mech.* **2000**, *94*, 57–66. [CrossRef]
95. Vasquez, P.A.; McKinley, G.H.; Cook, L.P. A network scission model for wormlike micellar solutions: I. Model formulation and viscometric flow predictions. *J. Non-Newton. Fluid Mech.* **2007**, *144*, 122–139. [CrossRef]
96. Samuel, M.M.; Dismuke, K.I.; Card, R.J.; Brown, J.E.; England, K.W. Methods of Fracturing Subterranean Formations. Google Patents US6306800B1, 23 October 2001.
97. Van Santvoort, J.; Golombok, M. Sweep enhancers for oil recovery. *J. Pet. Explor. Prod. Technol.* **2016**, *6*, 473–480. [CrossRef]
98. Zana, R.; Kaler, E.W. *Giant Micelles: Properties and Applications*; CRC Press: Boca Raton, FA, USA, 2007; Volume 140.
99. Štern, P.; Myška, J. Viscous and elastic properties of a cationic and a zwitterionic drag reducing surfactant. *Colloids Surf. A Physicochem. Eng. Asp.* **2001**, *183*, 527–531. [CrossRef]
100. Zakin, J.L.; Lu, B.; Bewersdorff, H.-W. Surfactant drag reduction. *Rev. Chem. Eng.* **1998**, *14*, 253–320. [CrossRef]
101. Lin, Z.; Chou, L.; Lu, B.; Zheng, Y.; Davis, H.T.; Scriven, L.; Talmon, Y.; Zakin, J.L. Experimental studies on drag reduction and rheology of mixed cationic surfactants with different alkyl chain lengths. *Rheol. Acta* **2000**, *39*, 354–359. [CrossRef]
102. Lu, B.; Li, X.; Scriven, L.; Davis, H.; Talmon, Y.; Zakin, J. Effect of chemical structure on viscoelasticity and extensional viscosity of drag-reducing cationic surfactant solutions. *Langmuir* **1998**, *14*, 8–16. [CrossRef]
103. Virk, P.S. Drag reduction fundamentals. *AIChE J.* **1975**, *21*, 625–656. [CrossRef]
104. Zakin, J.L.; Myska, J.; Chara, Z. New limiting drag reduction and velocity profile asymptotes for nonpolymeric additives systems. *AIChE J.* **1996**, *42*, 3544–3546. [CrossRef]
105. Brostow, W. Drag reduction in flow: Review of applications, mechanism and prediction. *J. Ind. Eng. Chem.* **2008**, *14*, 409–416. [CrossRef]
106. Wang, Y.; Yu, B.; Zakin, J.L.; Shi, H. Review on drag reduction and its heat transfer by additives. *Adv. Mech. Eng.* **2011**, *3*, 478749. [CrossRef]
107. Hong, C.; Jang, C.; Choi, H. Turbulent drag reduction with polymers in rotating disk flow. *Polymers* **2015**, *7*, 1279–1298. [CrossRef]
108. Nesyn, G.V.; Sunagatullin, R.Z.; Shibaev, V.P.; Malkin, A.Y. Drag reduction in transportation of hydrocarbon liquids: From fundamentals to engineering applications. *J. Pet. Sci. Eng.* **2018**, *161*, 715–725. [CrossRef]
109. Kotenko, M.; Oskarsson, H.; Bojesen, C.; Nielsen, M.P. An experimental study of the drag reducing surfactant for district heating and cooling. *Energy* **2019**, *178*, 72–78. [CrossRef]
110. Bewersdorff, H.-W.; Ohlendorf, D. The behaviour of drag-reducing cationic surfactant solutions. *Colloid Polym. Sci.* **1988**, *266*, 941–953. [CrossRef]
111. Myska, J.; Zakin, J.L. Differences in the flow behaviors of polymeric and cationic surfactant drag-reducing additives. *Ind. Eng. Chem. Res.* **1997**, *36*, 5483–5487. [CrossRef]
112. Utracki, L.A.; Jamieson, A.M. *Polymer Physics: From Suspensions to Nanocomposites and Beyond*; John Wiley & Sons: Hoboken, NJ, USA, 2011.
113. Elbing, B.R.; Solomon, M.J.; Perlin, M.; Dowling, D.R.; Ceccio, S.L. Flow-induced degradation of drag-reducing polymer solutions within a high-Reynolds-number turbulent boundary layer. *J. Fluid Mech.* **2011**, *670*, 337–364. [CrossRef]

114. Shibaev, A.V.; Tamm, M.V.; Molchanov, V.S.; Rogachev, A.V.; Kuklin, A.I.; Dormidontova, E.E.; Philippova, O.E. How a viscoelastic solution of wormlike micelles transforms into a microemulsion upon absorption of hydrocarbon: New insight. *Langmuir* **2014**, *30*, 3705–3714. [CrossRef]
115. Shibaev, A.V.; Kuklin, A.I.; Philippova, O.E. Different responsiveness to hydrocarbons of linear and branched anionic/cationic-mixed wormlike surfactant micelles. *Colloid Polym. Sci.* **2019**, *297*, 351–362. [CrossRef]
116. Fogang, L.T.; Sultan, A.S.; Kamal, M.S. Understanding viscosity reduction of a long-tail sulfobetaine viscoelastic surfactant by organic compounds. *RSC Adv.* **2018**, *8*, 4455–4463. [CrossRef]

fluids

MDPI

Article

Natural Convection in a Non-Newtonian Fluid: Effects of Particle Concentration

Chengcheng Tao [1], Wei-Tao Wu [2] and Mehrdad Massoudi [3,*]

1 University of Florida, Engineering School of Sustainable Infrastructure and Environment, Gainesville, FL 32611, USA; chengchengtao@gmail.com
2 School of Mechanical Engineering, Nanjing University of Science and Technology, Nanjing 210094, China; weitaowwtw@njust.edu.cn
3 U.S. Department of Energy, National Energy Technology Laboratory (NETL), Pittsburgh, PA 15236, USA
* Correspondence: mehrdad.massoudi@netl.doe.gov; Tel.: +1-412-386-4975

Received: 1 October 2019; Accepted: 23 October 2019; Published: 1 November 2019

Abstract: In this paper we study the buoyancy driven flow of a particulate suspension between two inclined walls. The suspension is modeled as a non-linear fluid, where the (shear) viscosity depends on the concentration (volume fraction of particles) and the shear rate. The motion of the particles is determined by a convection-diffusion equation. The equations are made dimensionless and the boundary value problem is solved numerically. A parametric study is performed, and velocity, concentration and temperature profiles are obtained for various values of the dimensionless numbers. The numerical results indicate that due to the non-uniform shear rate, the particles tend to concentrate near the centerline; however, for a small Lewis number (Le) related to the size of the particles, a uniform concentration distribution can be achieved.

Keywords: non-linear fluids; variable viscosity; natural convection; convection-diffusion; buoyancy force

1. Introduction

Fluid flow can occur for various reasons such as applications of external forces, presence of pressure or temperature gradients, natural convection (buoyancy driven flow), etc. The latter type is when the density of the fluid is a function of temperature and as a result due to a temperature dependent buoyancy (body) force the fluid can move (see Turner (1979) [1]). Natural convection and heat transfer in a suspension composed of solid particles and a fluid occur in thermal storage systems, chemical industry or food industry [2,3]. Studying the natural convection and flow of suspensions can provide better understanding of the complex mechanisms involved in these flows [4,5]. Particulate suspensions usually show some of the non-Newtonian features, such as shear-thinning, yield stress, thixotropy, dilatancy, normal stress effects, and even anisotropic thermal or momentum diffusivity. Metivier et al. (2017) [6] experimentally studied the onset of the Rayleigh-Bénard convection of a concentrated suspension of microgels subject to a temperature gradient. They focused their studies on the no-slip condition and found that the main control parameters for this flow is the ratio between the yield stress and the buoyancy force. Sun et al. (2019) [3] investigated the natural convection and heat transfer of a ferro-nanofluid with anisotropic thermal conductivity under a magnetic field. The numerical results show that the isotherms become elliptic and deviate from the circular pattern which is the typical pattern with isotropic thermal conductivity.

In general, due to certain effects such as the presence of lift force or drag force, the suspension can exhibit certain multi-component features, such as particle migration or particle sedimentation; moreover, in many situations, due to the presence of gravity or some other body forces (such as electro-magnetic forces) the solid particles can redistribute themselves and cause a change in the rheological properties of the suspension. Okada and Suzuki (1997) [7] experimentally investigated the natural convection of particulate suspension in a rectangular cell where the central part of the lower wall was heated. They found that the suspension forms different layers during the sedimentation of particles, and these layers disappear as the flow evolves; they attributed this phenomenon to the double diffusive convection caused by the volume fraction and the temperature gradient. Using Particle Tracking Velocimetry (PTV), Chen et al. (2005) [2] measured the velocity and the particle distributions in a square section with the bottom wall heated; they noticed that the flow patterns of the particulate suspension, such as sedimentation driven convection, is distinct from the flow of fluid with no particles.

Natural convection problems related to meteorology (see Batchelor (1954) [8]) and non-Newtonian fluids have been studied extensively (see Shenoy and Mashelkar (1982) [9]). For example, Rajagopal and Na (1985) [10] studied the natural convection of grade fluids between two vertical walls. Massoudi and Christie (1990) [11] considered the flow due to natural convection of a thermodynamically compatible third grade fluid between two vertical cylinders. Later, Massoudi et al. (2008) [12] studied the natural convection of a generalized second grade fluid with a temperature dependent and shear-rate dependent viscosity. In these studies, the fluid was not considered to be a suspension of particles in a fluid and as a result the effect of volume fraction was ignored.

In this paper we do consider the effect of volume fraction of the particles and we will look at the buoyancy driven flow of a particulate suspension between two inclined walls with variable transport properties. In Sections 2 and 3 we present the governing equations and the constitutive relations, respectively. In Section 4, we look at the simplified equations for the natural convection flow and present the governing equations and the boundary conditions along with our assumptions. In Section 5, the results are analyzed. Finally, in Section 6 we present the conclusions.

2. Governing Equations

As mentioned earlier, in general, most suspensions behave as multi-component fluids. They can be modeled using the techniques of suspension rheology or the techniques of multi-component materials (mixture theory). While the former method is easier to handle computationally (fewer equations), it also has the disadvantage that it cannot predict many of the interesting phenomena observed in multicomponent flows, such as the various possible interactions between different components, such as lift forces, drag forces, etc. For example, for a two-component system, the governing equations are written for each component (phase) and constitutive relations are needed for the two stress tensors, the interaction forces, the flux vectors, etc. Clearly, this approach, while more accurate, will be computationally more intensive. For a recent discussion of the multi-component approach we refer the reader to Rajagopal and Tao (1995) [13]; Massoudi (2003, 2008, 2010) [14–16]. As a compromise, one can look at the suspension which does have some type of structure (in this case solid particles which can be re-arranged and move with the velocity of the suspension), as a single component non-linear fluid, allowing for the presence of the particles through the introduction of a concentration (volume fraction) field ϕ. In this paper, we take this approach and model the suspension as a (single component) non-linear fluid; in this case the governing equations of motion are the conservation of mass, linear and angular momentum, and the energy equations. These equations are (see for example, Slattery (1999) [17]):

2.1. Conservation of Mass

$$\frac{\partial \rho}{\partial t} + \mathrm{div}(\rho \boldsymbol{v}) = 0 \tag{1}$$

where $\rho = \phi\rho_s + (1 - \phi)\rho_f$ is the density of the suspension, ϕ is the concentration of the particles, ρ_s and ρ_f are the density of particles and the fluid, respectively, $\partial/\partial t$ is the partial derivative with respect to time, "div" is the divergence operator, and $\boldsymbol{v}$ is the velocity vector. For an incompressible fluid,

$$\text{div}\ \boldsymbol{v} = 0 \tag{2}$$

2.2. Conservation of Linear Momentum

$$\rho \frac{d\boldsymbol{v}}{dt} = \text{div}\ \boldsymbol{T} + \rho \boldsymbol{b} \tag{3}$$

where $\boldsymbol{b}$ is the body force vector, $\boldsymbol{T}$ is the Cauchy stress tensor, and d/dt is the total time derivative, given by $d(.)/dt = \partial(.)/\partial t + [\text{grad}(.)]\boldsymbol{v}$. The balance of angular momentum indicates that in the absence of couple stresses, the stress tensor is symmetric.

2.3. Conservation of Energy

$$\rho \frac{d\varepsilon}{dt} = \boldsymbol{T} : \boldsymbol{L} - \text{div}\ \boldsymbol{q} + \rho r \tag{4}$$

where ε is the specific internal energy, $\boldsymbol{L}$ is the velocity gradient, $\boldsymbol{q}$ is the heat flux vector, and r is the radiant heating. Thermodynamical considerations require the application of the second law of thermodynamics (the entropy inequality); in this paper, we do not consider the entropy law (see Liu (2002) [18]). The specific internal energy, ε, is related to the specific Helmholtz free energy (Dunn and Fosdick, 1974 [19]) $\varepsilon = \Psi + \theta\eta$ where η is the specific entropy. The internal energy may in some ways depend on other parameters such as concentration. Nevertheless, in this paper, due to the nature of the kinematical assumptions for the flow field, ε drops out of the energy equation. Furthermore, we do not consider the effects of radiation in this paper.

2.4. Convection-Diffusion Equation for Particles

Here we assume that the particles do not have their own independent velocity, as is the case in two-phase flows; instead we assume that they flow with the velocity of the suspension where a convection-diffusion equation is used to describe the volume fraction field ϕ (see Probstein (2005) [20]):

$$\frac{\partial \phi}{\partial t} + \boldsymbol{v} \cdot \text{grad}\phi = \text{div}\ \boldsymbol{N} \tag{5}$$

where $\boldsymbol{N}$ is the flux determining the motion of the particles. In this approach as the particles are re-distributed, through ϕ, they influce the fluid motion via the shear viscosity of the fluid (which depends on ϕ).

3. Constitutive Relations

In looking at Equations (1)–(5), we can see that we need constitutive relations for $\boldsymbol{T}$, $\boldsymbol{q}$, $\boldsymbol{N}$ and the body force $\rho\boldsymbol{b}$. We will now discuss the constitutive relations needed for the closure in this problem.

3.1. Stress Tensor

Primarily, what distinguishes a non-Newtonian fluid from a Newtonian fluid, is its ability to exhibit one or many of the following characteristics: (1) shear-thinning or shear-thickening effects; (2) yield-stress; (3) normal stress effects; (4) creep; (5) relaxation; (6) thixotropy, etc. (see Macosko (1994) [21]; Schowalter (1978) [22]). In this paper, we focus on the shear-thinning (or shear-thickening) aspects and assume that the Cauchy stress tensor for the suspension is given by,

$$\boldsymbol{T} = -p\boldsymbol{I} + \mu(\phi, \boldsymbol{A}_1)\boldsymbol{A}_1 \tag{6}$$

where p is the pressure (the mean normal stress), $\boldsymbol{I}$ is the identity tensor, $\boldsymbol{A}_1 = \boldsymbol{L} + \boldsymbol{L}^T$ ($\boldsymbol{L} = grad\ \boldsymbol{v}$) and the shear viscosity is assumed to be given by

$$\mu(\phi, \boldsymbol{A}_1) = \mu^*(\phi)\left(1 + \alpha tr\left(\boldsymbol{A}_1{}^2\right)\right)^n \tag{7}$$

where "tr" is the trace of a 2nd order tensor and n determines whether the fluid is shear-thinning ($n < 0$), or shear-thickening ($n > 0$). The second law of thermodynamics indicates that the constant $\alpha \geq 0$ [Bridges and Rajagopal (2006) [23]]. In this paper, the viscosity is assumed to also depend on ϕ. Following the works of [24,25], we assume,

$$\mu^*(\phi) = \mu_r(1 - \phi/\phi_i)^{-1.82} \tag{8}$$

where ϕ_i is the volume fraction at which the relative viscosity μ^* tends to infinity. This value is around 0.68 for hard spheres [24,25]. For a recent discussion of a more general model of this type, see Tao, et al. (2019) [26]. Substituting Equations (7) and (8) in (6), we obtain the expression for $\boldsymbol{T}$:

$$\boldsymbol{T} = -p\boldsymbol{I} + \mu_r(1 - \phi/\phi_i)^{-1.82}\left(1 + \alpha tr\left(\boldsymbol{A}_1{}^2\right)\right)^n \boldsymbol{A}_1 \tag{9}$$

where μ_r is constant (also referred to as the reference viscosity). We use this equation in our analysis.

3.2. Heat Flux Vector

For the heat flux vector, we use the traditional Fourier's assumption where,

$$\boldsymbol{q} = -k\text{grad}\theta \tag{10}$$

where θ is the temperature, k is the (constant) thermal conductivity. In general, thermal conductivity of a non-linear fluid (suspension) is not constant; it can be a function of shear rate, concentration, etc. (see Miao and Massoudi (2015) [27], Yang, et al. (2013) [28], Yang and Massoudi (2018) [29]). For a recent review of the heat flux vector for granular-type fluids, see Massoudi (2006a, b) [30,31] and Massoudi and Kirwan (2016) [32].

3.3. Body (Buoyancy) Force

The body (buoyancy) force is given by $\rho\boldsymbol{b} = \rho(\theta)\boldsymbol{g}$; in general, for a suspension composed of a fluid and particles, the density will also depend on the volume fraction. In this paper, we ignore this effect. Here we use the usual Boussinesq-assumption (see Rajagopal et al. (1996) [33] and Rajagopal et al. (2009) [34], for detailed discussion), where the density is expressed as

$$\rho = \rho_{ref}\left(1 - \zeta\left(\theta - \theta_{ref}\right)\right) \tag{11}$$

where $\zeta = -\frac{1}{\rho}\frac{\partial\rho}{\partial\theta}\Big|_{\theta_{ref}}$ is the coefficient of thermal expansion which is assumed to be a constant here, and ρ_{ref} is the density of the suspension at the reference temperature θ_{ref}.

3.4. Particle Flux

We assume that the particle transport flux is given by [25,35]:

$$\mathbf{N} = -a^2\phi K_c grad\left(\dot{\gamma}\phi\right) - a^2\phi^2\dot{\gamma}K_\mu grad(\ln\mu) - Dgrad\phi + \phi(1-\phi)t_p\left(1 - \frac{\rho_f}{\rho_s}\right)\boldsymbol{g} \tag{12}$$

where the terms on the right-hand side are fluxes due to particle collision, changes in viscosity, Brownian motion and gravity, respectively. The last term $\mathbf{N}_g = \frac{2}{9}\phi(1-\phi)\frac{a^2(\rho_s - \rho_f)}{\mu(\phi, \boldsymbol{A}_1)}\boldsymbol{g}$, is the particle flux attributed to gravity, and has been used in studying several different problems in flows of solid-fluid suspensions [36,37]. In the above equation, a is the particle radius, $\dot{\gamma}$ is the shear rate $[\dot{\gamma} = \sqrt{1/2\text{tr}(\boldsymbol{A}_1{}^2)}]$, μ is the viscosity, K_c, K_μ are empirical coefficients, D is the diffusivity of the Brownian motion and t_p is the particle response time. To model $\boldsymbol{N}$, a similar approach, although from a different perspective, was provided by Bridges and Rajagopal (2006) [23] for chemically reacting fluids (see also Massoudi and Uguz (2012) [38]).

4. Flow Due to Natural Convection between Two Walls

We assume that a fluid-partilces suspension with density ρ and viscosity μ (which is a function of concentration and shear-rate) is situated between two walls (which are at different temperatures) titled at an angle β from the vertical direction; the heated wall is at y = −H and the cooler wall is at y = H, i.e., $\theta_1 > \theta_2$. The physical setting of the problem is shown in Figure 1. Because of the temperature gradient and the assumption that the density depends upon the temperature, the momentum and the energy equations are coupled; as a result, we expect that the fluid near the warmer wall would rise (due to the buoyancy effects) and near the cooler wall, the fluid would descend.

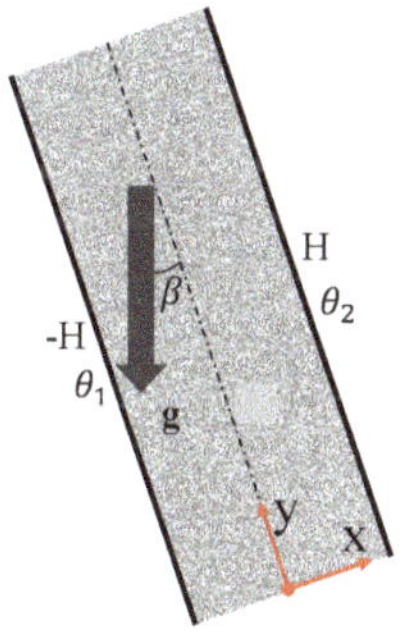

Figure 1. Physical sketch of the system.

This is the type of flow which can occur in double wall panels in buildings and in the operations of the Clusius-Dickel column, used for separating isotopes in liquid mixtures (see Bird et al. (2007) [39]). A relevant and related problem, though more complicated, is the natural convection in rectangular enclosures or cavities. Dawson and McTigue (1985) [40] provide a good overview of this problem, where they studied natural convection in fluid-saturated porous media.

For this idealized problem, we assume

$$\boldsymbol{v} = u(x)\boldsymbol{e}_y \tag{13}$$

$$\theta = \theta(x) \tag{14}$$

$$\phi = \phi(x) \tag{15}$$

With the above, Equation (12), conservation of mass is automatically satisfied. We should mention that an implicit assumption made in many buoyance driven flows, including our paper, is that while the fluid is mechanically incompressible, i.e., div $\boldsymbol{v} = 0$, thermally the fluid is assumed to be compressible, via the Boussinesq approximation. For an excellent discussion of this issue, see Prusa and Rajagopal (2013) [41]. Additionally, the linear momentum equation in component form in (*x*,*y*,*z*) direction reduces to

$$0 = -\frac{\partial p}{\partial x} - \rho_{ref} g sin\beta\left(1 - \varsigma\left(\theta - \theta_{ref}\right)\right) \tag{16}$$

$$0 = -\frac{\partial p}{\partial y} + \frac{d}{dx}\left(\mu^*(\phi)\left(1 + 2\alpha\left(\frac{du}{dx}\right)^2\right)^n \frac{du}{dx}\right) - \rho_{ref} g cos\beta\left(1 - \varsigma\left(\theta - \theta_{ref}\right)\right) \tag{17}$$

$$0 = -\frac{\partial p}{\partial z} \tag{18}$$

Let $\hat{p} = p + \rho_{ref} g sin\beta\left(1 - \zeta\left(\theta - \theta_{ref}\right)\right)$, then $\frac{\partial \hat{p}}{\partial x} = \frac{\partial \hat{p}}{\partial z} = 0$, then,

$$\frac{\partial \hat{p}}{\partial y} = \frac{\partial p}{\partial y} \tag{19}$$

We can now re-write Equation (18) as,

$$\frac{d\hat{p}}{dy} = \frac{d}{dx}\left[\mu^*(\phi)\left(1 + 2\alpha\left(\frac{du}{dx}\right)^2\right)^n \frac{du}{dx}\right] - \rho_{ref} g cos\beta\left(1 - \zeta\left(\theta - \theta_{ref}\right)\right) \tag{20}$$

Now, since the right-hand side of the above equation is not a function of y, we assume $\frac{d\hat{p}}{dy} = constant = C$. If we choose $C = -\rho_{ref} g cos\beta$, then the momentum equation in the y-direction reduces to

$$0 = \frac{d}{dx}\left[\mu^*(\phi)\left(1 + 2\alpha\left(\frac{du}{dx}\right)^2\right)^n \frac{du}{dx}\right] + \rho_{ref} g \zeta\left(\theta - \theta_{ref}\right) cos\beta \tag{21}$$

For the concentration flux, for steady-state condition, Equation (5) reduces to

$$\begin{aligned} 0 = \; & \mathrm{div}\left(-a^2 \phi K_c grad\left(\dot{\gamma}\phi\right) - a^2 \phi^2 \dot{\gamma} K_\mu grad(\ln \mu) - D grad\phi \right. \\ & \left. + \phi(1-\phi) t_p \left(1 - \tfrac{\rho_f}{\rho_s}\right) \boldsymbol{g}\right) \end{aligned} \tag{22}$$

Notice that at the solid boundaries, since we are assuming non-porous walls, we must ensure that there are no particles moving across the surfaces; this implies that the particle flux normal to the direction of flow should be zero [25]. That is,

$$\begin{aligned} 0 = \; & \boldsymbol{n}\cdot\left(-a^2 \phi K_c grad\left(\dot{\gamma}\phi\right) - a^2 \phi^2 \dot{\gamma} K_\mu grad(\ln \mu) - D grad\phi \right. \\ & \left. + \phi(1-\phi) t_p \left(1 - \tfrac{\rho_f}{\rho_s}\right) \boldsymbol{g}\right)|_{wall} \end{aligned} \tag{23}$$

Integrating Equation (22) and using Equation (23), we have,

$$\begin{aligned} 0 = \; & -a^2 \phi K_c grad\left(\dot{\gamma}\phi\right) - a^2 \phi^2 \dot{\gamma} K_\mu grad(\ln \mu) - D grad\phi \\ & + \phi(1-\phi) t_p \left(1 - \tfrac{\rho_f}{\rho_s}\right) \boldsymbol{g} \end{aligned} \tag{24}$$

The above equation implies that the total flux should be zero everywhere in the flow. For unsteady or multi-dimensional flows, this condition is not applicable. As a result, the expanded form of the convention-diffusion equation becomes,

$$\begin{aligned} 0 = \; & a^2 K_c\left(\phi^2 \tfrac{d}{dx}\left(\left|\tfrac{du}{dx}\right|\right) + \phi\left|\tfrac{du}{dx}\right|\tfrac{d\phi}{dx}\right) + a^2 K_\mu\left(\phi^2 \tfrac{1}{\mu}\tfrac{d\mu}{d\phi}\left|\tfrac{du}{dx}\right|\tfrac{d\phi}{dx}\right) + D\left(\tfrac{d\phi}{dx}\right) \\ & + \tfrac{2}{9}\phi(1-\phi)\tfrac{a^2\left(\rho_s - \rho_f\right)}{\mu} g sin\beta \end{aligned} \tag{25}$$

Using Equations (13) and (14), the energy equation, Equation (4), becomes

$$0 = \mu^*(\phi)\left(1 + 2\alpha\left(\frac{du}{dx}\right)^2\right)^n \left(\frac{du}{dx}\right)^2 + k\frac{d^2\theta}{dx^2}. \tag{26}$$

We now make the equations dimensionless by using the following reference quantities,

$$\overline{x} = \frac{x}{H}; \; \overline{u} = \frac{uH}{k}; \; \overline{\theta} = \frac{\theta - \theta_{ref}}{\Delta\theta} = \frac{\theta - \theta_{ref}}{\theta_1 - \theta_2} \tag{27}$$

where H is half the distance between the two walls, k is the thermal conductivity, $\nu = \mu_r/\rho_{ref}$ is the kinematic viscosity. The mean value of the two temperatures at the walls is taken as the reference temperature; i.e., $\theta_{ref} = 0.5(\theta_1 + \theta_2)$. The resulting dimensionless parameters are,

$$Pr = \frac{\nu}{k};\ \delta = \frac{2\alpha k^2}{H^4}; Ra = \frac{g\zeta\Delta\theta H^3}{\nu k};\ Br = \frac{\mu_r k}{\Delta\theta H^2}\Gamma_c = \frac{a^2 K_c}{H^2};\ \Gamma_\mu = \frac{a^2 K_\mu}{H^2}; \Gamma_g = \frac{a^2\left(\rho_s - \rho_f\right)gH}{k};\ Le = \frac{k}{D} \tag{28}$$

where Pr and Ra are the Prandtl and the Rayleigh numbers, Le is known as the Lewis number which is a measure of the ratio of thermal diffusivity to mass diffusivity and Br is the Brinkman number which is a measure of the ratio between heat produced by viscous dissipation and heat transported by molecular conduction. Notice that the Pr number can be canceled out without affecting Equation (29) below.

The dimensionless governing equations are then given as, (dropping the overbar symbol for simplicity),

$$0 = Pr\frac{d}{dx}\left((1-\phi/\phi_i)^{-1.82}\left(1+\delta\left(\frac{du}{dx}\right)^2\right)^n\frac{du}{dx}\right) + PrRa\theta cos\beta \tag{29}$$

$$\begin{aligned} 0 = \ & \Gamma_c\left(\phi^2\frac{d}{dx}\left(\left|\frac{du}{dx}\right|\right) + \phi\left|\frac{du}{dx}\right|\frac{d\phi}{dx}\right) + \Gamma_\mu\left(\phi^2\frac{1}{\mu}\frac{d\mu}{d\phi}\left|\frac{du}{dx}\right|\frac{d\phi}{dx}\right) \\ & + \frac{2\phi(1-\phi)}{9(1-\phi/\phi_i)^{-1.82}\left(1+\delta\left(\frac{\partial u}{\partial x}\right)^2\right)^n}\Gamma_g cos\beta + \frac{1}{Le}\frac{d\phi}{dx} \end{aligned} \tag{30}$$

$$0 = Br(1-\phi/\phi_i)^{-1.82}\left(1+\delta\left(\frac{du}{dx}\right)^2\right)^n\left(\frac{du}{dx}\right)^2 + \frac{d^2\theta}{dx^2} \tag{31}$$

Looking at the above equations, we can see that we need two boundary conditions for u, one for ϕ, and two for θ. The non-dimensional forms of the boundary conditions are given by

$$u(\pm 1) = 0 \tag{32}$$

$$\theta(-1) = 0.5; \theta(1) = -0.5 \tag{33}$$

where we have used the no-slip boundary condition for the velocity. Also, Equation (33) indicates that the temperature is higher at the left wall. For particle concentration the appropriate boundary condition may be given as an average value in an integral form (See Massoudi (2007) [42]):

$$\phi_{avg} = \frac{1}{2}\int_{-1}^{1}\phi dx \tag{34}$$

The above equations can be solved for the three field variables, namely, velocity, volume fraction and temperature.

5. Results and Discussions

In this paper, the system of the non-linear ordinary differential Equations (29)–(31) with the boundary conditions (32)–(34) are solved numerically using the MATLAB solver bvp4c, which is a collocation boundary value problem solver [43]. The step size is automatically adjusted by the solver. The default relative tolerance for the maximum residue is 0.001. The boundary conditions for the average/bulk concentration is numerically satisfied by using the shooting method.

Table 1 lists the values of the dimensionless numbers and other parameters used in Sections 5.1 and 5.2.

Table 1. The dimensionless parameters used in our study.

Section 5.1		Section 5.2	
Ra	0.1, 1.0, 2.0	$Ra.$	1.0, 2.0, 3.0
Γ_c	0.1, 1.0, 2.0	Γ_c	0.1, 1, 2.5
Γ_μ	1, 10, 100	Γ_μ	1, 10, 100
Le	0.1, 2.5, 10	Le	0.1, 10, 30
n	0.5, 0.0, 1	n	NA
δ	0.1, 0.5, 0.8	δ	NA
Br	1, 3, 4	Br	NA
ϕ_{avg}	0.05, 0.1, 0.2	ϕ_{avg}	0.05, 0.1, 0.15
β	NA	β	$0^\circ, 15^\circ, 30^\circ$
Γ_g	NA	Γ_g	0, 2.5, 5

5.1. Natural Convection with Neutrally Buoyant Particles

We first perform a parametric study for the case of natural convection of a suspension composed of neutrally buoyant particles in a fluid in a vertical channel; in this case, $\Gamma_g = 0$ and $\beta = 0^\circ$. Notice that according to Equation (12) the small size particles can lead to a negligible Γ_g. Figure 2 shows the effect of the buoyancy force term, Ra. We can observe two approximately parabolic velocity profiles where near the hotter wall the velocity is positive and near the colder wall the velocity is negative; the particles tend to concentrate near the region with the maximum and minimum velocity (low shear rates) due to the effect of the particle flux term Γ_c; the temperature shows higher values in the interior of the flow due to the effect of viscous dissipation. As the buoyancy force (Ra) increases, the magnitude of the velocity seems to increase, resulting in an increase in temperature. We also notice that more particles accumulate near the region with the maximum and minimum velocity, perhaps due to the higher values of the shear rate, see Equation (12). Figure 3 shows the effect of the shear-dependent viscosity. From Figure 3a, we can see that as the fluid changes from shear-thinning to shear-thickening (n changing from −1 to 1), the magnitude of the velocity tends to decrease; and the temperature and volume fraction profiles change a little for the range of parameters studied here. From Figure 3b, we notice that as δ increases, implying that the shear-thinning effect is stronger (notice $n = -0.5$), the magnitude of the velocity increases, while the concentration and temperature profiles do not change that much.

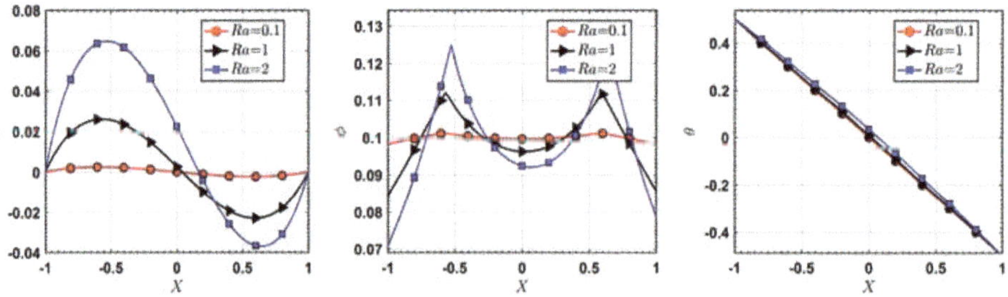

Figure 2. Effect of the buoyancy force term, the Rayleigh number (Ra) on the velocity, concentration and temperature profiles, when $\Gamma_c = 2.5$, $\Gamma_\mu = 0.1$, $Le = 10$, $n = -0.5$, $\delta = 1$, $Br = 5$ and $\phi_{avg} = 0.1$.

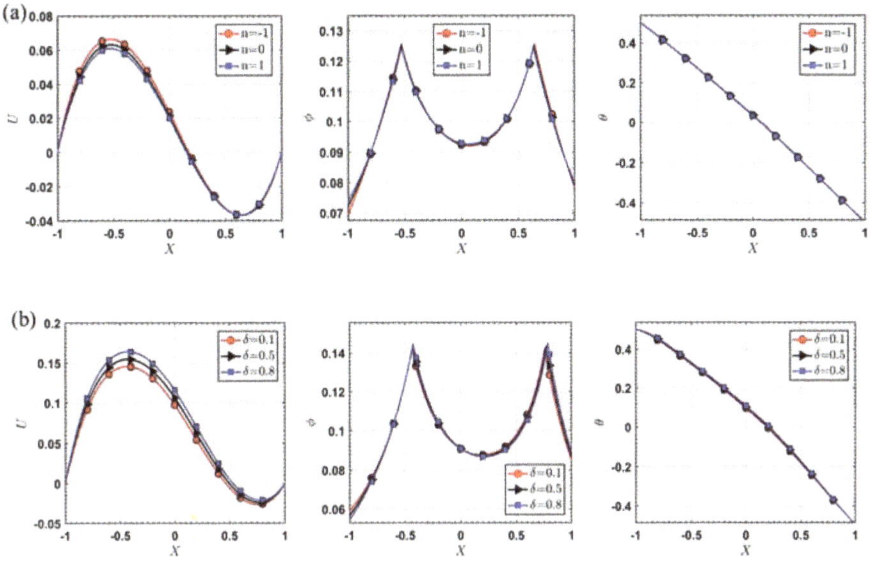

Figure 3. Parametric studies for shear-dependent viscosity. (**a**) Effect of n on the velocity, concentration and temperature profiles, when $\Gamma_c = 2.5$, $\Gamma_\mu = 0.1$, $Le = 10$, $\delta = 1$, $Ra = 2$, $Br = 5$ and $\phi_{avg} = 0.1$. (**b**) Effect of δ on the velocity, concentration and temperature profiles, when $\Gamma_c = 2.5$, $\Gamma_\mu = 0.1$, $Le = 10$, $n = -0.5$, $Ra = 3$, $Br = 5$ and $\phi_{avg} = 0.1$.

In Figure 4, we can see the effects of particle flux terms, Γ_c and Γ_μ, and the dimensionless number, Le. Recall that Γ_c represents the particle flux responsible for variable shear rates. As Figure 4a indicates, increasing Γ_c causes the particles to move towards the region with low shear rate, and a small value of Γ_c (0.1) leads to a uniform distribution of the particles; the velocity seems to increase as Γ_c increases, since the particle concentration near the region with low shear rate seems to produce a "lubrication" region near the wall. Figure 4b indicates that the effect of Γ_μ is opposite to that of Γ_c, implying that for the type of suspension considered here, Γ_μ tends to make the particles to be distributed more uniformly. Notice that the viscosity is proportional to the particle concentration, while according to Equation (12) Γ_μ forces the particle to move towards the region with lower viscosity. $1/Le$ is proportional to the coefficient of the flux due to the Brownian effects, therefore from Figure 4c, we see that a small value of Le leads to a uniform distribution of the particles; overall the effect of Le is similar but opposite to the effect of Γ_μ.

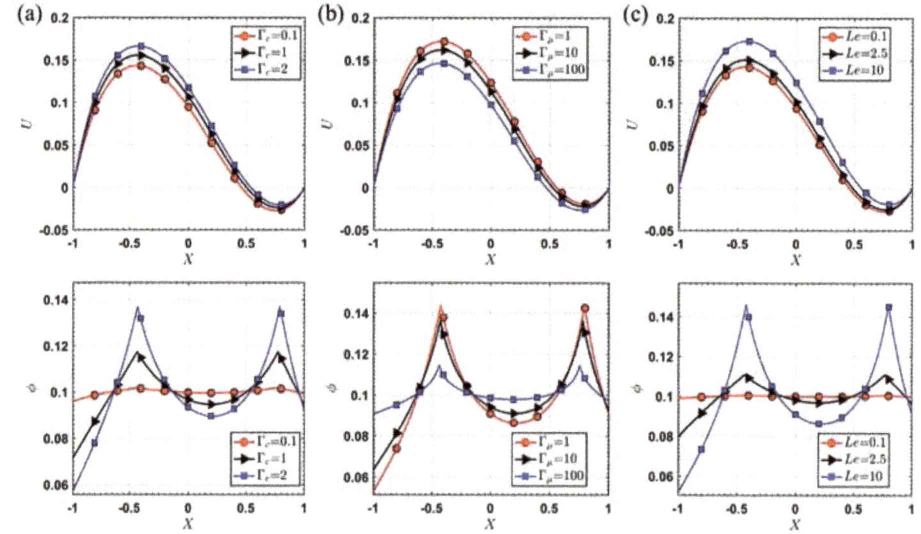

Figure 4. (**a**) Effect of Γ_c on the velocity, and concentration, when $\Gamma_\mu = 0.1$, $Le = 10$, $n = -0.5$, $\delta = 1$, $Ra = 3$, $Br = 5$ and $\phi_{avg} = 0.1$. (**b**) Effect of Γ_μ on the velocity, and concentration, when $\Gamma_c = 2.5$, $Le = 10$, $n = -0.5$, $\delta = 1$, $Ra = 3$, $Br = 5$ and $\phi_{avg} = 0.1$. (**c**) Effect of Le on the velocity, and concentration, when $\Gamma_c = 2.5$, $\Gamma_\mu = 0.1$, $n = -0.5$, $\delta = 1$, $Ra = 3$, $Br = 5$ and $\phi_{avg} = 0.1$.

Figure 5 shows the effect of the Brinkman number (Br). A larger value of Br indicates an increase in the temperature in the interior region; as a result, the velocity seems to increase indicating an increase in the buoyancy force. The effect of Br on the concentration profile is moderate, but we see that the position of the maximum concentration moves slightly. Figure 6 shows that as the bulk (average) concentration of the particles, ϕ_{avg}, increases, the magnitude of the velocity decreases, perhaps due to an increase in the viscosity; for particle concentration, a smaller ϕ_{avg} leads to a more uniform distribution of the particles.

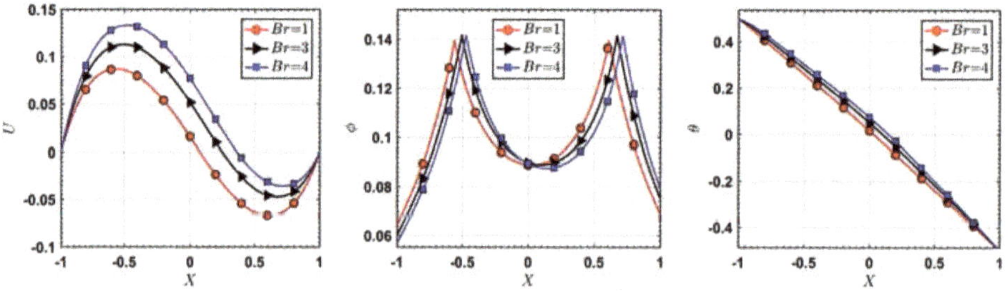

Figure 5. Effect of Brinkman number (Br) on the velocity, concentration and temperature profiles, when $\Gamma_c = 2.5$, $\Gamma_\mu = 0.1$, $Le = 10$, $n = -0.5$, $\delta = 1$, $Ra = 3$ and $\phi_{avg} = 0.1$.

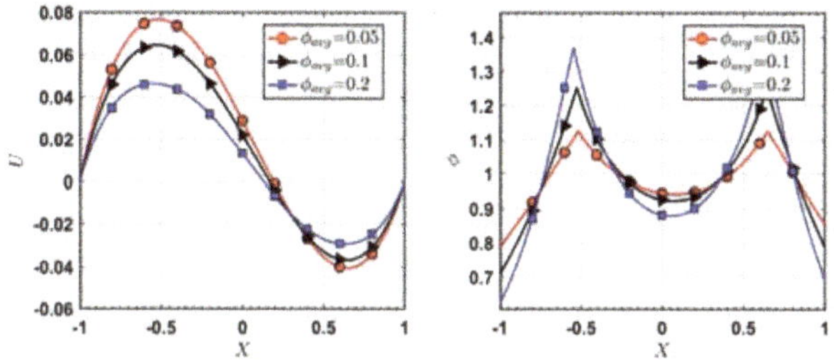

Figure 6. Effect of bulk (average) concentration (ϕ_{avg}) on the velocity and concentration profiles, when $\Gamma_c = 2.5$, $\Gamma_\mu = 0.1$, $Le = 10$, $n = -0.5$, $\delta = 1$, $Ra = 2$ and $Br = 5$.

5.2. Natural Convection with Particle Sedimentation

Now we look at a more general situation by considering two parallel walls which are tilted at an angle, giving rise to the possibility that particles may deposit. Figure 7 shows the effect of the particle flux due to gravity. Figure 7a indicates that as Γ_g increases, more particles tend to move and concentrate near the left wall (X = −1, see Figure 1); the particle concentration near the right wall decreases faster as Γ_g increases, and when $\Gamma_g = 5$ there are almost no particles at the right wall. For the velocity profile, the position of the maximum velocity tends to move slightly toward the left wall, perhaps due to an increase in the particle concentration in that region. Figure 7b shows that as β increases, indicating an increase or decrease in the X and Y component of the gravity, the velocity decreases and the particles tend to concentrate near the left wall; the temperatures seem to decrease a little.

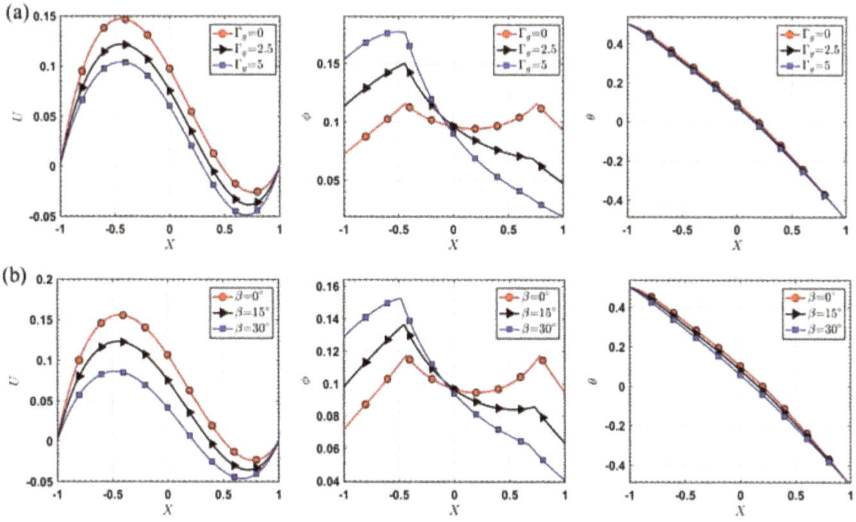

Figure 7. Effect of particle flux due to gravity. (**a**) Effect of Γ_g on the velocity, concentration and temperature profiles, when $\Gamma_c = 1.0$, $\Gamma_\mu = 0.1$, $\beta = 10^\circ$, $Le = 10$, $n = -0.5$, $\delta = 1$, $Ra = 3$, $Br = 5$ and $\phi_{avg} = 0.1$. (**b**) Effect of β on the velocity, concentration and temperature profiles, when $\Gamma_c = 1.0$, $\Gamma_\mu = 0.1$, $\Gamma_g = 1$, $Le = 10$, $n = -0.5$, $\delta = 1$, $Ra = 3$, $Br = 5$ and $\phi_{avg} = 0.1$.

Figure 8 shows the effect of Γ_c. Unlike the case with the neutrally buoyant particles, when Γ_c has a small value (0.1), the concentration now seems to decrease almost linearly in the X-direction. As Γ_c increases, the pattern of high particle concentration near the region with larger magnitude of velocity re-appears. Similar to the previous section, the effect of Γ_μ is opposite to that of Γ_c, as shown in Figure 8b. From Figure 8c, we can see that when *Le* is small, that is when the Brownian motion is strong, the particles are more uniformly distributed.

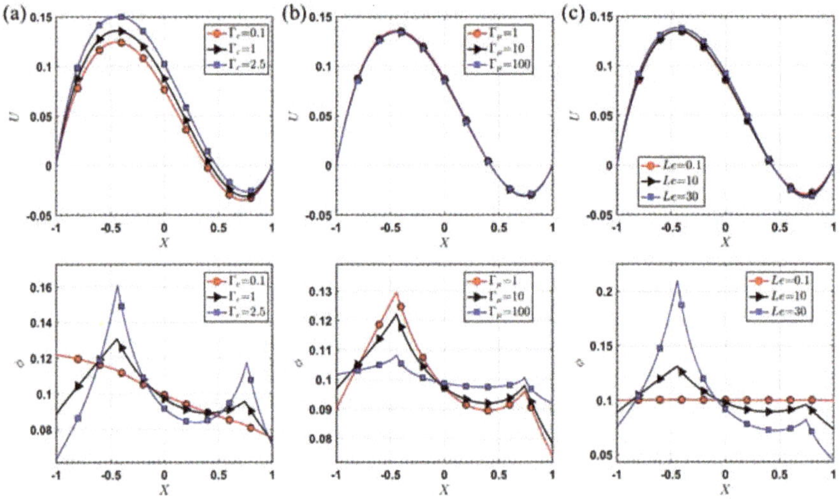

Figure 8. Effect of particle fluxes, Γ_c, Γ_μ and *Le*. (**a**) Effect of Γ_c on the velocity and concentration profiles, when $\Gamma_\mu = 0.1$, $\Gamma_g = 1$, $\beta = 10^\circ$, $Le = 10$, $n = -0.5$, $\delta = 1$, $Ra = 3$, $Br = 5$ and $\phi_{avg} = 0.1$. (**b**) Effect of Γ_μ on the velocity and concentration profiles, when $\Gamma_c = 1.0$, $\Gamma_g = 1$, $\beta = 10^\circ$, $Le = 10$, $n = -0.5$, $\delta = 1$, $Ra = 3$, $Br = 5$ and $\phi_{avg} = 0.1$. (**c**) Effect of *Le* on the velocity and concentration profiles, when $\Gamma_c = 1.0$, $\Gamma_\mu = 0.1$, $\Gamma_g = 1$, $\beta = 10^\circ$, $n = -0.5$, $\delta = 1$, $Ra = 3$, $Br = 5$ and $\phi_{avg} = 0.1$.

Figure 9 indicates that as *Ra* decreases, that is as the effect of the buoyancy force becomes less noticeable, particle sedimentation under gravity becomes more significant; meanwhile the values of the velocity and the temperature decrease. It should be noticed that the parametric studies of the Brinkman number (*Br*), and the terms related to the shear-dependent viscosity (n and δ) are not shown in this section, because the effects are similar to the Section 5.1. Figure 10 shows the effect of the bulk (average) concentration (ϕ_{avg}). With a small value of ϕ_{avg} (0.05), the concentration profile decreases monotonically along the X-direction, indicating that the particle distribution is dominated by the flux term due to gravity. We can also notice that increasing ϕ_{avg} results in a higher viscosity, causing a decrease in the velocity, viscous dissipation and the temperature.

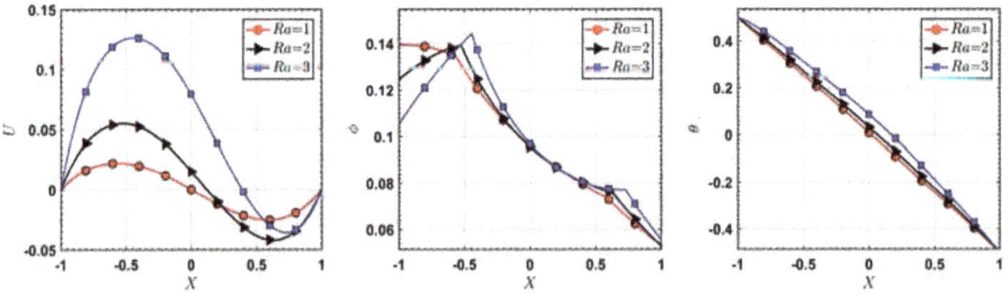

Figure 9. Effect of buoyancy force term, Rayleigh number (*Ra*) on the velocity, concentration and temperature profiles, when $\Gamma_c = 1.0$, $\Gamma_\mu = 0.1$, $\Gamma_g = 1$, $\beta = 10^\circ$, $Le = 10$, $n = -0.5$, $\delta = 1$, $Br = 5$ and $\phi_{avg} = 0.1$.

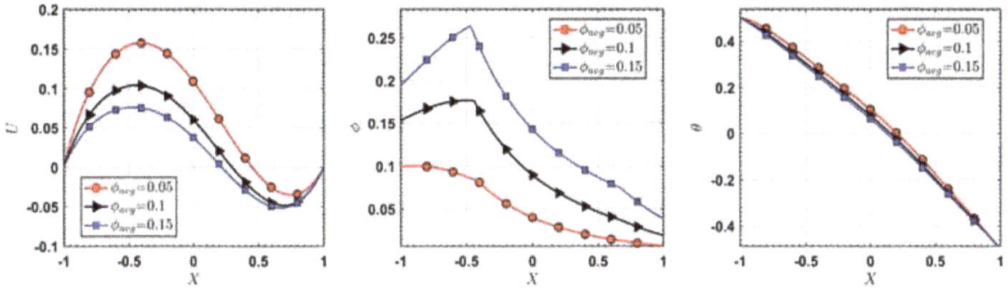

Figure 10. Effect of bulk (average) concentration (ϕ_{avg}) on the velocity, concentration and temperature profiles, when $\Gamma_c = 1.0, \Gamma_\mu = 0.1, \Gamma_g = 1, \beta = 10^\circ, Le = 10, n = -0.5, \delta = 1, Ra = 2$ and $Br = 5$.

6. Conclusions

In this paper we study the buoyancy driven flow of a suspension between two long vertically inclined walls. The suspension is modeled as a non-linear fluid, where the viscosity depends on the shear rate and the particle concentration. The motion of the particles is modeled by a convection-diffusion equation, where the particle transport flux is assumed to depend on the body force (gravity), and the variation of the shear rate and viscosity. The numerical results indicate that natural convection flow shows certain multi-component features noticed in flow of solid-fluid suspensions where the solid particles tend to move and concentrate near the region with low shear rate. Furthermore, under the effect of gravity, the particles tend to move and concentrate near the lower (left) wall; however, a small Lewis number (stronger Brownian diffusion) can generate a more uniform concentration distribution.

Author Contributions: C.T. and W.-T.W. did the numerical simulations. W.-T.W. and M.M. derived all the equations. M.M. supervised this work. All of the authors have provided substantial contributions to the manuscript preparation.

Funding: This research received no external funding.

Conflicts of Interest: The authors declare no conflict of interest.

Nomenclature

Symbol	Explanation
ρ	Density
ε	Specific internal energy
r	Radiant heating
η	Specific entropy density
g	Gravity
H	Characteristic length
k	Thermal conductivity
ζ	Coefficient of thermal expansion
$\dot{\gamma}$	Shear rate
K_c, K_μ	Coefficients of particle flux
t_p	Particle response time
n	Power-law index
a	Particle radius
D	Diffusion coefficient
μ	Viscosity
θ	Temperature
ϕ	Volume fraction
p	Pressure
β	Inclination angle
x	Spatial position

$\boldsymbol{v}$	Velocity
$\boldsymbol{b}$	Body force vector
$\boldsymbol{q}$	Heat flux vector
$\boldsymbol{N}$	Particle flux
$\boldsymbol{T}$	Cauchy stress tensor
$\boldsymbol{L}$	Gradient of the velocity vector
$\boldsymbol{D}$	Symmetric part of the velocity gradient
$\boldsymbol{I}$	Identity tensor
grad or ∇	Gradient symbol
div	Divergence operator
tr	Trace operator

References

1. Turner, J.S. *Buoyancy Effects in Fluids*; Cambridge University Press: Cambridge, UK, 1979.
2. Chen, B.; Mikami, F.; Nishikawa, N. Experimental studies on transient features of natural convection in particles suspensions. *Int. J. Heat Mass Transf.* **2005**, *48*, 2933–2942. [CrossRef]
3. Sun, X.; Massoudi, M.; Aubry, N.; Chen, Z.; Wu, W.-T. Natural convection and anisotropic heat transfer in a ferro-nanofluid under magnetic field. *Int. J. Heat Mass Transf.* **2019**, *133*, 581–595. [CrossRef]
4. Kang, C.; Okada, M.; Hattori, A.; Oyama, K. Natural convection of water–fine particle suspension in a rectangular vessel heated and cooled from opposing vertical walls. *Int. J. Heat Mass Transf.* **2001**, *44*, 2973–2982. [CrossRef]
5. Bustos, M.C.; Paiva, F.; Wendland, W. Control of continuous sedimentation of ideal suspensions as an initial and boundary value problem. *Math. Methods Appl. Sci.* **1990**, *12*, 533–548. [CrossRef]
6. Metivier, C.; Li, C.; Magnin, A. Origin of the onset of Rayleigh-Bénard convection in a concentrated suspension of microgels with a yield stress behavior. *Phys. Fluids* **2017**, *29*, 104102. [CrossRef]
7. Okada, M.; Suzuki, T. Natural convection of water-fine particle suspension in a rectangular cell. *Int. J. Heat Mass Transf.* **1997**, *40*, 3201–3208. [CrossRef]
8. Batchelor, G.K. Heat convection and buoyancy effects in fluids. *Q. J. R. Meteorol. Soc.* **1954**, *80*, 339–358. [CrossRef]
9. Shenoy, A.V.; Mashelkar, R.A. Thermal convection in non-Newtonian fluids. In *Advances in Heat Transfer*; Elsevier: Amsterdam, The Netherlands, 1982; Volume 15, pp. 143–225.
10. Rajagopal, K.R.; Na, T.-Y. Natural convection flow of a non-Newtonian fluid between two vertical flat plates. *Acta Mech.* **1985**, *54*, 239–246. [CrossRef]
11. Massoudi, M.; Christie, I. Natural convection flow of a non-Newtonian fluid between two concentric vertical cylinders. *Acta Mech.* **1990**, *82*, 11–19. [CrossRef]
12. Massoudi, M.; Vaidya, A.; Wulandana, R. Natural convection flow of a generalized second grade fluid between two vertical walls. *Nonlinear Anal. Real World Appl.* **2008**, *9*, 80–93. [CrossRef]
13. Rajagopal, K.R.; Tao, L. *Mechanics of Mixtures, Series on Advances in Mathematics for Applied Sciences*; World Scientific: Singapore, 1995; Volume 35.
14. Massoudi, M. Constitutive relations for the interaction force in multicomponent particulate flows. *Int. J. Non. Linear. Mech.* **2003**, *38*, 313–336. [CrossRef]
15. Massoudi, M. A note on the meaning of mixture viscosity using the classical continuum theories of mixtures. *Int. J. Eng. Sci.* **2008**, *46*, 677–689. [CrossRef]
16. Massoudi, M. A Mixture Theory formulation for hydraulic or pneumatic transport of solid particles. *Int. J. Eng. Sci.* **2010**, *48*, 1440–1461. [CrossRef]
17. Slattery, J.C. *Advanced Transport Phenomena*; Cambridge University Press: Cambridge, UK, 1999.
18. Liu, I.-S. *Continuum Mechanics*; Springer Science & Business Media: Berlin, Germany, 2002.
19. Dunn, J.E.; Fosdick, R.L. Thermodynamics, stability, and boundedness of fluids of complexity 2 and fluids of second grade. *Arch. Ration. Mech. Anal.* **1974**, *56*, 191–252. [CrossRef]
20. Probstein, R.F. *Physicochemical Hydrodynamics: An Introduction*; John Wiley & Sons: Hoboken, NJ, USA, 2005.
21. Macosko, C. *Rheology: Principles, Measurements and Applications*; Wiley-VCH Inc.: New York, NY, USA, 1994.
22. Schowalter, W.R. *Mechanics of Non-Newtonian Fluids*; Pergamon Press: Oxford, UK, 1978.

23. Bridges, C.; Rajagopal, K.R. Pulsatile Flow of a Chemically-Reacting Nonlinear Fluid. *Comput. Math. Appl.* **2006**, *52*, 1131–1144. [CrossRef]
24. Krieger, I.M. Rheology of monodisperse latices. *Adv. Colloid Interface Sci.* **1972**, *3*, 111–136. [CrossRef]
25. Phillips, R.J.; Armstrong, R.C.; Brown, R.A.; Graham, A.L.; Abbott, J.R. A constitutive equation for concentrated suspensions that accounts for shear-induced particle migration. *Phys. Fluids A Fluid Dyn.* **1992**, *4*, 30–40. [CrossRef]
26. Tao, C.; Kutchko, B.G.; Rosenbaum, E.; Wu, W.-T.; Massoudi, M.; Tao, C.; Kutchko, B.G.; Rosenbaum, E.; Wu, W.-T.; Massoudi, M. Steady Flow of a Cement Slurry. *Energies* **2019**, *12*, 2604. [CrossRef]
27. Miao, L.; Massoudi, M. Effects of shear dependent viscosity and variable thermal conductivity on the flow and heat transfer in a slurry. *Energies* **2015**, *8*, 11546–11574. [CrossRef]
28. Yang, H.; Aubry, N.; Massoudi, M. Heat transfer in granular materials: Effects of nonlinear heat conduction and viscous dissipation. *Math. Methods Appl. Sci.* **2013**, *36*, 1947–1964. [CrossRef]
29. Yang, H.; Massoudi, M. Conduction and convection heat transfer in a dense granular suspension. *Appl. Math. Comput.* **2018**, *332*, 351–362. [CrossRef]
30. Massoudi, M. On the heat flux vector for flowing granular materials—Part I: Effective thermal conductivity and background. *Math. Methods Appl. Sci.* **2006**, *29*, 1585–1598. [CrossRef]
31. Massoudi, M. On the heat flux vector for flowing granular materials—part II: Derivation and special cases. *Math. Methods Appl. Sci.* **2006**, *29*, 1599–1613. [CrossRef]
32. Massoudi, M.; Kirwan, A. On Thermomechanics of a Nonlinear Heat Conducting Suspension. *Fluids* **2016**, *1*, 19. [CrossRef]
33. Rajagopal, K.R.; Ruzicka, M.; Srinivasa, A.R. On the Oberbeck-Boussinesq approximation. *Math. Model. Methods Appl. Sci.* **1996**, *6*, 1157–1167. [CrossRef]
34. Rajagopal, K.R.; Saccomandi, G.; Vergori, L. On the Oberbeck–Boussinesq approximation for fluids with pressure dependent viscosities. *Nonlinear Anal. Real World Appl.* **2009**, *10*, 1139–1150. [CrossRef]
35. Subia, S.R.; Ingber, M.S.; Mondy, L.A.; Altobelli, S.A.; Graham, A.L. Modelling of concentrated suspensions using a continuum constitutive equation. *J. Fluid Mech.* **1998**, *373*, 193–219. [CrossRef]
36. Acrivos, A.; Mauri, R.; Fan, X. Shear-induced resuspension in a Couette device. *Int. J. Multiph. flow* **1993**, *19*, 797–802. [CrossRef]
37. Wu, W.T.; Aubry, N.; Antaki, J.F.; Massoudi, M. A non-linear fluid suspension model for blood flow. *Int. J. Non. Linear Mech.* **2019**, *109*, 32–39. [CrossRef]
38. Massoudi, M.; Uguz, A.K. Chemically-reacting fluids with variable transport properties. *Appl. Math. Comput.* **2012**, *219*, 1761–1775. [CrossRef]
39. Bird, R.B.; Stewart, W.E.; Lightfoot, E.N. *Transport Phenomena*; John Wiley & Sons: Hoboken, NJ, USA, 2007.
40. Dawson, P.R.; McTigue, D.F. A numerical model for natural convection in fluid-saturated creeping porous media. *Numer. Heat Transf.* **1985**, *8*, 45–63. [CrossRef]
41. Pruša, V.; Rajagopal, K.R. On models for viscoelastic materials that are mechanically incompressible and thermally compressible or expansible and their Oberbeck–Boussinesq type approximations. *Math. Model. Methods Appl. Sci.* **2013**, *23*, 1761–1794. [CrossRef]
42. Massoudi, M. Boundary conditions in mixture theory and in CFD applications of higher order models. *Comput. Math. Appl.* **2007**, *53*, 156–167. [CrossRef]
43. *MATLAB User's Guide*; The Mathworks Inc.: Natick, MA, USA, 1998.

MDPI
St. Alban-Anlage 66
4052 Basel
Switzerland
Tel. +41 61 683 77 34
Fax +41 61 302 89 18
www.mdpi.com

Fluids Editorial Office
E-mail: fluids@mdpi.com
www.mdpi.com/journal/fluids

www.ingramcontent.com/pod-product-compliance
Lightning Source LLC
LaVergne TN
LVHW071942080526
838202LV00064B/6654
* 9 7 8 3 0 3 9 2 8 3 0 8 8 *